# The biology of high-altitude peoples

# THE INTERNATIONAL BIOLOGICAL PROGRAMME

The International Biological Programme was established by the International Council of Scientific Unions in 1964 as a counterpart of the International Geophysical Year. The subject of the IBP was defined as 'The Biological Basis of Productivity and Human Welfare', and the reason for its establishment was recognition that the rapidly increasing human population called for a better understanding of the environment as a basis for the rational management of natural resources. This could be achieved only on the basis of scientific knowledge, which in many fields of biology and in many parts of the world was felt to be inadequate. At the same time it was recognized that human activities were creating rapid and comprehensive changes in the environment. Thus, in terms of human welfare, the reason for the IBP lay in its promotion of basic knowledge relevant to the needs of man.

The IBP provided the first occasion on which biologists throughout the world were challenged to work together for a common cause. It involved an integrated and concerted examination of a wide range of problems. The Programme was co-ordinated through a series of seven sections representing the major subject areas of research. Four of these sections were concerned with the study of biological productivity on land, in freshwater, and in the seas, together with the processes of photosynthesis and nitrogen-fixation. Three sections were concerned with adaptability of human populations, conservation of ecosystems and the use of biological resources.

After a decade of work, the Programme terminated in June 1974 and this series of volumes brings together, in the form of syntheses, the results of national and international activities.

INTERNATIONAL BIOLOGICAL PROGRAMME 14

# The biology of high-altitude peoples

EDITED BY

## P. T. Baker
*Professor of Anthropology*
*The Pennsylvania State University*

CAMBRIDGE UNIVERSITY PRESS

CAMBRIDGE
LONDON · NEW YORK · MELBOURNE

CAMBRIDGE UNIVERSITY PRESS
Cambridge, New York, Melbourne, Madrid, Cape Town, Singapore, São Paulo, Delhi

Cambridge University Press
The Edinburgh Building, Cambridge CB2 8RU, UK

Published in the United States of America by Cambridge University Press, New York

www.cambridge.org
Information on this title: www.cambridge.org/9780521111959

First published 1978
This digitally printed version 2009

*A catalogue record for this publication is available from the British Library*

*Library of Congress Cataloguing in Publication data*
Main entry under title:
The Biology of high-altitude peoples.
(International Biological Programme; 14)
1. Altitude, Influence of – Addresses, essays, lectures. 2. Man –
Influence of environment – Addresses, essays, lectures. 3. Physical anthro-
pology – Addresses, essays, lectures. I. Baker, Paul Thornell, 1927–.
II. Series. [DNLM: 1. Altitude. WD710 B613] GF57.B56    612′.01441′5
76-50311

ISBN 978-0-521-21523-7 hardback
ISBN 978-0-521-11195-9 paperback

# Table des matières

# Contents

# Содержание

# Contenido

# Contributors

P. T. Baker, Department of Anthropology, The Pennsylvania State University, University Park, PA 16802, USA

E. R. Buskirk, Human Performance Research Laboratory, The Pennsylvania State University, University Park, PA 16802, USA

E. J. Clegg, Department of Human Biology and Anatomy, University of Sheffield, Sheffield S10 2TN, UK

R. Cruz-Coke, Genetic Section, Universidad de Chile, Hospital Clinico José Joaquin Aguirre, Santiago, Chile

A. R. Frisancho, Center for Human Growth and Development, University of Michigan, Ann Arbor, Michigan 48104, USA

J. M. Hanna, Department of Physiology, University of Hawaii, Honolulu, Hawaii 96822

C. Jest, Musée de l'Homme, Place de Chaillot, 75016 Paris, France

M. A. Little, Department of Anthropology, State University of New York, Binghamton, New York 13901, USA

M. M. Mirrakhimov, State Medical Institute, 50-year October St 92, Frunze, Kirghizia, USSR

I. G. Pawson, G. W. Hooper Foundation, University of California, San Francisco, California 94143, USA

E. Picón-Reátegui, Universidad Nacional Mayor de San Marcos, Apartado 5073, Lima, Peru

J.-C. Quilici, Centre National de la Recherche Scientifique, Centre d'Hémotypologie, Toulouse, France

H. Vergnes, Centre National de la Recherche Scientifique, Centre d'Hémotypologie, Toulouse, France

# Foreword

The investigations of human populations within the Human Adaptability Section of the IBP (1964–74) were planned on a worldwide scale. One objective was to obtain comparable data on population characteristics over a wide range of ecosystems. Another was to examine and compare different ethnic and economic groups within similar biomes. A major effort was directed at a comprehensive coverage of the high-altitude and circumpolar regions of the world. Yet another aim was to direct international biomedical teams to 'threatened' or disappearing groups – hunter-gatherers and simple agriculturists – to study their often highly distinctive ecological characteristics (and at the same time to provide biomedical help). Yet another objective of interest to many participating countries was the biological condition of populations undergoing migration or living under man-made urbanized conditions.

When in 1974 the IBP was brought to a close, the Human Adaptability Section had largely fulfilled this global programme. Over 50 countries had mounted some 250 projects. Several thousands of scientific papers have already been published along with some 30 detailed monographs dealing with particular national projects. Details of all the contributing countries, their projects, the team personnel and their publications and reports (as known in 1975) are to be found in *Human Adaptability: a history and compendium of research within the IBP* by K. J. Collins & J. S. Weiner (Taylor & Francis, London 1977).

Over and above the purely local interest of a national project – and this is always important – the data obtained from different HA investigations can be utilized in many different ways; the exploitation of the findings will certainly go on for a long time. However, the Special Committee for the IBP decided that an immediate effort should be made to put together a significant proportion of the material in a series of readily accessible 'synthesis' volumes. Within this series of some 30 volumes, six have been planned to cover some of the major approaches mentioned above within the HA Section. The present volume under the editorship of Professor Paul Baker is the third of these to appear.

The first volume (Phyllis Eveleth & J. M. Tanner) deals in a systematic and comprehensive manner with a fundamental characteristic of human beings the world over, namely, development and physique. In this volume, children's growth patterns in particular are examined on a world scale in relation to many factors including climate, disease, nutrition and genetic constitution. The extensive tabulations make this an invaluable survey of base-line growth data for some hundreds of population samples.

Like this volume on growth two further volumes will deal with major response systems of the body. Professor R. J. Shepard has prepared a detailed and intensive survey of physiological work capacity of communities the world over. A synthesis volume on HA investigations into the tolerance to high temperatures and functioning of the respiratory system is also in preparation. In these volumes the significance and causes of variations in these physiological parameters are analysed in detail and a large amount of original data is brought together.

The present volume edited and prepared by Professor Baker and that by Professor Milan, now in press, each comprise a comparative survey of the demographic, genetic and biomedical characteristic of communities living respectively in high-altitude and circumpolar biomes. Each is aimed at understanding how communities differing in their economic base, in population size, and in genetic make-up, come to terms with the particular stresses of life in these environments.

The sixth volume, edited by Professor Ainsworth Harrison, and already published, complements all the other volumes in an interesting way. It comprises a series of case studies each of which provides a vivid illustration of special aspects of population biological structure which enter into the comparative surveys in the other volumes.

The appearance of this synthesis volume affords an opportunity for me as Convener of the HA Section to pay tribute to the vital contribution made by Professor Baker to this major aspect of the HA Programme. Without his enthusiastic leadership in his capacity as Coordinator of the high-altitude population theme and his inspiring field work in the High Andean region carried out year after year within the IBP, this major success of the programme would not have been achieved. His strategic understanding of the problems and opportunities of field work in the high mountain regions of the world is very well displayed in his contributions to this volume. As Theme Coordinator he was in close touch with the teams working in all the other high-altitude regions and he visited many of these locations. The present volume, which gives a worldwide conspectus of high-altitude communities, and the monograph *Man in the Andes: a multidisciplinary study of high-altitude Quechua* co-edited by Michael A. Little (Dowden, Hutchinson & Ross, Stroudsburg, Pennsylvania, 1976) which deals very largely with the work carried out and supervised by Professor Baker, represent major landmarks in our understanding of the mechanisms and limits of adaptation of human beings living and working at high altitudes.

J. S. WEINER

# IBP Human Adaptability section publications

J. S. Weiner (1969).
*A guide to human adaptability proposals*
Blackwell Scientific Publications, Oxford. 88 pp.

J. S.Weiner & J. A. Lourie (1969).
*A guide to field methods*
Blackwell Scientific Publications, Oxford. 652 pp.

S. Biesneuvel (ed.) (1969).
*Methods for the measurements of psychological performance*
Blackwell Scientific Publications, Oxford. 110 pp.

P. B. Eveleth & J. M. Tanner (1976).
*Worldwide variation in human growth*
Cambridge University Press, London 498 pp.

G. A. Harrison (ed.) (1977).
*Population structure and human variation*
Cambridge University Press, London. 342 pp.

F. Milan (ed.).
*The biology of circumpolar peoples*
Cambridge University Press, London. (In press.)

P. Baker (ed.) (1978).
*The biology of high-altitude peoples*
Cambridge University Press, London.

J. S. Weiner & J. E. Coles.
*Components of human physiological function: thermal responses and respiratory function*
Cambridge University Press, London. (In preparation.)

R. J. Shephard.
*Human physiological work capacity*
Cambridge University Press, London. (In press.)

A detailed guide to all the IBP projects contributing to the theme of this volume is given in *Human adaptability: a history and compendium of research within the IBP* (1977) by K. J. Collins & J. S. Weiner, published by Taylor & Francis Ltd, 10–14 Macklin Street, London WC2B 5NF. A collection of HA reports, reprints and archival material is held in the Library of the British Museum (Natural History), Cromwell Road, London SW7 5BD.

# 1. IBP high-altitude research: development and strategies

P. T. BAKER

The peoples of the high-altitude regions of the world have long fascinated both scientists and laymen from the lowlands. Early explorers from Europe discovered in the high-altitude regions of South America, north-eastern Africa and Central Asia highly complex cultures with substantial populations and often great wealth. The explorers also discovered that while these high areas caused them discomfort and often sickness the native populations seemed healthy and by the standards of their own physical capabilities considerably stronger and more fit. They continued over the centuries to carry back to the lowlands romantic stories of the health and wealth of these peoples. It is surely no accident that even today many laymen dream of a utopian Shangri-la hidden in the vastness of the Himalayas and even serious scientists still suggest that somewhere in the Andes and the mountains of the USSR there are human beings who live with a high degree of physical fitness to ages far exceeding those found for lowlanders.

With the beginnings of such organized sports as mountain climbing and balloon ascents a more serious scientific interest in the effects of altitude emerged. By the 1800s some of the more dramatic physical effects of high altitude on lowland man had been documented (Ward, 1975) and by the early 1900s physiological research on the causes of these effects were seriously underway. Interest in how the hypoxia of high altitude affected lowland man accelerated during the first half of the century so that by 1965 a review of the literature produced a bibliography of more than 5,000 published articles and books on the topic (Wulff, Braden, Shillito & Tomashefski, 1968).

Despite this great interest, knowledge about the biology of populations native to the world's high-altitude areas remained fragmentary. Beginning in the 1920s some research on the high-altitude peoples of the Peruvian Andes began. Under the strong leadership of Carlos Monge M., Peruvians, along with European and North American scientists, inaugurated studies on physiological responses to hypoxia of the Peruvian highland Quechua. Some early studies on the growth of natives were followed in the late 1950s and early 1960s by efforts to broaden the knowledge base concerning reproduction and specific altitude-related disease.

Information about the biology of Ethiopia's high-altitude people was essentially non-existent in the early 1960s while information on the biology

1

of the Asiatic high-altitude peoples was limited to a few observations made by members of English and New Zealand mountain climbing expeditions (Pugh, 1966).

## The International Biological Programme project

When the International Biological Programme (IBP) was formulated in the early 1960s it was realized that such a broad-based biological program whose theoretical foundations lay in ecology would have to include a section devoted to the study of the effect of environment on biological man (Worthington, 1975). The project entitled Human Adaptability was convened by Professor Joseph Weiner who first developed and later co-ordinated the project for the life of the IBP. As part of the early development process Weiner was able, with the help of the Wenner–Gren Foundation for Anthropological Research, to hold a conference on our state of knowledge in human adaptability. I am sure that the mere twenty of us who contributed to the state-of-knowledge book which developed from this conference (Baker & Weiner, 1966) were inadequate to the task, but the information and research directions suggested in the book offer a base on which the information in this book may be judged.

As Weiner (1966, 1969) noted it was hoped that a significant segment of the Human Adaptability project efforts could be devoted to a few 'intensive multidisciplinary regional studies'. These were to be based on life in very stressful environments and on habitat and population contrasts. It was believed that such studies could progress beyond traditional descriptive information and provide insights into how our species had adapted biologically to such a broad diversity of environments, many of which were totally alien to the tropical lowlands in which we evolved.

Harrison, Pugh and myself (Baker & Weiner, 1966) writing on different aspects of man at high altitude all emphasized what we saw as the unique opportunities for understanding fundamental processes by the study of man at high altitude. Among the various environmental stresses to which man is exposed the hypoxia of high altitude is the one probably least subject to behavioral or technological amelioration. Thus, Harrison (1966) suggested that high altitude might offer a situation to which classic experimental design might be applied. Basically he suggested that a natural experiment situation should be found where four populations lived under the same conditions with the exception of the altitude of residence. Two should live at high altitude and two at low altitude. Of the two at high altitude, one should consist of migrants from the native low-altitude population. One of the low-altitude populations should be migrants who are high-altitude natives. The structure of this design and what could be deduced from studying these populations is outlined as follows:

2

(1) HAN: a group living at high altitude and native to altitude, in the sense of having a long generational history of high-altitude residence.

(2) HAN↓: a subdivision of group (1) who live under identical conditions but who have migrated down to sea level.

(3) LAN: a group living at low altitude with other conditions similar to those of group (1).

(4) LAN↑: a subdivision of group (3) which migrated up to the same altitude and conditions of group (1).

With these populations, intergroup comparisons would provide the following information.

HAN–LAN = total differences caused by altitude

HAN–LAN↑ = genetic features of altitude adaptation

LAN–LAN↑ = ontogenetic and physiological adaptations to altitude (acclimatization) plus detrimental effects of altitude

HAN–HAN↓ = ontogenetic and physiological adaptation (acclimatization) to downward migration plus detrimental effects of downward migration

LAN–HAN↓ = differences in response to sea-level pressure produced by genetic differences

Unfortunately, but perhaps not surprisingly, populations fulfilling these exacting design requirements were not found, but the IBP program in Ethiopia (Harrison *et al.*, 1969; Clegg, Pawson, Ashton & Flinn, 1972) and to some extent the studies in Peru (Baker, 1976*b*) were able to follow modified forms of this design.

The World Health Organization became interested in the development of the high-altitude project and co-operatively with the Pan American Health Organization and the United States IBP committee sponsored a major conference in 1967 to assess our state of knowledge and research needs in relation to man at high altitude. The final research recommendations of this conference were divided into three topical areas. They were: (1) physiological adaptation and acclimatization to altitude, (2) the study of human biology at high altitudes, and (3) the health aspects of altitude. The recommendations for research under these three topics were indeed comprehensive but in many instances it developed that IBP was not the appropriate vehicle for supporting the needed research.

While research on physiological acclimatization to high altitude and the health problems of high altitude continued after this conference they were generally not incorporated in the IBP-supported research of any country with the exception of the USSR. It is my further impression that research on these topics declined in the time after this conference as compared to the five preceding years. The reasons for a diminished interest in these significant scientific and practical problems is not clear to me.

3

## The biology of high-altitude peoples

While the promising start of IBP research on high altitude thus suffered a significant setback, the interest generated in the human biology aspects developed and formed the major segment of research specifically undertaken with IBP designation. Even this development was not without difficulties and some history of these problems may be useful for placing the studies reviewed in this book in perspective.

A significant problem for IBP high-altitude project was the inability of the scientists in most high-altitude countries to form IBP committees in their own countries. Although individual scientists in many Andean countries worked on IBP projects they received very limited support from within. In the Asian mountains the USSR was the only country to carry out a substantial research program on high-altitude peoples under the label of IBP. Since it was deemed appropriate to limit this book to the primary topic of high-altitude peoples, not all of the USSR research has been reviewed and Dr Mirrakhimov's chapter has been limited to the physiological aspects.

India did have an IBP committee but was unable to devote significant funds to high-altitude research so that the work completed on high-altitude natives in India was supported from sources other than IBP. The Tibetan area, as part of China, was not part of IBP and the studies in the Himalayan countries of Nepal and Bhutan were limited in part by the lack of native scientific participants.

Of the individual projects completed all suffered to some extent from funding limitations; nevertheless, the sum total of these projects were great enough so that a quite new and more synthetic picture of the biology of high-altitude peoples may now be developed. Principle among these long-term and multidisciplinary efforts were the following:

(1) WHO and Chilean sponsored studies of the Aymara in northeastern Chile.

(2) French and Bolivian studies on the Aymara and Quechua in the region of La Paz in Bolivia.

(3) US and Peruvian studies on the Quechua of southern Peru.

(4) British and Norwegian studies of the Amhara in northwestern Ethiopia.

(5) British, French, and US studies of the Sherpa, Tibetan migrants and other high-altitude natives of Nepal.

(6) Specialized Italian supported studies on the Quechua of Peru and Sherpa of Nepal.

(7) USSR studies on the native peoples in the Tien Shan and Pamir mountains.

(8) Selected studies of US populations in the high-altitude areas of Colorado.

4

In this book the authors have attempted to integrate the information gained from these projects with the previous information available and indicate some of the research needs growing from these results.

## Research designs and methods

The conferees at the WHO/PAHO/IBP meeting in recommending the nature of the research needed on the biology of high-altitude peoples suggested a somewhat dualistic approach. On the one hand it was suggested that descriptive information was needed. The categories of information needed were structured in a manner which would provide a comprehensive description of the biological fitness of each population studied.

The recommendations of the human biology group as printed in the report (PAHO, 1967) were as follows:

It was agreed to recommend that, with the use of compositive methods of approach and standardized procedures, information should be obtained in the following categories:

(1) *Fertility and the components of fertility*
  (*a*) by demographic needs,
  (*b*) using methods in the reproductive physiology of man and of animals which could be applied to human population studies.

(2) *Growth, development and aging*
  (*a*) age changes and variability in characteristics thought to be of adaptive value at high altitude,
  (*b*) age changes and variability in characteristics in relation to the somatic fitness of individuals,
  (*c*) such studies should not be divorced from the psychological and intellectual changes which occur during development.

(3) *Nutrition*
In all cases the nutritional assessment of the populations studied should be made in as detailed a manner as possible, commensurate with the resources available. Such assessments should include:
  (*a*) the nutritional status of individuals,
  (*b*) detailed nutritional surveys where possible,
  (*c*) biochemical studies related to nutrition.

(4) *Special problems relating to work capacity*
Both physiological and psychological methods should be used.

(5) *Epidemiology*
  (*a*) In all cases the pattern of disease distribution in populations should be studied. Where additional demographic information is available it is

highly important that more vigorous epidemiological studies should be made.

(*b*) It is of great importance that demographic methods should be developed which would enable the relationships between age, disease and morbidity to be ascertained.

## (6) *Genetics*

Further information is required on:

(*a*) the distribution of polymorphic systems in high-altitude populations,

(*b*) the hereditability of quantitatively varying traits, particularly those presumed to be adaptive in nature,

(*c*) congenital defects, especially those presumed to have a genetic component.

## (7) *General*

In all these studies the following are essential:

(*a*) There is as precise an analysis as possible of all aspects of the environment.

(*b*) Adequate precautions must be taken to insure statistical representation and control situations. This will often mean the study of lowland populations.

(*c*) The demographic background of the populations under study must be ascertained in as great a detail as possible.

While the descriptive data were considered essential there was a strong concern that the data should be collected whenever possible within a research design which would permit the development of conclusions on how the high-altitude populations had adapted to their environment. At that time short-term acclimatization, developmental biological responses and genetic adaptations were all considered possible mechanisms of adaptation and adjustment to high-altitude environments. In addition, it was clear that many biological responses in these populations had significance in relation to multiple environmental stresses such as reduced oxygen pressure, cold and aridity. It was, therefore, suggested that studies of a broad and multidisciplinary nature should be undertaken in an intensive manner on appropriate population samples.

These ideas appeared to be highly compatible with the desires of most investigators since almost all of the subsequent studies were in fact developed using a variety of population comparison strategies. Because of the financial and other problems cited earlier only a few of the studies were able to develop the disciplinary breadth and sampling procedures which were recommended.

A review of the research strategies utilized in the various projects is enlightening as it allows the reader of the subsequent chapters to evaluate the relative informational and theoretical advances which each design produced.

## Descriptive studies

The number of purely descriptive studies were quite limited but were of considerable value when they provided information on previously unknown populations or on fitness indicators which were unknown in partially described populations. Examples of this type of work include the study of Donoso, Apud, Sanudo & Santalaya (1971) on the work capacity of the high-altitude Aymara in Chile, Andersen's (1973*a, b, c*) similar studies of Ethiopia's Amhara, and Lang & Lang's (1971) general description of the Sherpa.

### Population modeling

Another type of descriptive research which was only rarely attempted in the high-altitude projects was process or flow modeling. Recent research has suggested the great potential of this approach (Baker, 1976*c*) but the only attempts in relation to the high-altitude project were small studies of gene flow (Dutt, 1976*a*) and a community study of energy flow in the Peruvian Andes (Thomas, 1973, 1976).

### Intra-group comparisons

Comparison within given population and community units was a research strategy used quite often to solve specific problems in high-altitude regions. Weitz (1973) compared the aerobic capacity of active porters in highland Nepal with less active males in the community. By contrasting these results with a similar study on Tibetans at low altitudes he was able to demonstrate that age and activity effects on aerobic capacity were identical for the two groups. Mazess (1969) used a similar strategy when studying students in a physical education school in southern Peru. By comparing these students with the rather inactive students in a nearby university he assessed the effects of short-term training on the aerobic capacity of the Quechua. Bouloux (1968) compared well-cared-for children in an orphanage to lower class children of the same genetic background to demonstrate that improved child care increased growth rates in high-altitude Bolivian children. In still a different type of intra-population comparison Baker (1969) divided individuals in a high-altitude Peruvian community into two groups in order to demonstrate that the more acculturated individual showed more age increase in systemic blood pressure than the traditional native.

## The biology of high-altitude peoples

### Altitude contrasts

The most popular research design was probably the traditional one of contrasting native populations at different altitudes. In several of these studies populations at various altitudes were used to detect the relative effect while in others only very high-altitude populations were compared to near sea level ones of similar genetic history.

The altitude gradient approach was used to study altitude effects on a number of fitness indicators. Scientists in the USSR used this approach to study both oxygen transport (see Chapter 10) and the growth of children (Miklashevskaya, Solovyeva & Godina, 1972, 1974). Cruz-Coke, Cristoffanini, Aspillaga & Biancani (1966) used the altitude gradient in northern Chile to examine the effects of altitude on both genetic structure and demography. Dutt (1976b) also utilized data on populations at three altitude levels to examine the possible relationship between altitude and fertility.

In Peru a number of gradient designed studies were undertaken. Hoff *et al.* (1972) by studying the growth of Quechua children and adults at several altitudes were able to reconfirm that the large chests of the Quechua were basically genetic in origin and little affected by altitude. Garruto (1976) by the study of Quechua at several altitudes found that their hematology was apparently much less affected by altitude than that of lowlanders. One of the smallest significant altitude gradients was demonstrated by Frisancho & Baker (1970) who showed that the difference between residence at 4,000 and 4,500 m appeared to affect general body growth rates.

While several attempts to demonstrate altitude effects were made with comparisons of high- and low-altitude natives (Boyce, Haight, Rimmer & Harrison, 1974; Ruffie, Larrouy & Vergnes, n.d.) the most elaborate design was the one used by Haas (1976). He developed a design containing multiple intra-population comparisons of social class, urban versus rural residence, and ethnic background, all contrasted by altitude to demonstrate that in southern Peru altitude is the major factor affecting infant growth. His ideal design for sampling is presented in Fig. 1.1. In fact he was able to fill only eight of the twelve sample cells projected but these proved adequate for the demonstration.

### Lowland migrants to high altitude

The use of migrant–native comparisons proved as popular as Harrison projected, although the entire design he recommended was never completely fulfilled. Of the migrant design models the most common one was a comparison of individuals brought to altitude for short times with

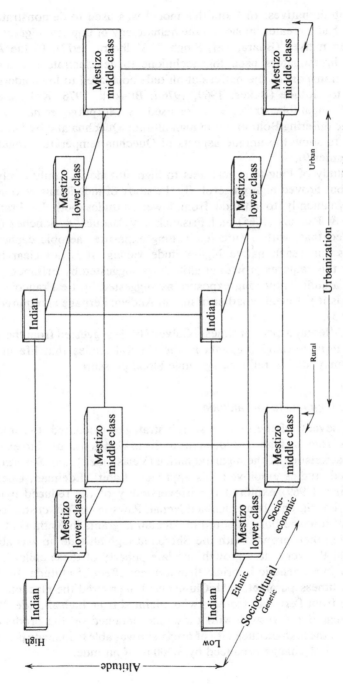

Fig. 1.1. Sampling design structure for analyzing the effects of altitude and other variables on infant growth and development in southern Peru.

9

high-altitude natives. In India this model was used to demonstrate that altitude had a greater impact on the hematology of upward migrants than it did on natives (Bharadwaj, Singh & Malhotra, 1972). In the Andes several investigators used this technique to demonstrate the superior oxygen transport of the natives at altitude compared to lowlanders who moved to altitude (Baker, 1969, 1976a; Buskirk, 1976; Kollias *et al.*, 1974). Comparable strategies were used by Morpurgo *et al.* (1970) to show the differing Bohr effect in high-altitude Quechua and by US investigators to show the unique aspects of Quechua temperature regulation (see Chapter 9).

The study of long-term migrants to high altitude was only rarely employed but proved highly useful. By the study of individuals who moved semi-permanently to altitude from lower altitudes both Le Francois, Gautier & Pasquis (1965) and Frisancho, Velásquez & Sanchez (1973) concluded that, with a sufficiently long exposure, aerobic capacity in migrants can reach native high-altitude values. It is not clear to me whether this requires growth at altitude as suggested by Frisancho *et al.* or only a sufficiently long exposure as suggested by Le Francois *et al.* Neither is it yet clear whether or not an Andean heritage is required (see Chapter 5).

On a different aspect of fitness, Galvez (1966) suggested from the study of long-term lowland migrants in the Central Andes that life at high altitude may lead to reduced systemic blood pressure.

### Highland migrants to low altitude

As with several of the other research strategies discussed, the study of migrants from the high-altitude areas to the lowlands was used to examine the characteristics of the highland native (Velásquez, 1966). Several IBP-sponsored studies employed this approach. Lahiri, Edelman, Cherniak & Fishman (1969) explored the irreversibility of the reduced hypoxic drive in the migrants from highland Nepal. Pawson (1976b, c) studied the growth of lowland-born children of Tibetan migrants in Kathmandu. By contrasting their growth with the Sherpa at high altitude he was able to show that the very slow growth and late puberty of these children was probably more genetic in origin than a direct effect of altitude. In still a different fitness parameter, McClung (1969) measured the placentas and newborn from Peruvian mothers who migrated from high altitude. When she compared the results with those she obtained on the products of mothers in the high-altitude city of origin she was able to indicate precisely the amount of change produced by 3,300 m of altitude.

*Complex migration designs*

The research design which came closest to Harrison's original proposal was the one carried out by Harrison himself, Clegg and others in Ethiopia (Clegg, Pawson, Ashton & Flinn, 1972; Harrison *et al.*, 1969). In these studies they were able in a number of instances to compare not only natives but migrants to and from communities located at approximately 1,500 and 3,000 m. The altitude differential was not as great as one would like for detecting some of the more subtle effects of altitude and financial restrictions limited both sample size and the variety of fitness parameters studied. Nevertheless, as the analysis in this book demonstrates, the design proved highly valuable.

In a different migrant design Baker and others studied migrants from high altitude to the coast of Peru, migrants from low altitude and sedentes, all of whom had a reasonably similar genetic history (Baker, 1976b). This study was undertaken late in IBP and many of the results are yet to be published. Even so this design has also proved useful in demonstrating more precisely how altitude affects such variables as fertility (Abelson, Baker & Baker, 1974), general health (Dutt & Baker, 1977) and adolescent growth (Beall, Baker, Baker & Haas, 1977).

**Book structure**

Given the broad variety of topics covered by IBP high-altitude studies and the diversity of research designs implemented, it was felt that the most useful synthesis which could be attempted was one which adhered to the population fitness parameters outlined at the beginning of IBP. This was made possible in part by the fact that most investigators adhered to the measurement techniques suggested in an early IBP manual on methods (Weiner & Lourie, 1969). The structure of the topics deviates somewhat from the original structure for several reasons. First, some areas of knowledge have been more thoroughly explored than others and second, despite attempts on the part of all participants, the language barriers have sometimes been difficult. It was, therefore, felt that a separate chapter on the studies in the USSR was justified. Finally, the subject of human population biology has itself been somewhat better defined by the existence of IBP, leading to a greater sophistication in structure (Baker, 1976c; MAB, 1971, 1973). While the following chapters discuss in some detail the results of the IBP-sponsored studies no attempt has been made to limit the topical review to these results and the authors have attempted to summarize in the most complete fashion possible all the available information.

# The biology of high-altitude peoples

## References

Abelson, A. E., Baker, T. S. & Baker, P. T. (1974). Altitude, migration and fertility in the Andes. *Social Biology*, **21**, 12–27.

Andersen, K. Lange. (1973a). The effect of altitude variation on the physical performance capacity of Ethiopian men. I. Maximal oxygen uptake and some related circulatory and respiratory functions in relation to aging. In *Physical Fitness*, ed. V. Seliger, pp. 13–33. Prague: Universita Karlova.

Andersen, K. Lange. (1973b). The effect of altitude variation on the physical performance capacity of Ethiopian men. II. Development of physical performance capacity during adolescence. In *Physical Fitness*, ed. V. Seliger, pp. 34–46. Prague: Universita Karlova.

Andersen, K. Lange. (1973c). The effect of altitude variation on the physical performance capacity of Ethiopian men. III. Vital capacity and forced expiratory volume of adult men in relation to age, height and maximal oxygen uptake. In *Physical Fitness*, ed. V. Seliger, pp. 47–54. Prague: Universita Karlova.

Baker, P. T. (1969). Human adaptation to high altitude. *Science, Washington*, **163**, 1149–56.

Baker, P. T. (1976a). Work performance of highland natives. In *Man in the Andes: A Multidisciplinary Study of High Altitude Quechua*, ed. P. T. Baker & M. A. Little, pp. 300–14. Stroudsburg, Pa.: Dowden, Hutchinson & Ross.

Baker, P. T. (1976b). Research strategies in population biology and environmental stress. In *The Measures of Man: Methodologies in Biological Anthropology*, ed. E. Giles & J. Friedlaender, pp. 230–59. Cambridge, Mass.: Peabody Museum Press.

Baker, P. T. (ed.). (1976c). *Human Population Problems in the Biosphere: Some Research Strategies and Designs*. MAB Technical Note No. 3. Paris, France: UNESCO.

Baker, P. T. & Weiner, J. S. (ed.). (1966). *The Biology of Human Adaptability*. Oxford: The Clarendon Press.

Beall, C. M., Baker, P. T., Baker, T. S. & Haas, J. D. (1977). The effects of high altitude on adolescent growth in southern Peru. *Human Biology*, **49**, 109–24.

Bharadwaj, N., Singh, A. P. & Malhotra, M. S. (1972). Body composition of the high altitude natives of Ladakh. A comparison with sea level residents. *Human Biology*, **45**, 423–34.

Bouloux, C. J. (1968). *Contribution à l'Etude Biologique des Phénomènes Pubertaires en très haute Altitude (La Paz)*. Toulouse, France: Centre Regional de Transfusion Sanguine et d'Hematologie.

Boyce, A. J., Haight, J. S., Rimmer, D. B. & Harrison, G. A. (1974). Respiratory function in Peruvian Quechua Indians. *Annals of Human Biology*, **1**, 137–48.

Buskirk, E. R. (1976). Work performance of highland natives. In *Man in the Andes: A Multidisciplinary Study of High Altitude Quechua*, ed. P. T. Baker & M. A. Little, pp. 283–99. Stroudsburg, Pa.: Dowden, Hutchinson & Ross.

Clegg, E. J., Pawson, I. G., Ashton, E. H. & Flinn, R. M. (1972). The growth of children at different altitudes in Ethiopia. *Philosophical Transactions of the Royal Society of London, Series B*, **264**, 403–37.

Cruz-Coke, R., Cristoffanini, A. P., Aspillaga, M. & Biancani, T. (1966). Evolutionary forces in human populations in an environmental gradient in Arica, Chile. *Human Biology*, **38**, 421–38.

Donoso, H., Apud, E., Sanudo, M. & Santalaya, R. (1971). Capacidad aerobica como indice de adecuidad fisica en muestras de poblaciones (urbanas y nativas de la altura) y un grupo de atletas de seleccion. *Revista Medica de Chile*, **99**, 719–31.

Dutt, J. S. (1976a). Population movement and gene flow. In *Man in the Andes: A Multidisciplinary Study of High Altitude Quechua*, ed. P. T. Baker & M. A. Little, pp. 115–27. Stroudsburg, Pa.: Dowden, Hutchinson & Ross.

Dutt, J. S. (1976b). Altitude and Fertility: The Bolivian Case. Doctoral Dissertation in Anthropology, The Pennsylvania State University.

Dutt, J. S. & Baker, P. T. (1977). Environment, migration and health in Southern Peru. *Social Science & Medicine* (in press).

Frisancho, A. R. & Baker, P. T. (1970). Altitude and growth: A study of the patterns of physical growth of a high altitude Peruvian Quechua population. *American Journal of Physical Anthropology*, **32**, 279–92.

Frisancho, A. R., Velásquez, T. & Sanchez, J. (1973). Influence of developmental adaptation on lung function at high altitude. *Human Biology*, **45**, 583–94.

Galvez, J. (1966). Presion arterial en el sujeto de nivel del mar con residencia prolongada en las grandes alturas. *Archivos del Instituto de Biologia Andina*, **1**, 238–43.

Garruto, R. M. (1976). Hematology. In *Man in the Andes: A Multidisciplinary Study of High Altitude Quechua*, ed. P. T. Baker & M. A. Little, pp. 261–82. Stroudsburg, Pa.: Dowden, Hutchinson & Ross.

Haas, J. D. (1976). Prenatal and infant growth and development. In *Man in the Andes: A Multidisciplinary Study of High Altitude Quechua*, ed. P. T. Baker & M. A. Little, pp. 161–79. Stroudsburg, Pa.: Dowden, Hutchinson & Ross.

Harrison, G. A. (1966). Human adaptability with reference to the IBP proposals for high altitude research. In *The Biology of Human Adaptability*, ed. P. T. Baker & J. S. Weiner, pp. 509–20. Oxford: The Clarendon Press.

Harrison, G. A., Küchemann, C. F., Moore, M. A. S., Boyce, A. J., Baju, T., Mourant, A. E., Godber, M. S., Glasgow, B. G., Kopeć, A. C., Tills, D. & Clegg, E. J. (1969). The effects of altitudinal variation in Ethiopian populations. *Philosophical Transactions of the Royal Society of London, Series B*, **256**, 147–82.

Hoff, C. J., Baker, P. T., Haas, J. D., Spector, R. & Garruto, R. (1972). Variaciones altitudinales en el crecimiento y desarollo fisico del Quechua Peruano. *Revista del Instituto Boliviano de Biologia Altura*, **4**, 5–20.

Kollias, J., Buskirk, E. R., Akers, R. F., Prokop, E. K., Baker, P. T. & Picón-Reátegui, E. (1968). Work capacity of long time residents and newcomers to altitude. *Journal of Applied Physiology*, **24**, 792–9.

Lahiri, S., Edelman, N. H., Cherniak, N. S. & Fishman, A. P. (1969). Blunted hypoxic drive to ventilation in subjects with life-long hypoxemia. *Federation Proceedings*, **28**, 289.

Lang, S. D. R. & Lang, A. (1971). The Kunde hospital and a demographic survey of the upper Khumbu, Nepal. *New Zealand Medical Journal*, **74**, 1–8.

Le Francois, R., Gautier, H. & Pasquis, P. (1965). Ventilatory oxygen drive in high altitude natives. Proceedings of the 23rd International Congress of Physiological Sciences. Tokyo, No. 434.

MAB (1971). *International Co-ordinating Council of the Programme on Man and the Biosphere. First Session, Paris, 9–19 November 1971.* MAB Report No. 1. Paris: UNESCO.

MAB (1973). *Working Group on Project 6: Impact of Human Activities on Mountain and Tundra Ecosystems, Lillehammer, 20–23 November 1973.* MAB Report No. 14. Paris: UNESCO.

Mazess, R. B. (1969). Exercise performance of Indian and white high altitude residents. *Human Biology*, **41**, 494–518.

McClung, J. (1969). *Effects of High Altitude on Human Birth*. Cambridge, Mass.: Harvard University Press.

Miklashevskaya, N. N., Solovyeva, V. S. & Godina, E. Z. (1972). Growth and development in high altitude regions of S. Kirghizia, U.S.S.R. *Vopros Anthropologii*, **40**, 71–91.

Miklashevskaya, N. N., Solovyeva, V. S. & Godina, E. Z. (1974). New anthropological investigations in the Pamirs. In *Racial–Genetic Processes in Ethnic History, A Collection in Memory of G. F. Debets*, pp. 188–200. Moscow: Science Publishers.

Morpurgo, G., Battaglia, P., Bernini, L., Paolucci, A. M. & Modiano, G. (1970). Higher Bohr effect in Indian natives of Peruvian highlands as compared with Europeans. *Nature, London*, **227**, 387–8.

PAHO (1967). Report on WHO/PAHO/IBP Meeting of Investigators on Population Biology of Altitude, Pan American Health Organization, Washington, DC (unpublished).

Pawson, I. G. (1977a). Growth and development in high altitude populations: a review in Ethiopian, Peruvian, and Nepalese studies. *Proceedings of the Royal Society of London, Series B*, **194**, 83–98.

Pawson, I. G. (1977b). Growth characteristics of populations of Tibetan origin in Nepal. *American Journal of Physical Anthropology* (in press).

Pugh, L. G. C. (1966). A programme for physiological studies of high altitude peoples. In *The Biology of Human Adaptability*, ed. P. T. Baker & J. S. Weiner, pp. 521–32. Oxford: The Clarendon Press.

Ruffie, J., Larrouy, G. & Vergnes, H. (n.d.). Hematologie comparée des populations amerindiennes de Bolivie et phénomènes adaptifs. In *Definition et Analyse Biologique des Populations Amerindiennes: Etude de leur Environment*, ed. J. C. Quilici, Chapter 15. Recherche cooperative sur Programme No. 87. Toulouse, France: Centre d'Hemotypolgie, Centre National de la Recherche Scientifique.

Thomas, R. B. (1973). *Human Adaptation to a High Andean Energy Flow System. Occasional Papers in Anthropology No. 7*. The Pennsylvania State University.

Thomas, R. B. (1976). Energy flow at high altitude. In *Man in the Andes: A Multidisciplinary Study of High Altitude Quechua*, ed. P. T. Baker & M. A. Little, pp. 379–403. Stroudsburg, Pa.: Dowden, Hutchinson & Ross.

Velásquez, T. (1966). Acquired acclimatization to sea level. In *Life at High Altitude, Scientific Publication No. 140*, pp. 58–63. Washington, DC: Pan American Health Organization.

Ward, M. (1975). *Mountain Medicine: A Clinical Study of Cold and High Altitude*. London: Crosby Lockwood Staples.

Weiner, J. S. (1966). Major problems in human population biology. In *The Biology of Human Adaptability*, ed. P. T. Baker & J. S. Weiner, pp. 1–43. Oxford: The Clarendon Press.

Weiner, J. S. (1969). *A Guide to the Human Adaptability Proposals*, 2nd edn, *IBP Handbook No. 1*. Oxford: Blackwell Scientific Publishers.

Weiner, J. S. & Lourie, J. A. (1969). *Human Biology: A Guide to Field Methods. IBP Handbook No. 9*. Oxford: Blackwell Scientific Publishers.

Weitz, C. A. (1973). The Effects of Aging and Habitual Activity Pattern on

Exercise Performance among a High Altitude Nepalese Population. Doctoral Dissertation in Anthropology, The Pennsylvania State University.

Worthington, E. B. (ed.). (1975). *The Evolution of IBP.* London: Cambridge University Press.

Wulff, L. Y., Braden, I. A., Shillito, F. H. & Tomashefski, J. F. (1968). *Physiological factors relating to terrestrial altitudes: a bibliography. The Ohio State University Libraries Publications No. 3.* Columbus, Ohio: Ohio State University Press.

# 2. The high-altitude areas of the world and their cultures

I. G. PAWSON & CORNEILLE JEST

Of the world's major ecosystems, mountainous areas offer a variety of opportunities to examine the ways in which the structure of human populations is conditioned by the physical environment. One of the reasons for this potential is the diversity of high-altitude environments that exists in different parts of the world. Although reduced atmospheric pressure is common to all highland areas, surface features, geographical location and climatic variables all contribute to the individual characteristics of a particular zone (Clegg, Harrison & Baker, 1970). As a result, the environmental variables that interact to produce a given set of conditions in one area may be different, either in a functional or quantitative sense, from those that exist elsewhere. The main purpose of this chapter is to describe this physical diversity of the world's major highland areas, and to provide a descriptive background for the populations whose biocultural characteristics are described in subsequent chapters.

To date, most publications dealing with the biological or cultural characteristics of highland peoples have contained descriptions of the physical and social environment of study populations which would render simple repetition or précis of this material superfluous. For this reason, it seemed more appropriate to present a synthesis of this material in a comparative framework so that differences and similarities between highland areas could be readily observed. With this aim in view, the structure of this chapter will consist of sections dealing with the geographical location and principal physical characteristics of highland regions. The climatic conditions in these zones are discussed in Chapter 9. Following these sections, regional summaries of the areas in which IBP researchers have worked will present descriptions of the social and cultural milieu of each population. These summaries will include historical data and brief demographic profiles to set the stage for the more detailed descriptions that follow.

## The 2,500-m line as a high-altitude 'delimiter'

In deciding the physical limits of 'high' and 'low' altitude, a number of variables must be considered. The transition from lower, through moderate, to high altitude cannot be described as a series of quantum changes between geographical zones since the changes in environmental

17

conditions are in fact a continuum. In addition, variables that bear upon the body's ability to withstand these conditions must also be taken into account in deciding a high-altitude delimiter line. Factors such as physical fitness, activity patterns and thermal tolerance are conspicuous modifiers of an individual's ability to withstand the stress of living in conditions of reduced barometric pressure (Thomas, 1975). In addition, changes in some physiological parameters, such as those associated with night vision, become apparent at altitudes of only 1,500 m (McFarland, 1969) while the sedentary individual may not experience respiratory stress until he is at about twice this altitude. However, most individuals experience some high-altitude effects at altitudes of 2,500 m and over (Clegg *et al.*, 1970) and for the purposes of describing the physiological effects of altitude, the 2,500-m contour line may be taken as a general divider between 'high' areas of land as opposed to 'low'. What follows is a general description of these mountainous environments.

## Geographical location of the world's major highland zones

The location of the principal areas of land encompassed by the 2,500-m contour is shown in Fig. 2.1. The regions that support sizable and permanent human populations are the Andes mountains of South America, the highlands of northern Ethiopia and the area of the Tibetan plateau. With the exception of the Tibetan plateau, these areas are usually separated by well-defined climatic and geographical boundaries from the lowlands. In South America, the highland zone which extends from Colombia in the north to central Chile in the south, is mostly flanked by an area of arid desert which extends the length of its western side and by a deeply eroded escarpment on the side which adjoins the Amazon basin (Bowman, 1916; Troll, 1968). Much of the biosocial history of the Andean region has been shaped by its isolation from the lowlands.

Although more than 25% of the total area of Ethiopia is situated above 1,800 m, it is not easy to define a circumscribed highland zone as it is in the Andes. Nevertheless, the highland plateau to the north has long been a focus for human settlement and much of the extensive history of Ethiopia centers in this zone (Simoons, 1960; Sellassie, 1972). Both geographical and cultural boundaries separate this area from the southern highlands, and from Ethiopia's neighbor, the Sudan (Hess, 1970). Our description of the physical characteristics of Ethiopia will therefore emphasize this area of the country.

In terms of absolute size and elevation, the Tibetan plateau probably represents the world's largest and highest land mass. Tibet itself consists of a single high plateau with an average elevation in excess of 4,500 m. To the north, this plateau is bordered by the Kunlun mountain mass which

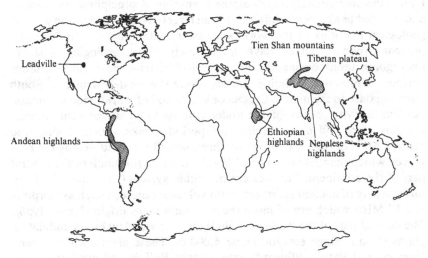

Fig. 2.1. Areas of the world above 2,500 m.

extends from longitude 77° to 93° E. The western and southern borders of the central Tibetan plateau are fringed by the Himalayan range which contains the world's highest peak, Mount Everest (8,848 m). Despite the relative isolation of the Tibetan plateau, this region and the mountainous masses that surround it are part of an extensive interconnecting system of valleys and plateaus that extend eastwards to Tien Shan and the Kirghiz highlands, and to the Siberian uplands to the north (Gansser, 1964). The physiographical and cultural diversity of this region makes it difficult to present a balanced description of each area within the context of the present work. Partly because access to much of this region has long been restricted, there are no up-to-date ethnographic summaries upon which cultural comparisons with other highland areas can be based. With these considerations in mind, our attention in this chapter will be confined to the highlands of Nepal, and the populations of Tibetan origin that live immediately adjacent to the Sino–Nepalese border.

*Geological characteristics and history of the world's highland zones*

Although each of the three principal high-altitude areas of the world could be described as 'mountainous', their individual topographies differ to such an extent that elevation above sea level remains the single common denominator in the geological characteristics of these regions. For example, the *altiplano* of Peru and the highlands of Ethiopia seldom convey the impression of high mountain scenery while inhabited areas of the

19

Himalayas immediately recall alpine scenarios of precipitous mountains and isolated peaks. Such differences can largely be ascribed to the diverse geological histories of the world's major mountain masses. In Peru, the present Andean ranges have a relatively short geological history, undergoing uplift during the Pliocene and Pleistocene periods. Earlier periods of mountain-building activity along the western fringe of South America date to the middle Cretaceous, a period of world-wide mountain-building activity. Many of the underlying rocks of the present Andean systems were laid down during this period (Jenks, 1956). Today, the central Andean region consists of three broadly defined geological provinces which parallel the coast of South America for much of the central part of the continent. The western mountain system (*Cordillera Occidental*) consists of ancient sediments with volcanic outcrops such as Corpuna and El Misti which are of more recent, Quaternary origin (Jenks, 1956). The central part of the Andean chain is characterized by broad undulating plains at an average elevation near 4,000 m. These plains, which extend from central Peru southwards into eastern Bolivia and northern Chile, are collectively known as the *altiplano*. The underlying structure of the *altiplano* consists of accumulations of Cretaceous deposits that fill much of the Titicaca trough. However, superficial deposits of alluvial and lacustrine material, associated with a large lake that covered much of the *altiplano* during the early Pleistocene, and the outwash from Pleistocene glaciation, have leveled the *altiplano* to a considerable extent (Jenks, 1956).

On the western slopes of the Andes, a deeply eroded escarpment separates the high-level plains of the *altiplano* from the Amazon basin. Known as the *Cordillera Oriental*, this region is composed predominantly of ancient faulted rock formations which have become intersected by the action of alluvial and glacial erosions (Jenks, 1956).

The topography of each of these geological horizons is continually being modified by the action of glacial and tectonic activity and, recently, by man himself (Thomas & Winterhalder, 1976).

In contrast to the relatively well-defined geological horizons of Peru, the topography of Ethiopia presents a more confused picture of numerous plateaus separated by deeply incised valley systems that are often extremely difficult to traverse (Buxton, 1970). Ethiopia rests on the Pre-Cambrian crystalline block that underlies much of Africa. The country is intersected by a number of rift-valley systems that establish the connection between the African Rift Valley in the south and the syncline of the Red Sea to the north (King, 1962). This rift system is responsible for the division of the country into three more-or-less well-defined relief regions, the western highlands, the eastern highlands, and a third zone that comprises the Rift Valley itself and the western lowlands (Hess, 1970). How-

20

ever, the topography of the western highlands differs considerably in a north–south direction. Almost without exception, the northern highlands achieve greater mean elevations than those in the south. Lying at altitudes of between 2,400 and 3,700 m, this region has been termed the 'Amhara Highlands' by Büdel (1954). It consists of basalt plateau blocks of varying size from which rise volcanic outcrops such as Ras Dashan (4,620 m), the highest peak in Ethiopia. At the center of the Amhara highlands lies the basin of Lake Tana, from which emanates the Blue Nile river gorge that in places cuts as much as 2,100 m below the level of the plateau. The valleys in this region present considerable obstacles when traveling from north to south (Simoons, 1960). A road built during the Italian occupation of Ethiopia runs southwards from Eritrea across several large river valleys towards Lake Tana. Along this route, altitudinal differences of more than 1,000 m in just a few kilometers are not uncommon. The eastern highlands are less difficult of access from the north than those in the western part of the country. The Simien escarpment which presents a formidable barrier to access from Eritrea and the north has no counterpart in this region.

Division of the eastern highlands into plateau blocks has not occurred so extensively in the west resulting in conditions that are favorable for the development of agriculture. The topography of the lowland regions of Ethiopia comprising the rift zone and the southern highlands consists of numerous lakes surrounded by extensive lacustrine and fluvian deposits (Hess, 1970). Superimposed on these plains are a number of volcanoes of fairly recent origin, including the 3,000-m-high Zakwula near the capital, Addis Ababa. In the Rift Valley region, these volcanoes are frequently associated with hot springs and other indications of current geological activity.

The topographical complexity of the Himalayan region makes it difficult to present balanced descriptions of the main geological zones in this part of the world. In the northwestern region, for example, the junction of the westward extensions of the Kunlun mountain mass with the Himalayan ranges of Kashmir presents a confused array of mountain systems and valleys. Yet despite the geological complexity of the Himalayas as a whole, it is possible to differentiate within our area of primary interest, Nepal, three parallel zones. These are the main Himalayan range, immediately adjacent to the Sino–Nepalese border and containing many peaks over 7,000 m in altitude, the middle ranges, or lesser Himalayas, which form a network of ranges in the north-central part of the country, and the outer Himalayas, consisting mainly of foothills rising from the Indian plains, with average elevations of 1,000 to 1,500 m (Gansser, 1964).

During the Eocene period, a large sea called the Tethys covered the area now known as Tibet and the southern geosyncline which coincided with

21

this ocean covered the area now occupied by the Himalayan ranges listed above. Although these mountains contain all the principal geological horizons from PreCambrian to present, the principal elevation of the central Himalayan ranges took place in Eocene and Oligocene periods. The successive orogenies that formed the Himalayas were caused by the northward movement of the Indian subcontinent that continued through the Miocene epoch to post-Pleistocene times. Continued geological activity in the area of Assam bears witness to the fact that this process of mountain-building activity is still underway. Because of the steep gradients encountered almost everywhere in the central Himalayas, the erosive power of rivers and glacial runoffs is strong. As a result, the ranges are frequently intersected by deep transverse gorges that rarely permit the development of agriculture except at their peripheries. The lack of sizable areas of arable plateau makes the Himalayas unique in the world's major mountain systems. The windswept Tibetan plateau, at an average elevation of about 4,570 m, is mainly unsuitable for agriculture except in the southern and southeastern part of the country (Gansser, 1964; Wadia & West, 1964).

## The Ethiopian highlands

### History of Ethiopia

Of the African states, Ethiopia has an extraordinarily rich history that belies its recent position as one of the poorer countries of the world. With the exception of Egypt, no other country in Africa can trace its history back to antiquity and draw on such a rich cultural heritage that includes the period of the Aksumite Empire and the succession of kings that ruled from the imperial city of Gondar. The northern highlands, which will be our focus here, have long been the population center of the country (Simoons, 1960). This area is inhabited by the largest of Ethiopia's many population groups, the Amharas and the Tigreans. Together, these two groups represent the descendants of peoples who migrated from southern Arabia before 1,000 B.C. (Sellassie, 1972). The history of Ethiopia has been influenced in part by its geographical position at the junction of Asian and African continents. Because of this location, it has served as a focal point for the introduction of Asian influences into Africa and as a center for exchange for the countries that surround the Red Sea (Hess, 1970). Yet, despite its commanding position on the Horn of Africa, Ethiopian culture has remained remarkably stable (Ullendorf, 1973), even in the face of constant diffusion of external influences through its borders. Before he was deposed in a military coup in 1975, the last reigning emperor of Ethiopia, Haile Sellassie, symbolized this permanency as the world's longest ruling monarch.

22

The earliest period of recorded history in Ethiopia consists of biblical traditions surrounding the visit of the Queen of Sheba to the court of King Solomon, as recounted in the Book of Kings. During the first millennium B.C. the indigenous inhabitants of Ethiopia became mixed with Semitic immigrants from the southern parts of the Arabian peninsula, center of the Sabaean culture. These peoples brought with them many aspects of Sabaean culture including astral religion, sacred royalty, agriculture and architectural styles (Sellassie, 1972). In addition, the Sabaean language was adapted to form a blend of native and Arabian tongues that has been preserved to this day by the Ethiopian Church. Towards the end of the first millennium B.C. the influences of hellenistic Egypt were felt in Ethiopia in the form of trading posts that were established along the coasts and to which came Ptolemaic fleets for the purpose of transporting valuable products such as spices, incense and myrrh (Pankhurst, 1961). During the first century B.C., the Aksumite Empire developed in the northern highlands of Ethiopia. At its greatest extent, this empire held sway over a large part of eastern Africa and Arabia. The origins of Aksumite culture may be traced to new immigrants from the Arabian peninsula and beyond who intermingled with the descendants of the earlier Sabaeans. The center of this civilization was Aksum just north of the Amhara highlands. Archeological remains of this region attest to the importance of Aksum as a focal point for much of the commercial and religious life of that part of the ancient world. The Aksumite emperors developed commerce along both coasts of the Red Sea and beyond and instituted a system of coinage that symbolized the financial prosperity of the period (Buxton, 1970). The golden age of Aksum is associated with the reign of Ezana in the fourth century A.D. It was at this time that the empire reached its greatest extent and exercised political influence throughout the eastern world (Pankhurst, 1961). Ezana adopted Christianity as the official religion of his empire and instituted a modified form of the ancient Sabaean alphabet as Ethiopia's earliest definitive script. After a succession of emperors, a series of invasions by Persia heralded the decline of Aksum's influence (Sellassie, 1972). In the seventh and eighth centuries the spread of Christianity throughout the world brought about substantial declines in the value of Ethiopia's principal exports, myrrh and incense, which had been used by numerous pagan cults for the purposes of embalming. As a result of declining revenues, Aksum ceased to issue coinage although the activities of Aksumite trading ships on the Red Sea continued till the twelfth century (Sellassie, 1972).

During the early Middle Ages, the capital of the Aksumite Empire moved southwards to the present site of Adana. A series of incursions by Judaic peoples to the south contributed to the declining influence of Aksum's culture.

23

## The biology of high-altitude peoples

Towards the end of the tenth century A.D., a group of princes, known as the Zague, from central Ethiopia and not associated with the Aksumite emperors, usurped control of the empire and ruled for about 300 years during which time they fostered relations with Egypt and re-established Christianity as the country's official religion (Sellassie, 1972). The most famous of the Zague emperors, Lalibela, constructed a series of churches carved out of solid rock which can be seen today at the town that bears his name (Buxton, 1970).

In the late thirteenth century, the Zague dynasty was replaced by adherents to the ancient Aksumite traditions who represented descendants of the original emperors of Aksum. These rulers did much to restore the economic prosperity and spiritual leadership of the sovereign that had become eclipsed during the Zague dynasty (Sellassie, 1972).

The modern period of Ethiopian history dates from the early sixteenth century which saw the 'discovery' of Ethiopia by Portugal. The years that followed represent an unsettled period in the history of the country, with conflicts between Turks and Christians that culminated in the Muslim invasions that lasted from 1523 to 1543 (Buxton, 1970). Peace was not established throughout the country until the latter part of the fifteenth century or early part of the sixteenth. A national reaction to the persistent efforts of Portugese Catholic missionaries to convert the country to their faith resulted in the expulsion of most Europeans and the start of a period of Ethiopian history that resembles the European Renaissance (Pankhurst, 1961). The founding of the city of Gondar in the center of the Amhara highlands served as a focus for much of the political and religious life of the country in this period of transition. In the years that followed, Gondar became the second largest city in Africa (after Cairo) with a population of more than 100,000 (Buxton, 1970). Despite several crises including foreign intervention and internal struggles between rival groups of Christian (Amhara–Tigrean) and pagan (Galla) princes, the position of Gondar as capital of Ethiopia continued uninterrupted until the mid-nineteenth century when Addis Ababa, the present capital, was founded by the emperor Menelik II. Except for a brief period of foreign domination during the Italian occupation of 1935 to 1941, Ethiopia has retained its independence throughout the years of colonialism. The most recent emperor of Ethiopia, Haile Sellassie, represented a lineage that survived almost unbroken from the earliest period of Ethiopian history. Shortly before his death in 1975, Haile Sellassie was deposed in a military coup in Ethiopia's first departure from a constitutional monarchy.

Because there are few reliable census data from Ethiopia, the demographic characteristics of the population are only imperfectly known. (More detailed demographic profiles for Ethiopian IBP study populations will be presented in a later section of this chapter.) Although no national census has ever been undertaken, the United Nations (1975) estimated that in 1974 there were approximately 27.2 million inhabitants, making the country one of the most populous states in tropical Africa. Apart from the principal urban centers of Amhara and Addis Ababa, the most heavily populated regions are the southern provinces where hoe cultivation of *ensete*, a type of banana belonging to the Musaceae family, permits densities of up to 200 persons per square kilometer (Schaller, 1972). In areas of plow cultivation, such as the northern, or Amhara, highlands, population density is typically lower. In the province of Begemdir, for example, the estimated density in 1967 was eighteen persons per square kilometer (Schaller, 1972). Within a particular zone, it is possible to differentiate a vertical as well as a horizontal displacement of the population.

In the northern agricultural provinces, as well as in the areas of hoe cultivation to the south, the moderate altitude of the *woina dega* (hill country) at altitudes of 2,000–3,000 m seems to be more favored than the *dega* (highlands) (Hess, 1970). In the Simien mountains, the upper limit of human habitation is about 3,800 m. The estimates of Staszewski (1957), although made in 1940, provide a general indication of the vertical distribution of Ethiopia's population. Excluding the northern province of Eritrea, approximately 50% of the total population live at elevations in excess of 2,000 m and only 9.3% live in the *k'olla*, which are those lands that border on the hill country, at altitudes of up to 1,000 m.

For the purposes of classification, the partitioning of Ethiopia's population into racial groups is of little value when one considers the extensive degree of ethnic diversity seen throughout the country. Linguistic identification allows a simpler method of description for the overall population, but with approximately 70 languages and 200 dialects (Hess, 1970) only general distinctions are possible. The most important group of languages from an historical, political and cultural point of view are those of Semitic origin (Hess, 1970). Within this group of languages, two major subdivisions can be discerned. The first group of northern Semitic languages includes Ge'ez, the Ethiopian liturgical and literary language and Tigrinya (Tigrean), spoken by the inhabitants of Eritrea. The other group of Semitic languages includes a number of tongues of which Amarinya (Amharic) is the most prominent (Hess, 1970). Today, Amharic is the official language of Ethiopia and is spoken as a first language by approximately six million people in the provinces of Begemdir and Simien, Gojam, Welo and Shewa.

## The biology of high-altitude peoples

The term 'Amharic', which is often applied to the inhabitants of these provinces, refers more to a linguistic identification than to biological relationships. Indeed, 'Amharic' has come to be synonymous with 'Ethiopia' and reflects a focal point for the assimilation of widely diverse cultural groups into a more homogeneous Ethiopian society.

The peoples of the northern Ethiopian plateau, who have been the focus for studies conducted under IBP, are usually referred to as the Amharas and the Tigreans, although, as we have seen, linguistic groupings such as these may be of little use in determining biological relationships between peoples. However, the two are distinguished by a number of physical characteristics from their neighbors in the Sudan.

### Economic and political structure

Despite a continuing shift of the population towards the major urban centers, the mainstay of the Ethiopian economy continues to be agriculture. As we indicated earlier in this chapter, two types of agricultural practice may be distinguished. The first is plow cultivation which is centered mainly in the northern part of the country in the highlands inhabited by Amharic- and Tigrean-speaking peoples (US Army, 1966). The second group of agriculturalists lives in the southern part of the country. Here the economy is based on the hoe cultivation of *ensete* (US Army, 1966). In the north, a variety of cereals, legumes and oil seeds are cultivated, but by far the most important crop is *teff* (*Eragrostis abysinnica*), a type of grass that produces an extremely small seed (Simoons, 1960). The seed is ground into a flour from which *injera*, an unleavened, porous bread is made. *Injera* forms the staple food of the inhabitants of the northern and central regions of Ethiopia. It is usually eaten with a stew (*wet*, pronounced 'what') which may contain meat, vegetables, or a combination of ingredients (US Army, 1966). *Teff* is grown at altitudes of up to 3,000 m although in the higher elevations it is usually cultivated with other crops such as barley or wheat. In addition, millet is grown mainly for the purpose of brewing beer. Legume crops of the northern highlands include a variety of peas, horse beans and lentils, and the oil seeds grown include linseed and *nug* (*Guizotia abysinnica*). *Teff*, however, remains the mainstay of the agriculturally based economy in almost all areas of the northern highlands (Simoons, 1960).

The principal method of cultivation is based on fields which are plowed by a hook prior to planting. Plowing, therefore, requires a substantial expenditure of labor on the part of the farmer, and for this reason, is usually done by the men. However, after the crop has been planted, the fields must be kept free from weeds, a task that is normally undertaken by women. In most areas where *teff* cultivation is practiced, stock farming

26

forms an important adjunct to the economy. In the Simien, most families keep at least one or two oxen, although these are kept primarily for use in plowing and not as a source of food (US Army, 1966).

Traditionally, political and economic power in Ethiopia was in the hands of a titled nobility whose influence was based on the ownership of land (Ullendorf, 1973). Tenants were under a series of obligations to provide goods and services that included military service. Mainly because most arable land tended to be under the control of relatively few families, there were few gradations of social class in traditional Ethiopiean society. Those that did exist consisted of the ruling landowners, an intermediate class of clergy and small landowners, who were followed in turn by the tenant farmers. In addition, certain laborers such as the potters and blacksmiths were not considered part of Ethiopian society and frequently came from outcast ethnic groups (US Army, 1966). Social mobility in traditional Ethiopia was easiest in times of war when outstanding services might be rewarded by a grant of land. Similarly, the favor of the emperor or other high government official could also place an individual in a higher social class (US Army, 1966).

In traditional Amharic society, kinship was traced through the male line although names were generally not retained for longer than a generation (US Army, 1966). Residence patterns were determined more by the economic circumstances of the couple in question than by the dictates of long-standing traditions although it was considered ideal for a newly married couple to live near or with the bridegroom's family until they had become sufficiently economically independent to move to the immediate vicinity of fields designated for their use. Marriage was based on the economic arrangements made by the prospective marriage partners' families (US Army, 1966). In Amharic and Tigrean society the divorce rate was very high, estimates ranging from 60 to 90% of all marriages. Although barrenness was one of the principal causes for divorce, the extreme youth of the bride and groom (in the Simien, brides of seven to ten years of age were not uncommon) may contribute to the economic instability of the marriage during its early years (US Army, 1966).

Today, the entire structure of Ethiopian society is in a period of transition. The ruling military council of Ethiopia is instituting a series of reforms which is bound to alter the traditional basis for political and economic influence in the country. At the time of writing it is too early to speculate what the nature of these changes might be.

## The biology of high-altitude peoples

### The Peruvian Andes

*History of Andean civilization in Peru*

A major difficulty in reconstructing the history of human civilization in the Andean region is that the pre-Hispanic peoples did not develop a written language (Willey, 1971). The Incas, for example, passed down historical records by word of mouth, a practice which was no doubt influenced by the desire of successive generations of nobles and priests whose task it was to portray Inca civilization in as favorable a light as possible. Following the conquest of Peru by Pizarro, Spanish chroniclers recorded many of these traditions, so that the historical accuracy of pre-colonial events is determined as much by the Spaniards' interpretations of these oral traditions as by the accuracy of the traditions themselves. Archeology provides a valuable adjunct to these orally transmitted records, particularly for the early stages of Andean civilization that preceded the emergence of the Inca state.

The Incas occupy a place in the history of Andean civilization even though the period of their major influence only lasted for about 100 years, from the first part of the fifteenth century until 1532, the year of the Spanish invasion. Before that time, their influence was confined to the region immediately surrounding Cusco and did not extend more than a few miles from their capital (Lanning, 1967). Despite this relatively short period of domination, the Incas established a sphere of influence which is remarkable considering the time involved. At its greatest extent, during the reign of Huayna Capac (1483–1525) the Inca Empire extended from what is now the northern border of Ecuador to the Maule River, south of Chile's capital, Santiago. In an east–west direction, the empire extended from the Pacific to the Gran Chaco of Argentina (Willey, 1971). Peoples that came under the sovereignty of the Inca emperors still inhabit the Andean region and retain many of the cultural and economic elements of the Inca rule. The cultural conservatism of Andean peoples may be partly explained by the failure of the Spanish Government to develop effective control over this region, a failure caused in part by the natural ecology of the area itself.

The earliest archeological evidence for human occupation in the Andean region has been found in a cave at Ayacucho, Peru, located at an altitude of 2,900 m. The next to lowest level in this cave has been dated to approximately 20,000 years before the present (MacNeish, 1971). Other early finds dating from 16,000 to 11,000 B.P. have been found in Peru (Lanning & Patterson, 1967; Lynch & Kennedy, 1971), central Chile (Montané, 1968), Argentina (Bird, 1938; Haynes, 1965) and Venezuela (Rouse & Cruxent, 1963). The pre-Incan civilizations consisted of a series of cultures which, until the empires of Huari and Tihuanaco in the ninth

28

century A.D., were mainly along the coast (Lanning, 1967; Mosely, 1975), although settlements existed in the highlands long before this (Willey, 1971). Archeological remains on the coast dating to the middle of the fourth millennium B.C. suggest that these peoples subsisted primarily on sea food (Sanders & Marino, 1970). Settlements were scattered and tended to occur close to patches of sparse vegetation, or *lomas*, in the coastal valleys where river drainages from the mountains provided sufficient moisture to support plant life (Lanning, 1965). During the period that extended from 2,500 to 2,000 B.C., a resource exploitation based principally on marine resources supported permanent villages (Patterson, 1971; Mosely, 1975), particularly in the region of Peru between the Casma and Huarmey rivers. Little is known about the population of the highlands at this time except for the temple of Kotosh, near Huanuco (Izumi & Sono, 1963).

Pottery was introduced at the beginning of the second millennium B.C. at various sites along Peru's central coast (Lanning, 1967; Willey, 1971), although its use spread slowly. At the same time, agriculture became more widespread though a general reliance on sea food continued as the primary means of subsistence (Lanning, 1967). Because of the extremely low rainfall in the coastal regions at all times of the year, cultivation was restricted to the few valleys that were irrigated by natural springs or rivers.

Early pottery styles show a lack of uniformity which points to the absence of widespread cultural traditions that one would expect to emerge with a centralized political and economic structure. The earliest such system to develop was associated with the Chavin civilization, whose influence extended over much of Peru except for the southern highlands and parts of the southern coastal region (Lanning, 1967). Dating of this period is uncertain although the limits of Chavin culture probably fall within the period 5,000–1,400 B.C. The characteristic motif of Chavin art is a series of feline-like representations, a style that may indicate belief in a series of cat-like deities (Rowe, 1967). Metal working and pottery traditions established during the period of Chavin influence continued in the coastal region for some time and developed by the early first millennium A.D. into the cultures of Moche on the north coast of Peru, and Nazca in the south (Lanning, 1967). Both exhibited advanced metalworking techniques including the *cire-perdu* (lost wax) method of casting and a great variety of artistic styles including polychrome decoration on pottery.

At the same time, the cultures of Pucara and Tihuanaco were growing in the southern highlands (Willey, 1971). Tihuanaco, which became important as a ceremonial center, developed some time after Pucara and had a sphere of influence that extended southwards from the Titicaca basin to the Chilean coast (Willey, 1971). Characteristic Tihuanaco artistic styles, consisting of stiff formalized figures and motifs, were adopted by

29

the Huari culture to the north, near the present town of Ayacucho. The influence of Huari probably extended over most of central and northern Peru (Willey, 1971), suggesting a highly developed political structure to cope with the administration of such a large area. However, by about 1,000 A.D., both Tihuanaco and Huari were abandoned and any unit that had developed quickly disappeared.

The Chimu Empire, the last major pre-Incan civilization in Peru, grew up at the town of Chan-Chan (Lanning, 1967) on the coast near the present-day city of Trujillo. The Chimu rulers were responsible for large-scale irrigation projects which expanded the area of cultivable land around the coastal valleys (Sanders & Marino, 1970). This heavy reliance on irrigation may have contributed to the overthrow of the Chimu Empire by the Incas, since whole towns frequently received their water from a neighboring valley, making them particularly vulnerable during a siege.

As indicated earlier, the Incan civilization only achieved a position of major importance in South America during the 100 years that preceded the Spanish invasions of 1532. The rapid rise of Inca influence is remarkable not only because of the geographical extent of the empire from the north of Ecuador to central Chile, but also because of the diversity of peoples that came under Inca rule. At its greatest extent, the Inca Empire contained between 6 million (Willey, 1971; Sanders & Marino, 1970) and 30 million people (Dobyns, 1966) who spoke a variety of unrelated languages. The key to the success of Inca colonial expansion lies in their understanding of the ecological diversity of the regions they conquered. Instead of a centralized authority that administered the economy of wide areas, the Inca established a system of semi-independent local resource areas (Lanning, 1967). In the coastal region, for example, these areas consisted of river valleys separated by expanses of barren desert, while on the steeper slopes of the inland hills, these areas were established in linear fashion along the altitude gradient, according to the type of crop that could be grown (Brush, 1976*b*). In this way, the economic structure of a given area was linked to its ecology.

Another reason for the success of Inca civilization was the custom of using local officials to run the political and economic affairs of a given district (Murra, 1961). The Inca state consisted of a pyramid-like hierarchy at the point of which was the emperor, and at whose base were units of 100, 500, 1,000, 5,000, and 10,000 individuals. Each unit was governed by a *kuraka* who was frequently a member of the community he governed. To ensure loyalty, the *kuraka*'s sons were often taken to Cusco for the duration of his governorship. In this way, the Incas gained a series of guarantees against insurrection in outlying regions of the empire, as well as a steady supply of labor for the government bureaucracy (Lanning, 1967). Another example of how the Incas maintained effective control over an area is their tolerance of local religious customs (Lanning, 1967).

Instead of insisting on rigid adherence to the official state religion, they allowed the inhabitants of a certain area to continue in the worship of local deities, so long as they satisfied the demands of the central government in the form of taxes. Frequently, a local totem or idol was removed to Cusco where it was kept in one of the temples, as a further guarantee for the good behavior of the group to which it belonged. In this way, the effigy served as another link between the central government and the inhabitants of a subjugated area.

At the time of the Spanish invasion of Peru in 1532, the Inca Empire was in a civil war that arose as a result of a disputed succession between the rival Inca brothers, Huascar and Atahuallpa. Prescott's (1908) account of the Spanish intervention in this dispute and the conquest of Peru that followed during the next two years tells how the Spanish managed to gain control over much of the territory formerly administered by the Incas, although it was not until 1569 that control of the country was firmly established.

Spanish settlement of the highlands appears to have been hindered by the ecological constraints imposed by high altitude and the nature of the terrain itself. A classic account of the difficulties encountered by the Spaniards in these regions is contained in Monge's (1948) book, *Acclimatization in the Andes*. After the consolidation of Spanish rule Peru remained under colonial domination for about 300 years and achieved independence in 1824, being one of the last strongholds of Spanish power in South America. Monge believed this may have been related to the difficulties lowland troops had in fighting at high altitude. The next fifty years were characterized by a general instability in Peru's political and economic structure. The Spanish attempted to reassert their authority in an extended campaign that lasted from 1864 to 1869 and in the so-called War of the Pacific, Chile won from Peru and Bolivia valuable nitrate fields in the Atacama Desert, a region still controlled by the former country.

In recent years, Peruvian society has undergone numerous changes, the most recent being an active policy of social and economic reforms.

*General population characteristics*

The focus for IBP-related research into the biocultural characteristics of high-altitude populations in the Andes has been in the *altiplano* areas of Peru, Bolivia and Chile. The District of Nuñoa in southern Peru is reasonably representative of the rural portions of this area and since it was the site of a major IBP effort an outline of the ecology and culture may be useful (Baker, 1976). This district, situated in the Department of Puno, was originally chosen because of its elevation (4,000 m and above), its relative isolation and the low level of European (Spanish) admixture in its population. The study population itself lives in and around the town

31

of Nuñoa. Detailed descriptions of the demographic characteristics of this population are contained in Baker & Little (1976) and only a summary will be presented here.

Archeological surveys indicate that the region of Nuñoa has supported human settlements since pre-Inca times (Baker, 1969). Records dating from the period of the Spanish conquest show that tribute was paid from this area in the form of a wide variety of products, suggesting a rather extensive area of jurisdiction that extended well into the *montaña* or eastern escarpment. This contrasts with the linear arrangement of resource areas, mentioned earlier, which characterized the Incaic economic system in the highland zone. As Thomas (1973) has pointed out, the fact that the economic structure of Nuñoa is based on the agricultural produce of several different ecozones must have contributed to the stability of the population as a whole. Since the Spanish conquest the growth of the *hacienda* system of land ownership has displaced the former pattern of community ownership, although present agricultural reforms by Peru's military government may reverse this trend.

In 1961, the District of Nuñoa contained a population of 7,750 and a total area of 1,600 km². Twenty-eight percent of the total population lived in the town of Nuñoa while the remainder were dispersed throughout the surrounding district (Baker, 1976). These figures yielded a mean population density of 3.5 persons per square kilometer. Three more-or-less well-defined levels of social stratification exist in the Nuñoa region: these are defined primarily on an individual's degree of 'westernization' and adherence to the national culture and practices of the Spanish-speaking population of Peru's major urban centers (Baker, 1976; Escobar, 1976). By far the most numerous of these social classes are the Indígenas, or Indians, a group which in the Nuñoa region at least are almost free from European admixture and cultural influences. Of the three groups, the Indians have departed the least from the traditional economic pursuits of the Andean native. They practice a combined agricultural and pastoral economy, the adaptiveness of which is suggested by the absence of environmental degradation in the ecozone they inhabit (Thomas, 1973). Another group, the Cholos, serve as intermediaries between the Indian population and the upper class, or Mestizos, who control much of the production and distribution of goods and services in the district. The Cholos frequently speak both Spanish and Quechua (the local Indian language) and have knowledge of both the national as well as the local economy. The Mestizos are less than 3% of the total population but own land and economic resources (Baker 1976). This disparity of wealth has become the object of recent attempts at reform by the Peruvian Government but it is too early to tell what the outcome will be. Spanish is the primary language of the Mestizo class, although most have at least a rudimentary knowledge of Quechua.

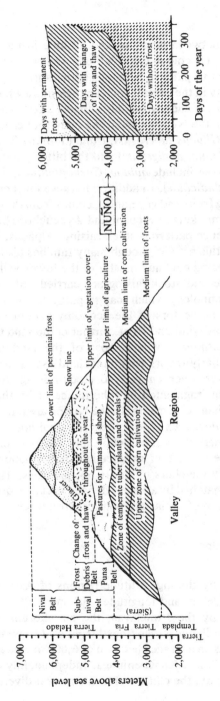

Fig. 2.2. Vertical zonation of agricultural vegetation types and frost climates in the Andes of southern Peru. The zonation of frost climates is based on information from Mt. El Misti in southern Peru. Modified from Troll (1968).

33

*Land use*

The relationship between land use and altitude in the central Andes is shown in Fig. 2.2. Of course exact latitude and regional aridity produce some variations in this pattern. Although agriculture is by far the dominant subsistence activity in the central Andean region as a whole, the high, level plains, or *altiplano*, of the region only support a few frost-resistant crops (Thomas, 1973). In the Nuñoa region, the main root crops are *oca* (*Oxalis crenata*), *ulloco* (*Ullucus tuberosus*), *isano* (*Tropaeolum tuberosum*), the sweet potato (*Solanum andigenum*) and the bitter potato (*Solanum curtilobaum*). Grain crops include *quinoa* (*Chenopodium quinoa*) and *cañihua* (*Chenopodium pallidicaule*). In addition to these local crops, a variety of produce, including fruits and vegetables, comes from the eastern *montaña*, on its way to the markets at Sicuani and Ayacucho. However, the most important subsistence pattern is stock-raising. Alpacas, llamas and sheep form the basis of this activity though many families also keep horses and cattle, the latter being confined mostly to the lower valley floors.

Both agriculture and stock-raising are carried out within a generally pre-Columbian technology which may in part account for the rather low product yields in all sectors of the economy (Thomas, 1973). Climatic constraints, such as wind and frost, further compound the problem. The ecology of the *altiplano*, where most of the land is unsuitable for agriculture particularly favors communal land use or the development of large ranches given over entirely to stock-raising. Such land tenure systems have been fragmented by inheritance or by the system of land grants to local Indian settlements. In the Nuñoa area, these native-owned settlements, or *ayllus*, coexist with the traditional *haciendas* or estates in which tenant farmers work a parcel of land and pay rent in the form of a proportion of the harvest, or an agreed amount of labor or services. The tenants are usually members of the Indígenas class. Their inferior economic position is matched by a clearly defined social inequality. For a more complete description of social and political structure see Escobar (1976).

## The Himalayan region

*Introduction*

The area covered by the high-altitude regions of Asia extends between latitudes 26° and 36° N and longitudes 70° and 110° E. A high plateau, Tibet, is rimmed by the Himalayas, the Karakorum and Kunlun and further west by the Pamir, cross-ribbed from east to west by high ranges. The main range has an average elevation of 6,000 m containing the highest peaks, Everest, K2, and Kanchenjunga. Independently of the enormous variety of topography, the climate presents a great diversity with a major contrast between the southern part of the mountain regions exposed to the summer monsoon, with a gradient of rainfall from east to west, and the aridity of the trans-Himalayan valleys and the Tibetan plateau with very low rainfall and extreme temperatures. In winter diurnal variation is as great as 50 deg C.

Rainfall is chiefly related to the Indian monsoon rainfall of which the Himalayas catch a large portion. The Ganges–Brahmaputra water system has a flow of 240,000 m³ s⁻¹ in the rainy season and 10,000 m³ s⁻¹ in the dry period. The rainfall diminishes westward (2,000 to 3,000 mm in Darjeeling, 1,500 mm in Kumaon). In the arid zone protected from the rainy winds the rainfall goes below 300 mm. In the eastern Himalayas, the winter limit of snow is 2,000 m; in the west (Kumaon) it reaches down to 1,600 m. The snow line or level to which the snow recedes in the course of the year ranges from 5,400 to 5,800 m on the southern exposures. The southern part of the Himalayan range, exposed to the monsoon, has a mild climate with a rich flora and fauna. On the northern face, altitude and climate combine to form a steppe with a few plants such as *Lonicera*, *Caragana*, and *Ephedra*. Wild animal life is scarce.

The ecological approach suggests that the high regions of Asia are utilized by man in two fashions.

(1) The few groups in permanent residence above 3,500 m are very individualized and mainly pastoralists. Some communities develop agriculture with the help of irrigation.

(2) The populations who have their permanent settlements in places lower than 2,500 m use the area for pasturage and summer residence.

*The populations*

The mountain region of central Asia is sparsely populated (estimated to be less than 25 million). In the northern part the population is Tibetan (2.5 million). Kara-Kirghiz and sedentary groups are found in Chinese Turkestan, and on the southern slopes of the Himalayas the populations are of Tibeto–Burman language plus populations belonging to the Indian culture. Agricultural and pastoral activities extend over a very large range of environmental zones thus involving a high degree of seasonal mobility.

The actual land available for agriculture is only about 5% of the geographical area of the regions considered. Fields are generally established on river terraces. The crops they bear are threatened not only by drought, hail, frost or snow, but also by erosion consequent to natural calamities. Irrigation is a necessity in the trans-Himalayan valleys. The main crops are barley, wheat and buckwheat. Agricultural practices are traditional and land holdings are very small.

The Himalayan region presents in its alpine zones extensive pasture

35

areas which are utilized by the nomadic and sedentary populations living in high-altitude areas (over 3,500 m), by inhabitants of the hills (1,000 to 2,500 m) and by nomadic populations such as the Bakarwal who spend the winter in Jammu and Kashmir (1,000 m) and the summer in the western parts of Ladakh (3,500 to 4,500 m).

Yak, sheep, goats, cattle and horses are the principal pastoral animals. The nomads at high altitude derive most of their needs from these animals. There is some evidence that the people of central Asia possessed domesticated common cattle before the yak. The domestication of the latter has now been fully developed and cross-breeds developed. The yak, also found in the wild state in Tibet, is kept in a state of domestication and is well adapted to the ecological conditions of extreme temperatures and scanty herbage.

The calendar of activities for the nomads (for example, the Drogpa of Tibet) is closely related to the natural vegetational cycle so that the population is in perpetual search for pasture areas.

Trade is essentially based on the exchange of rock salt from the lakes of Tibet and grain, rice or barley from the valleys. Because of trade the Himalayan range, far from being an obstacle, has been for centuries a zone of contact and circulation of ideas between the Indian subcontinent and central Asia (Snellgrove & Richardson, 1968).

### Problems of development

The high regions of Asia partly cover the Republic of Afghanistan, Pakistan, India, Nepal, Bhutan and the People's Republic of China. Man in the high regions of Asia is essentially a pastoralist having a way of life dependent on the natural conditions. Recent political developments are modifying the life of high-altitude populations by improving communication and transport, modifying social structures and the system of production and by encouraging sedentarization (in Tibet) and high-altitude tourism.

Economic development in central Asia is occurring as a consequence of the political situation. For example, the control of the political borders has accelerated the construction of a network of communications and improved the means of living of the mountain populations. The Border Roads Organization of the Government of India built nearly 8,000 km of roads between 1962 and 1970 over some of the most inaccessible areas of the Himalayas. In Bhutan the Indian Government has aided the construction of 1,200 km of roads for vehicles where none existed before 1960 (Mehta, 1974). While the achievements of this organization are remarkable in terms of the tremendous odds it had to face and as a result of the topology, it is doubtful if any thought was given during the road-building operation to the conservation of the Himalayan region.

*Case study: Nepal*

Nepal spans a complete cross-section of the Himalayas from the edge of
the Indo–Gangetic plain in the south to the edge of the Tibetan plateau
in the north. Between these two outer lines lie a wide variety of environments ranging from tropical to alpine. The terrain everywhere is of deep
branching V-shaped valleys aligned generally north–south and separated
by steep mountain ridges. This land, enclosed by longitudes 80° 15′ and
88° 10′ E and latitudes 26° 20′ and 30° 10′ N is bordered on the north by
the People's Republic of China (Tibet Autonomous Region) and on the
south by the Republic of India. It covers a surface of 141,577 km².

The land ranges from a mere 50 m above mean sea level at the southern
foothills to over 8,000 m on the northern crestline. Variations in altitude
and terrain are accompanied by variations in climate, natural vegetation,
soil, agricultural potential and the distribution of the population.

### The ecological zones

Due to the fact that the Himalayas lie between latitudes 26° and 30° N,
life remains possible at very high altitude. The natural timber line is at
4,000 m; herbaceous plants grow up to 5,500 m. In this important altitudinal transect we can differentiate no less than eleven vegetation zones
(Dobremez, 1975) and more than fifty types of forests (Stainton, 1972).
The characteristics of these zones are shown in Table 2.1.

In this zonation, vegetation varies from east to west. The eastern part
of the Himalayan range is influenced by the tropical monsoon climate
with dense forests; in the western part the tropical elements give way
to Mediterranean ones. In the trans-Himalayan zone, north of the main
crestline trees do not grow and vegetation is reduced to a steppe of
*Lonicera* and *Caragana* (Dobremez & Jest, 1971).

### Population distribution and human settlements

The uneven distribution of population by broad geographical regions needs
to be emphasized. Nepal has a population of 11.3 million (census of 1971).
The hills and mountains with 72% of the land area support only 53% of
the total population (density 58 km²). In the higher valleys of northwest
Nepal, the density falls to five inhabitants per square kilometer.

Nepal is a mosaic of more than thirty ethnic groups having their own
languages, social organizations, cultures and religions (Bista, 1972). Less
than 60,000 people live in the more than a third of the total surface of Nepal
above 3,000 m. The high mountain valleys are inhabited by populations of
Tibetan language and culture. Several trails link these regions with lower
Nepal and with the People's Republic of China via passes over 5,000 m.
Geographically, while the northern regions are cut off from the rest of
Nepal for the winter, free access to the Tibetan region, its pasture areas

The biology of high-altitude peoples

Table 2.1. *The vegetational zones of Nepal's natural plants*

| Zone | Natural plants | Altitude (m) | Human use |
|---|---|---|---|
| | Plant limit | 6,000 | |
| | Cushion plants | 5,500 | 5,500 m highest |
| Subnival | Shrubs/open grassland | | pasture areas |
| | | 5,000 | |
| Alpine upper | Alpine grassland | | |
| | | 4,500 | |
| Alpine lower | *Juniperus/Rhododendron* | | 4,200 m limit of |
| | | | agriculture with |
| | | 4,000 | irrigation |
| | Natural timber line | | |
| Subalpine upper | *Betula utius, Rhododendron* sp. | | |
| | | 3,600 | |
| Subalpine lower | *Abies spectabilis, Lari griffithii* | | |
| | | 3,000 | |
| Montane upper | *Cupressus torulosa* | | |
| | *Quercus semecampifolia* | 2,600 | |
| Montane lower | *Quercus lanata* | | |
| | | | 2,200 m upper limit |
| | | 2,000 | of rice cultivation |
| Subtropical upper | *Olea, Pinus roxb., Schima wallichii* | | |
| | *Olea, Castanopsis indica* | 1,500 | |
| Subtropical lower | | | |
| | | 1,000 | |
| Tropical upper | Tropical forests | | |
| | *Shorea robusta* | | 800 m lower limit of |
| | | | winter transhumance |
| | | 400 | for sheep from |
| | | | middle valleys |
| Tropical lower | Tropical forests | | |

and salt deposits, exists throughout the year. This situation symbolizes the mountain region's traditional twofold orientation.

The subsistence technology includes both agriculture and pastoral nomadism (very close to the agro-pastoral subsistence found throughout Tibet and Ladakh). Water is the limiting factor and fields have to be irrigated. Barley is the staple crop grown at altitudes between 3,500 m and 4,300 m. A little wheat and some potatoes are also grown. The high regions have large pasture areas where yak and hybrids (dzo), sheep, goats and horses are bred. Several patterns of husbandry are practiced (Fürer-Haimendorf, 1972; Jest, 1974; Goldstein, 1975). These are:

(1) Large-scale nomadism in which herds of yak and sheep are moved periodically to different pasture grounds. The herders live all the year in

38

traditional Tibetan tents. The animal movement cycle includes pasture areas in Tibet and Nepal with the animals migrating to Tibet in October for the winter and returning to Nepal in April–May for the summer.

(2) Nomadism as in (1) with animals wintering in the villages, where they are fed dry fodder.

(3) Summer pastoralism which alternates with winter grazing in lower areas (sheep and goats).

The maintenance of herds of sheep and yak is important not only for the transport of salt from Tibet to the middle valleys of Nepal but also for wool, meat and milk products. Trade with Tibet, the hills of Nepal and with India provides an important dimension to the high settlements' adaptation. In former times only the two first commodities were important. While rock salt and wool trade is now carried on in controlled trading centers in Tibet, new trade opportunities have developed in Kathmandu and India. Tibetan antiques (statues, jewelry), musk, etc. are now important trade items in these areas. However, this source of income is highly unstable and likely to decrease in the years to come.

The ecozone of the middle valleys of Nepal is able to produce two crops a year, with rice, maize and buckwheat as summer crops and wheat and barley as winter crops. The surplus is absorbed primarily by the high valleys and by Tibet. In return they obtain salt and wool. (Salt trade, on the basis of barter, is totally in the hands of the high-valley people.)

The populations of the high valleys of Nepal are Buddhist and had until recently strong ties with the religious centers in Tibet. There are many monasteries and temples of an unique architectural type and traditions of craftsmanship and design linked with religion. They are still vigorous and continue to develop.

In the trans-Himalayan valleys (3,500 to 4,200 m) the various economic strategies employed by the inhabitants generate surpluses which allow them to live above subsistence level. The situation is different in the middle valleys of Nepal (2,000 to 3,000 m). Because of population pressure on scarce arable land, intensified by erosion and the traditional technology (slash-and-burn cultivation), the decline of trade with Tibet and the opening to settlement through eradication of malaria of more fertile lowland (at 50 m to 300 m above sea level) with better social facilities in the Terai, a permanent migratory movement is occurring. This migratory movement is of two types: (1) a movement of people off the ridges to the valleys, and, (2) a southward movement from the hills and mountains to the Terai. In addition, a temporary migration also occurs in winter, from the highest settlements to the Terai.

Despite government efforts, these movements have resulted to some extent in uncontrolled encroachment on the forests. Human encroachment in forest areas, whether or not controlled, leads to the decline in wild life or the forced migration of indigenous jungle fauna to accommodate newcomers.

### Water and erosion

Although Nepal has very important resources, no proper action has been taken for their control and management. The potential for hydroelectric power is estimated at 83,000 MW (equivalent to the combined present hydro-electric capacity of Canada, USA and Mexico) (IBRD Report, 1975). Water is absolutely indispensable in the trans-Himalayan regions where rainfall is very low and agriculture impossible without irrigation. Only recently have irrigation, generation of power and drinking water projects been developed. Flood control and watershed management of the main rivers are planned but not yet implemented.

Not only the quantity but the characteristics of the rainfall in Nepal play a role in erosion. Precipitation averages 2,200 mm annually (a water weight of 24 tonnes per hectare), 70% of which falls between June and September. In the absence of forest cover or with poor agricultural contours such a concentration of rain rapidly erodes the hillsides, causing land-slides, blocking streams and inflicting flood damage in the lowlands. The loss of topsoil is one of the most dramatic consequences.

Other factors of major importance in erosion or loss of forest cover are shifting cultivation (primarily between 1,800 m and 2,700 m, especially in western Nepal), felling of forests for fire-wood, and overgrazing and lack of proper terracing. On steep hillsides the retaining walls of many terraced plots are thin and weak and often give way under the pressure of the water.

Efforts have been made to prevent land-slides and erosion in the program of the Trisuli Watershed project (UNDP/FAO, 1966–9), and in the Jiri project under Swiss Technical Cooperation.

### Pastoralism

Hillsmen of Nepal (mainly populations of Tibeto–Burman language) have long relied on animal husbandry as the dominant facet of their subsistence economy (Messerschmidt, 1975).

Pastoralism can be considered the main activity of populations living above 2,000 m, the upper limit of rice cultivation. (The cattle population of Nepal including water buffalo is estimated to be 4.4 million.)

Three types of pastoral activities can be defined in the central Himalayas:

(1) In the middle valleys there is breeding of common cattle and water buffalo with summer grazing in high alpine meadows (from 3,000 m to 4,000 m) and mountain forests. During the winter months the animals are kept on the unused fields near the villages.

(2) There is transhumance of goats and sheep covering a large altitudinal range. The herding cycle consists of a summer ascent to alpine meadows as high as 5,500 m and a winter descent to valleys as low as 600 m.

(3) In the trans-Himalayan valleys, above 3,500 m, yak (*Phoephagus grunniens*), dzo (its cross-breed with the Tibetan dwarf cattle) and sheep are bred.

In the high regions, the practice of yak and sheep herding remains a viable economic pursuit.

### Forest and plant products

In the central part of Nepal, forest resources have been reduced on a large scale by an increasing population in need of timber, fuel and more arable land. On the rim of the valley of Kathmandu some species – *Myricaria esculenta, Schimal wallichii, Spondias axilaris* – are in danger because their zone of growth corresponds to the belt of human settlements.

Timber is not exploited except on a local basis because of the total lack of transport facilities. In the trans-Himalayan valleys where forest resources are available, wood products are provided to the neighboring regions of Tibet.

The collection of medicinal plants is an important activity of the populations of the high regions. Plants such as *Picrorhiza* sp., *Swertia chirata, Nardostachys jatamensi, Rheum emodi, Orchis latifolia*, etc. are commercialized, transformed to drugs in Nepal, or exported to India.

### Wild life and recreational uses

Official reports state that the wild life has decreased in the past few years. In the case of a number of species, among which is the musk deer (*Moschus moschiferus*) whose musk pod is valuable in pharmacology and perfumery, this decrease is of alarming proportions. The marmot (*Marmota bobak*), a small lagomorph (*Ochotona roylei*), wild sheep (*Pseudois nahoor*), the Tibetan sheep (*Ovis ammon hodgsoni*) and very occasionally the gazelle (*Procapra picticaudate*) and the antelope (*Pantholops hodgsoni*) are found in the trans-Himalayan steppes (Blower, 1971–2). The birds, of which more than 700 species have been described, are one of Nepal's attractions (Fleming, 1975).

The attractions of Nepal's mountain ranges are apparent to tourists; the enthusiasm of thousands of trekkers shows this clearly (72,000 tourists in Nepal in 1974, among them 12,000 trekkers) (Stein, 1971). Although the increase of tourism has benefited the Nepal Government's foreign exchange earnings, the destructive effects to some Himalayan regions caused by vastly increased numbers of people threaten the natural and cultural resources upon which tourism depends. For example, the simultaneous presence in the Khumbu–Everest area of several mountaineering

## The biology of high-altitude peoples

expeditions, and/or groups of trekkers severely perturbates the regional trade, food becomes scarce due to the demand at inflated prices and there is an extensive consumption of wood for fuel.

If the present trend continues, the cost of tourism in terms of capital and cultural degradation will eventually have a negative impact on revenue and destroy the ethnic entity of the peoples of Nepal.

### Preservation of biota and maintenance of natural life systems

The general protection of biota and maintenance of life systems is a new idea in a developing country like Nepal. The Department of Wild Life has made extensive surveys of the different regions and localized zones where natural life systems should be preserved (Khumbu and Rolwaling, Langtang-Helambu, Dolpo-Shey, Jumla-Rara) (Blower, 1971-2). The Government of Nepal has taken action by creating national parks in the Khumbu region (Everest range), Langtang (north of Kathmandu) and is in the process of surveying another one in west Nepal around the Rara lake. These three park areas are inhabited and thus the local population will have to participate in the protection of nature and management of the park while continuing their traditional agro-pastoral activities.

## Overview

Geographers have long perceived that altitude zones have many floral and faunal similarities when they occur near the same latitude. What our brief survey of the cultures of the regions also suggests is that there may also be some cultural similarities occasioned by the similarities in potential land-use patterns. In a recent article Brush (1976a) argued that there are indeed cultural regularities among the high-altitude peoples of the world and that a search for these regularities would be a useful exercise in cultural ecology. While one must stress that there are, of course, regional differences as there are regional differences in the biology of high-altitude peoples, research leading to a more careful definition of the similarities and differences would aid us in our attempts to understand man's relationship with his environment.

## References

Baker, P. T. (1969). Human adaptation to high altitude. *Science, Washington,* **163**, 1149–56.

Baker, P. T. (1976). Evolution of a project: Theory, method, and sampling. In *Man in the Andes: A Multidisciplinary Study of High Altitude Quechua,* ed. P. T. Baker & M. A. Little, pp. 1–20. Stroudsburg, Pa.: Dowden, Hutchinson & Ross, Inc.

Baker, P. T. & Little, M. A. (eds.). (1976). *Man in the Andes: A Multidisciplinary*

Cultures of high-altitude areas

Study of High Altitude Quechua. Stroudsburg, Pa.: Dowden, Hutchinson & Ross, Inc.

Bird, J. B. (1938). Antiquity and migrations of the early inhabitants of Patagonia. *Geographical Review*, **28**, 250–75.

Bista, D. B. (1972). *People of Nepal.* Kathmandu: Ratna Pustak Bhandar.

Blower, J. (1971–2). *Reports on Wildlife Preservation.* Kathmandu: Ministry of Forests, Government of Nepal.

Bowman, I. (1916). *The Andes of Southern Peru.* New York: Henry Holt & Co.

Brush, S. B. (1976a). Introduction, cultural adaptations to mountain ecosystems symposium. *Human Ecology*, **4**, 125–33.

Brush, S. B. (1976b). Man's use of an Andean ecosystem. *Human Ecology*, **4**, 147–66.

Büdel, J. (1954). Klima-morphologische Arbeiten in Äthiopien im Frühjahr, 1953. *Erdkunde*, **8**, 139–56.

Buxton, D. (1970). *The Abyssinians.* New York: Praeger Publishers.

Clegg, E. J., Harrison, G. A. & Baker, P. T. (1970). The impact of high altitudes on human populations. *Human Biology*, **42**, 486–518.

Dobremez, J. F. (1975). *Le Népal, Ecologie et Phytogéographie.* Paris: Editions du CNRS.

Dobremez, J. F. & Jest, C. (1971). Carte écologique du Népal. Région Annapurna–Dhaulagiri. I. 250,000 ème. *Cahiers Népalais–Documents I.* Paris: Editions du CNRS.

Dobyns, H. F. (1966). An appraisal of techniques with a new hemispheric estimate. *Current Anthropology*, **7**, 395–416.

Escobar, G. (1976). Social and political structure of Nuñoa. In *Man in the Andes: A Multidisciplinary Study of High Altitude Quechua*, ed. P. T. Baker & M. A. Little, pp. 60–84. Stroudsburg, Pa.: Dowden, Hutchinson & Ross, Inc.

Fleming, R. (1975). *Birds of Nepal.* Bombay.

Fürer-Haimendorf, C. Von. (1972). *The Sherpa of Nepal: Buddhist Highlanders* (revised edition). London: John Murray.

Gansser, A. (1964). *Geology of the Himalayas.* London: Interscience Publishers.

Goldstein, M. C. (1975). Tibetan speaking agro-pastoralists of Limi: a cultural ecological overview of high altitude adaptation in the northwest Himalaya. *Objects et Mondes*, **16**, 259–68.

Haynes, C. V. (1965). Carbon-14 dates and early man in the New World. *Interim Research Report No. 9, Geochronology Laboratories.* Tucson: University of Arizona.

Hess, R. L. (1970). *Ethiopia.* Ithaca, New York: Cornell University Press.

IBRD Report (1975). A review of major issues related to Nepal's development prospects. Washington, DC (mimeograph).

Izumi, S. & Sono, T. (1963). *Andes 2: Excavations at Kotosh, Peru, 1960.* Tokyo: Kadokawa Publishing Company.

Jenks, W. F. (1956). Peru. In *Handbook of South American Geology.* Geological Society of America Memoir 65. Baltimore, Md.: Waverly Press.

Jest, C. (1974). *Tarap, une vallée de l'Himalaya.* Paris: Editions du Seuil.

King, L. C. (1962). *The Morphology of the Earth.* New York: Hafner Publishing Co.

Lanning, E. P. (1965). Early man in Peru. *Scientific American*, **213** (4), 68–76.

Lanning, E. P. (1967). *Peru Before the Incas.* Englewood Cliffs, NJ: Prentice-Hall.

Lanning, E. P. & Patterson, T. C. (1967). Early man in South America. *Scientific American*, **217** (5), 44–50.

43

Lynch, T. F. & Kennedy, K. A. R. (1970). Early human cultural and skeletal remains from Guitarrero Cave, Northern Peru. *Science, Washington,* **169**, 1307–9.

MacNeish, R. S. (1971). Early man in the Andes. *Scientific American,* **224** (4), 36–46.

McFarland, R. A. (1969). Review of experimental findings in sensory and mental functions. In *Biomedicine of High Terrestrial Elevations,* ed. A. H. Hegnauer, pp. 250–65. Natick, Mass.: US Army Research Institute of Environmental Medicine.

Mehta, G. N. (1974). *Bhutan, Land of the Peaceful Dragon.* Delhi: Bombay Vikas.

Messerschmidt, D. A. (1975). Gurung sheperds of Lamjung Himal. *Objets et Mondes,* **14**, 307–16.

Monge, M. C. (1948). *Acclimatization in the Andes.* Baltimore, Md.: The Johns Hopkins Press. (republished by Blaine Ethridge Books, Detroit, 1973.)

Montané, J. (1968). Paleo-Indian remains from Laguna de Tagua Tagua, central Chile. *Science, Washington,* **161**, 1137–8.

Mosely, M. E. (1975). *The Maritime Foundations of Andean Civilization.* Menlo Park, Calif.: Cummings Publishing Company.

Murra, J. (1961). Social structure and economic themes in Andean ethnohistory. *Anthropological Quarterly,* **39**, 47–59.

Pankhurst, R. (1961). *An Introduction to the Economic History of Ethiopia from Early Times to 1800.* London: Staples Printers Ltd.

Patterson, T. C. (1971). The emergence of food production in central Peru. In *Prehistoric Agriculture,* ed. S. Struever, pp. 181–207. Garden City, NY: The Natural History Press.

Prescott, W. H. (1908). *History of the Conquest of Peru.* New York: Everyman's Library.

Rouse, I. & Cruxent, J. M. (1963). *Venezuelan Archaeology. Yale University Caribbean Series, No. 6.* New Haven: Yale University.

Rowe, J. (1967). Form and meaning in Chavin art. In *Peruvian Archaeology,* ed. J. Rowe & D. Menzel, pp. 72–103. Palo Alto, Calif.: Peek Publications.

Sanders, W. T. & Marino, J. (1970). *New World Prehistory.* Englewood Cliffs, NJ: Prentice-Hall.

Schaller, K. (1972). *Äthiopien. Eine geographische-medizinische Landeskunde.* Berlin: Springer-Verlag.

Sellassie, S. H. (1972). *Ancient and Medieval Ethiopian History to 1270.* Addis Ababa, Ethiopia: United Printers.

Simoons, F. J. (1960). *Northwest Ethiopia.* Madison: The University of Wisconsin Press.

Snellgrove, D. L. & Richardson, H. E. (1968). *A Cultural History of Tibet.* London: Weidenfeld & Nicolson.

Stainton, J. D. A. (1972). *Forests of Nepal.* London: John Murray.

Staszewski, J. (1957). *Vertical Distribution of World Population. Polish Academy of Science, Geographical Studies, No. 14.*

Stein, J. A. (1971). Tourism, environment and development in the Himalayan region. The need for study of mountain environment and design. *Design,* January 1971. Delhi.

Thomas, R. B. (1973). *Human Adaptation to a High Andean Energy Flow System. Occasional Papers in Anthropology No. 7.* University Park, Pa.: The Pennsylvania State University.

Thomas, R. B. (1975). The ecology of work. In *Physiological Anthropology*, ed. A. Damon, pp. 59–79. London: Oxford University Press.

Thomas, R. B. & Winterhalder, B. P. (1976). Physical and biotic environment of southern highland Peru. In *Man in the Andes: A Multidisciplinary Study of High Altitude Quechua*, ed. P. T. Baker & M. A. Little, pp. 21–59. Stroudsberg, Pa.: Dowden, Hutchinson & Ross, Inc.

Troll, C. C. (1968). The cordilleras of the tropical Americas: aspects of climate, phytogeographical and agrarian ecology. In *Geo-ecology of the Mountainous Regions of the Tropical Americas*, ed. C. Troll. Bonn: Ferd. Dummlers Verland.

Ullendorf, E. (1973). *The Ethiopians*. London: Oxford University Press.

UNDP/FAO. (1966–9). Trisuli Watershed Project. Kathmandu: UNDP/FAO, Ministry of Forests, Government of Nepal.

UN (1975). *Demographic Yearbook: 1974*. New York: Department of Economic and Social Affairs, United Nations.

US Army. (1966). US Army Area Handbook for Ethiopia. Department of the Army Pamphlet No. 550–28.

Wadia, D. N. & West, W. D. (1964). Structure of the Himalayas. International Geological Congress, Twenty-second Session. New Delhi, India.

Willey, G. R. (1971). *An Introduction to American Archaeology*, vol. 2, *South America*. Englewood Cliffs, NJ: Prentice-Hall.

# 3. A genetic description of high-altitude populations

R. CRUZ-COKE

### Limitations of genetic studies

Our knowledge of the genetic structure of high-altitude populations is very limited for two principal reasons: the infancy of the science of population genetics and the paucity of formal genetic researchers in field studies.

Population genetics is a science that is concerned with understanding the nature and source of inherited characteristics of the human species. The mathematical foundations of the theory of population genetics were laid down between 1908 and 1931 and the first textbook appeared only in 1955 (Li, 1955). Moreover, the first text on the genetics of human populations is very recent (Cavalli-Sforza & Bodmer, 1971). Only a few mathematical models have been constructed and tested with human population data.

Formal genetic studies on high-altitude populations began during the last decade. So far only small samples have been examined and surveyed and these only for a few geographical areas. In general, human biological and demographic research on high-altitude populations are relatively recent (Baker & Weiner, 1966). Consequently, literature on the subject is scarce and this review will necessarily be very limited and preliminary (Clegg, Harrison & Baker, 1970).

This review will summarise the most pertinent genetic problems beginning with a description of the genetics of high-altitude populations around the world; i.e., distribution of genetic polymorphisms, differences between highlanders and lowlanders, detection of evolutionary forces affecting the genotypes of highlanders, and finishing with an analysis of some proposed mechanisms for genetic adaptation to high altitude.

### Distribution of genetic polymorphisms

According to Neel & Salzano (1964) a primary step for the analysis of a population's biology should be the intensive and extensive survey of the frequency of known genetic polymorphisms. A genetic polymorphism is defined as the occurrence in a given population of two or more genes at one locus, each with at least a frequency of one per cent. Polymorphism also usually refers to alternative hereditary forms that can be distinguished from each other and whose inheritance is clearly understood (Cavalli-Sforza & Bodmer, 1971).

47

Table 3.1. *Most significant distinctive genetic polymorphisms common to selected high-altitude populations in Africa, Asia and America*

| Genetic polymorphism | Africa (Simien)[a] | Asia (Himalayas)[b] | America (Andes)[c] |
|---|---|---|---|
| ABO | $I_{A_2} > 0.08$ | $I_B > 0.20$ | $I_O > 0.98$ |
| Rhesus | $Rh_0 > 0.20$ | $Rh_1 > 0.55$ | *rh* absent |
| Haptoglobin | Ahaptoglobulinaemia 0.06 | $Hp_2 > 0.66$ | $Hp_1 > 0.69$ |
| Most common | | | |
| HLA antigen | *HLA7* | *HLA9* | *HLA, W15* |
| HLA haplotype | *1, W17* | *11, W15* | *W31, W15* |

Data from   [a] Harrison *et al.* (1969).
       [b] Tiwari (1966); Bodmer & Bodmer (1973).
       [c] Post, Neel & Schull (1968); Cruz-Coke (1972).

For practical purposes it is possible to classify genetic polymorphisms in five groups: (1) red-cell antigens (ABO, Rhesus, MN, Kell, etc.), (2) white-cell antigens (HLA system), (3) plasma proteins (haptoglobin, group-specific component, immunoglobulin Gm, etc.), (4) red-cell proteins (haemoglobins, glucose-6-phosphate dehydrogenase, adenylate kinase, etc.), and (5) miscellaneous (PTC taste and colour blindness).

In relation to high-altitude populations systematic surveys of genetic polymorphisms have been performed only in the highlands of South America. The highland populations in Asia and Africa have been explored only sporadically. The extensive surveys have generally been based on the study of red-cell antigens, while the frequency of other polymorphisms have been measured in only a few regions. Consequently, it is difficult to establish comparisons in order to define a 'profile' or look for common features of genetic polymorphisms among the high-altitude populations of the world. Our description will therefore be necessarily tentative.

For the prime task of describing the genotype distribution of peoples living at high altitudes around the world, one must first consider the history of high-altitude peoples. These groups live in regions which were covered by ancient glaciers during the long periods of glaciations. We are living now in the fourth interglacial period, the last glaciation having ended some 10,000 years ago (MacNeish, 1971). Consequently, highlanders must have invaded high mountain areas after that time in all regions of the world, in Africa, Asia and America. The time period of possible micro-evolution and development of genotype diversity is thus relatively short. At first sight one must therefore conclude that it would have been difficult for these peoples to develop adaptive genes in order to establish a distinctive set of genotypes for living at high altitudes.

A first comparison of the polymorphic similarities and differences between high-altitude groups is presented in Table 3.1. From this table it may be observed that in the three most important high-altitude regions on earth there are distinct differences in the patterns of genetic polymorphisms. A complete list of the human biological differences among the populations of each area is substantial. Consequently, it is possible to support the idea that, at present, the high-altitude populations of our planet do not have in common a special set of genotypes.

### South America

The possibility of genetic differentiation of high-altitude populations is most easily studied in the Americas. We know that the ancestors of the present Amerindian population entered the continent some 70,000 years ago, before the Bering Sea land-bridge sank under the sea. The oldest paleo-Indian skeletons found in North America are estimated to be 48,000 years old (Bada, Schroeder & Carter, 1974). In South America, finds appear to be more recent, by the order of 22,000 years (MacNeish, 1971). We do not know the number of persons crossing the Bering Strait, the time span over which this occurred or the degree of heterogeneity of the invaders (Neel, 1973).

By the time the immigrants reached South America and the Andes ranges, the *altiplano* was already practically ice-free as a consequence of the retreating Wisconsin glaciation. Therefore, it is probable that the highland and lowland regions of South America were occupied during the same periods. Table 3.2 shows the distinctive features for each genetic polymorphism in the South America highlands. An analysis of the list of the most frequent alleles shows clearly that the people of the Andes have a relatively low degree of microdifferentiation. In all types of polymorphisms there are a few unusual allelic frequencies, an absence of abnormal types and in general a paucity of genotypes. Most advanced studies with red-cell enzymes showed also the absence of new polymorphic systems that could be special or distinctive in frequency for these peoples (Modiano, Bernini & Carter, 1972; Constans & Quilici, 1973).

Using data from extensive surveys which Post, Neel & Schull (1968) carried out along the Andes ranges it is possible to detect clines within the highlanders. Cruz-Coke (1972) studied the cline of the Rhesus system in the Andes from north to south. He discovered a divergent cline for the most common haplotypes of the Rhesus system. The segment CDe ($Rh_1$) decreases from north to south, while the segment cDE ($Rh_2$) increases toward the south reaching the highest frequency in the *altiplano* at the

Table 3.2. *Some genetic polymorphisms common to high-altitude populations in the Andes of South America*

| Genetic polymorphism | Special feature or most common allele |
|---|---|
| **Red-cell antigens** | |
| ABO | Isogenic for $I_O$ |
| Rhesus | *rh* absent |
| Kell | *K* absent |
| Duffy | $Fy^a > 0.79$ |
| Lewis | Only *Le* (*a-*) |
| Lutheran | Only *Lu* (*a-*) |
| Diego | $Di^{a+}$ present |
| **White-cell antigens** | |
| HLA antigen | *HLA* 15 |
| HLA haplotype | *W31, W15* |
| **Serum proteins** | |
| Haptoglobins | $Hp_1 > 0.69$ |
| Group specific component | $Gc_1 > 0.80$ |
| Immunoglobulins | $Gm(1, 17, 21) > 0.85$ |
| **Red-cell proteins** | |
| Haemoglobin | Absence of abnormal types |
| G-6-PD | Absence of abnormal types |
| Phosphoglucomutase | $PGM_1$ absent |
| Phosphogluconate dehydrogenase | $PGD^c$ absent |
| Adenyl kinase | $AK^2$ absent |
| Adenosine deaminase | $ADA^2$ absent |
| Acid phosphatase | $p^c$ absent |
| **Miscellaneous** | |
| PTC tasting | Rare or absent non-tasters |
| Colour blindness | Rare or absent defectives |

Data from Post *et al.* (1968); Cruz-Coke (1970); Quilici, Ruffie & Marty (1970); Modiano *et al.* (1972); Quilici (1975).

isolate of Huallatire with 0.843. Also gene $Di^a$ decreases significantly from north to south.

Clines in relation to altitude have been investigated in the sharp altitudinal gradient rising from Arica on the western edge of the Andes near the Pacific Ocean in Chile (Cruz-Coke, unpublished data; Rothhammer & Spielman, 1972). While genetic polymorphism did not show a clear relationship to altitude there was a significant change along the gradient in anthropometric characteristics.

Table 3.3. *Some distinctive genetic polymorphisms common to selected high-altitude populations of Asia*

| Genetic polymorphism | Tibet–Bhutan Himalayas[a] | Pakistan Karakoram[b] |
|---|---|---|
| ABO | $I_{A_2} < 0.02$ | $I_{A_2} > 0.04$ |
| Rhesus | $rh < 0.09$ | $rh > 0.20$ |
| MNSs | $L_{MS} > 0.40$ | $L_{MS} < 0.24$ |
| Kell | $K > 0.03$ | $K$ absent |
| Duffy | $Fy^b > 0.21$ | $Fy^b > 0.30$ |
| Lutheran | — | $Lu^a$ absent |

Data from: [a] Tiwari (1966); Glasgow *et al.* (1967).
          [b] Clegg *et al.* (1961).

### Asia

It is not yet possible to develop an overview of the general distribution of genetic polymorphisms in the highlands of the continent of Asia. The highlands of Central Asia, the Himalayas, Pamyr, Karakoram and Tien-shan mountains, are located at the crossroads of routes connecting the ancient Chinese, Indian and Mongoloid civilizations of the East. In these regions man has lived during the last million years and has survived three interglacial ages. Thus, the evolutionary potential is quite different from the American situation.

Using data from the few genetic studies available, it is possible to suggest the most distinctive features for the area (see Table 3.1). Nevertheless, as shown in Table 3.3 there are significant differences in some polymorphisms between the different mountain peoples. There is a great variability in the distribution of most common polymorphisms, which probably reflects the variable miscegenation of Mongoloid and Caucasoid populations. Isogenicity is rare and in every polymorphism all types of alleles are present, including all antigens of the HLA system (Bodmer & Bodmer, 1973).

### Africa

The genotypes of African highlanders are very different from those of the Asian and American continents. In fact, the African highlands are lower than those of Asia and glaciation was less important and extensive in the ice ages. Moreover, Ethiopia is very near to the Olduvai Gorge and Lake Rudolf where man emerged some million years ago. It is probable that the highlands of Ethiopia have been occupied by man longer than other highlands in the world.

51

# The biology of high-altitude peoples

Table 3.4. *Some distinctive genetic polymorphisms common to the high-altitude populations of Ethiopia, Africa*

| Genetic polymorphism | Special feature | |
|---|---|---|
| ABO | $I_{A_2}$ | 0.08 |
| Rhesus | $rh$ | 0.20 |
| | $Rh_o$ | 0.24 |
| MNSs | $Ms$ | 0.38 |
| Kell | $K$ | 0.03 |
| Duffy | $Fy^a$ | 0.55 |
| Haptoglobin | Ahaptoglobulinaemia | 0.06 |
| Immunoglobulin | $Gm(1,5,10,11)$ | 0.51 |
| Adenylate kinase | $AK_2$ | 0.02 |
| Acid phosphoglucomutase | $PGM_1{}^1$ | 0.69 |
| Phosphogluconate dehydrogenase | $PGD^c$ | 0.10 |
| G-6-PD | Absence of abnormal types | |

Data from Harrison *et al.* (1969); Ikin & Mourant (1962).

High-altitude populations in Africa are restricted almost exclusively to mountains in Ethiopia, particularly to the Simien highlands. Table 3.4 shows the most characteristic features of these highlanders. It may be noted that there is a great variability in most common genetic polymorphisms, particularly within the ABO and Rhesus systems, the latter showing ten different haplotypes. Red-cell proteins are also very rich. No abnormal G-6-PD or haemoglobins have been reported.

## Differences between highlanders and lowlanders

In order to test the hypothesis stated previously that the human species could not have reached a significant state of genetic differentiation in the highlands of the world because of the very short time which has elapsed after the end of the last glaciation, we shall analyse the genotypic differences between highlanders and nearby lowlanders, on each continent, and test the difference by the use of genetic distance measures.

### South America

Cruz-Coke (1972) has compared the mean of gene frequencies for eleven genetic systems. The information used was derived from the extensive data accumulated in a survey on the Andes (Post *et al.*, 1968). Only two significant differences were detected: the allele $Hp_1$ is very low in lowlands (0.55) and high (0.72) in the highlands. On the contrary, the allele $Di^a$ is significantly higher (0.16) in lowlands than in highlands (0.06). Constans

Table 3.5. *Genetic serological distances between selected very high-altitude peoples and other highland and lowland populations in South America*

| Country...<br>Place...<br>Language... | Chile<br>Huallatire<br>Aymara[a] | Bolivia<br>La Paz<br>Aymara[b] | Peru<br>Puno<br>Quechua[b] | Peru<br>Nuñoa<br>Quechua[c] |
|---|---|---|---|---|
| Genetic distances of<br>the above from | | | | |
| Highland neighbours<br>with same language | | | | |
| Population a | 0.241 | — | — | 0.028 |
| Population b | 0.316 | — | — | 0.110 |
| A distant highland<br>group with different<br>language | — | 0.288 | 0.288 | — |
| A lowland neighbour<br>with different language | 0.370 | 0.393 | 0.370 | 0.300 |
| Distant lowland groups<br>with different language | | | | |
| (Brazil) Xavantes | 0.449 | 0.374 | 0.336 | — |
| (Venezuela) Yanomama | 0.656 | 0.514 | 0.479 | — |

Data from: [a] Rothhammer & Spielman (1972).
[b] Neel (1973).
[c] Garruto & Hoff (1976).

& Quilici (1973) have also demonstrated a genetic difference between the two alleles of group specific component. $GC_1$ was found to be higher in highlands while $GC_2$ was higher in lowlands.

The other polymorphic distinctions common to South American highlanders (see Table 3.2) were also common in nearby lowland Amerindian populations. In fact, Table 3.2 is an extension of the one developed by Neel & Salzano (1964) to show the genetic characteristic of all South American Indians. Thus, there does not appear to be a clear-cut genetic frontier between highlanders and lowlanders in South America.

In order to obtain a multivariate overview of the genetic relationship between highlanders and lowlanders in South America, the model of phylogenetic analysis developed by Cavalli-Sforza & Edwards (1967) has been applied by some investigators. This method estimates genetic distances between all possible pairs of population groups. The distances generated by this technique are proportional to the chord length of a multidimensional sphere on whose surface the allele frequencies are plotted. The distances correspond roughly to fractions of a gene substitution separating the two groups compared, summed over all for the

53

## The biology of high-altitude peoples

independent alleles considered (Rothhammer & Spielman, 1972). Lalouel & Morton (1973) have criticised this measure on theoretical grounds; however, the technique does appear useful in general ways for estimating multivariate genetic affinities between human populations.

Table 3.5 summarises the major findings from studies on genetic distances in South America. Results show that genetic distances are small between highland neighbours speaking the same language, and relatively large between highland and distant lowland populations, as those of Xavantes in Brazil and Yanomamas in Venezuela. At first sight there appears to be a relationship between geographical and genetic distances, although the limited sample sizes and small number of groups studied suggest a cautious interpretation.

In spite of the relative genetic common features of the 'American Indian', there is a wide diversity within and between the classical tribes enumerated by Neel & Salzano (1964). Genetic distances between these tribes are not completely correlated with geographical distances or altitude. In a matrix of genetic distances between twelve tribes in South America (two highland and ten lowland), the mean pair-wise genetic distance is 0.385; the mean distance of the Aymara from eleven other tribes is 0.378; and that of Quechuas is only 0.340 (Neel, 1973). Consequently, altitude does not appear to be a significant factor in the process of microdifferentiation between the pre-Columbian populations in South America.

### Asia

In the Himalayas and Tibet, highland populations have some affinities with Indians and Chinese, but small differences have been pointed out by Tiwari (1966). $A_2$ gene is higher in Tibetans than in neighbouring Indians, Chinese and southeast Asians. In the Rhesus systems the incidence of $Rh_2$, $Rh_0$ and $Rh_z$ is very high compared with neighbours. In the Karakoram ranges in Baltistan (Pakistan), highlanders have numerous affinities with their lowland neighbours. The presence of $A_2$, rh genes and MS in the group shows a relationship to Caucasoid peoples of India, while the low Diego suggests a relationship to Mongoloids of China and southeast Asia. Nevertheless there are significant differences from both groups as shown by the very high cDE $(Rh_2)$ in Karakoram. Clegg, Ikin & Mourant (1961) concluded that the Baltis peoples in Karakoram were a Caucasoid population with a certain admixture of Mongoloid, and some affinities with the Turkomans and Uzbecks of the Soviet Union.

*Africa*

According to Harrison *et al.* (1969) highland populations in Ethiopia are not genetically different from the immediate lowland neighbouring groups. The comparative high frequencies of *MS* and *Fyª* genes suggest a rather high proportion of Caucasoid ancestry. The only genetic system for which a significant difference was found between highland and lowland populations was that of a phosphogluconate dehydrogenase variant *PGD$^c$*, which has lower values in lowlanders.

## A common highlander pattern

The review on the distribution of genotypes of highlanders around the world has shown that they have few genetic differences from lowlander neighbouring peoples, and that only sporadic genetic polymorphisms have significantly different frequencies between both groups of population. Nevertheless there is one general exception to this conclusion. In all highland populations surveyed, no examples of sickle-cell haemoglobins, thalassemia or abnormal G-6-PD have been found. One doubtful case was reported by Harrison *et al.* (1969) in Ethiopia and two cases of hetero-zygotes for Hb$_E$ by Glasgow *et al.* (1965) in Bhutan. These two cases probably originated in southeast Asia. In America, no abnormal type of Hb or G-6-PD has been found. Consequently, the only genetic pattern common to all high-altitude populations in the world seems to be the absence of abnormal types of the sickle-cell polymorphisms, classically described in lowland malarian areas (Allison, 1964).

Finally, it is important to stress the fact that, generally, all South American high-altitude populations surveyed have a very low degree of Caucasian admixture, less than 5%, indicating an important degree of isolation in relation with the European invaders.

## Evolutionary forces affecting the genotypes of highlanders

According to the genetic theory, the human species may be grouped into 'Mendelian populations', defined as populations of interbreeding individuals who share a common pool of genes which are transmitted according to Mendel's laws of inheritance. Mendelian populations are evolving under the influence of the systematic effects of mutation, natural selection and migration and the dispersive forces of random drift and inbreeding (Cavalli-Sforza & Bodmer, 1971).

Highland populations are generally distributed in small and isolated groups. This characteristic is an important factor in assessing the possibility for change in their genotypic structure. Small population isolates

## The biology of high-altitude peoples

may gradually become distinct from their neighbouring isolates through the processes of natural selection and random drift. This occurs unless the process is counteracted by gene mutation and gene flow between populations. A small amount of gene flow into an isolate can suffice to restrict the action of random drift. Consequently it has been suggested that high-altitude populations may be useful for detecting the action of evolutionary forces in shaping the genetic structure of human populations.

We shall review the studies which attempted to detect the effect of evolutionary forces on high-altitude populations in order to examine the possible causes for the specific and peculiar distribution of genetic polymorphisms in high-altitude groups.

### Mutation

Mutation is the raw material of evolution, since mutant genes in the population create diversity. Nevertheless this force is slow and has proved impossible to detect in most small populations such as those surveyed in the mountain isolates. This fact is unfortunate, because one suspects that, at high altitudes, populations may show increased mutation rates because of the increased environmental mutagenic forces in the form of natural radiation. Cosmic radiation, for example, is ten times higher at 4,000 metres than at sea level (Cruz-Coke, Cristoffanini, Aspillaga & Biancani, 1966).

### Natural selection

Natural selection in man is difficult to study because of the complex selective process. Since the process implies that some genotypes leave more offspring than others, it can only be deduced to have existed and its intensity can only be measured *ex post facto* (Schull, 1963).

According to Crow (1958), it is possible to measure potential total selection intensity if we are able to detect differential survival and fertility rates in succeeding generations. Crow has provided a measure of selection pressures by an Index called 'opportunity for selection (I)' and its two component parts: $I_m$ (mortality) and $I_f$ (fertility). Table 3.6 shows the estimation of selection pressure for two high-altitude populations of different size and language in the Andes (Cruz-Coke *et al.*, 1966; Garruto & Hoff, 1976). It appears that in both groups natural selection would operate mainly through mortality and that the fertility component is less important. Although it is probable that environmental factors such as hypoxia, cold, malnutrition, cosmic radiation and the lack of vitamins in soil, which are associated with life at high altitudes, are responsible for the high infant mortality rate which reaches 33% in some regions, one must assume that

56

Table 3.6. *Estimation of natural selection and random drift in two populations of different size and languages in very high-altitude areas of the Andean* Altiplano *in South America*

| | Peru | Chile |
|---|---|---|
| Country... | Nuñoa | Huallatire |
| Place... | 4,000 | 4,300 |
| Altitude (metres)... | Quechua | Aymara |
| Language... | | |
| Potential natural selection (indices) | | |
| Mean progeny number | 6.7 | 7.3 |
| Variance | 9.0 | 9.9 |
| Fertility index ($I_f$) | 0.200 | 0.185 |
| Mortality index ($I_m$) | 0.522 | 1.178 |
| Total selection index (I) | 0.827 | 1.581 |
| Potential random drift (indices) | | |
| Population breeding size (N) | 2635 | 67 |
| In-migration rate | 0.130 | 0.061 |
| Coefficient of breeding isolation | 325.8 | 3.4 |
| Deviation due to genetic drift per generation | ±0.007 | ±0.048 |

Data from Cruz-Coke *et al.* (1966) and Garruto & Hoff (1976).

only a small fraction of mortality and fertility are genetically effective. Neither mortality *per se* or fertility *per se* are necessarily selective. Thus, Crow's Index while a useful measure for demographic description must be interpreted with great caution in relation to selection.

The possibility of detecting selection pressure using data from the distribution of genetic polymorphisms has been proposed recently by Bodmer (1973). Such an approach may be applied to Andean populations and in fact there appears to be a linkage disequilibrium between the antigens of the HLA system. In the Aymara-speaking peoples in the Andes the frequency of the haplotype *W31, W15* is 0.211, while the two independent alleles *W31* and *W15* have frequencies of 0.360 and 0.430, respectively. The expected product of these two frequencies is 0.154 instead of the reported 0.211. The difference of 0.053 suggests a substantial and positive linkage disequilibrium. This difference might be interpreted to be a result of selective pressure. According to Fisher (1930) interactive natural selection would favour close linkage between interacting loci. There is evidence of a relationship between HLA and certain diseases and immune responses. If selection for certain haplotypes in the Andes, such as *W31* and *W15*, increases resistance to these diseases it would lead to the elimination of other haplotypes. In fact, there are only 16 alleles of the HLA system in the Andean population, against 25 to 30 in Europe and Asia.

Cruz-Coke (1972) has suggested that the high frequency of cDE ($Rh_2$)

## The biology of high-altitude peoples

haplotype of the Rhesus system in the *altiplano* in South America is a result of selection pressures acting on differential susceptibility to infectious diseases. Cruz-Coke (1972) discovered an association between haplotype CDf (*Rh₁*) and *Salmonella typhi* in vivo. Individuals with *Rh₁* have a *Salmonella typhi* infection risk 76% higher than non-*Rh₁* individuals. Typhoid fever appears to kill only individuals with *Rh₁*. Consequently, the relative low frequency of *Rh₁* in the *altiplano* could be due to the epidemics of typhoid fever which decimated highland populations after the Spanish invasion of America in the sixteenth century.

The common absence of sickle-cell polymorphisms in the highlands may also be explained by the action of natural selection. On the one hand, the presence of malaria which selects for the sickle-cell polymorphism is rare at high altitude, and on the other hand, the increased oxygen delivery problems associated with altitude may make the effects of sicklemia more serious. Harrison *et al.* (1969) compared the operation of natural selection on these polymorphisms in Sardinia and Ethiopia. In Sardinia, highland and lowland populations have similar blood-group frequencies but lowland populations, exposed as they are to malaria, show higher frequencies of sickle-cell polymorphisms than the highland ones. In Ethiopia natural selection in relation to malaria is also probably operating quite differently in the highlands and lowlands. Nevertheless it would be difficult to detect genetic differentiation in Ethiopia where the migration rates are very high unless the selection was very strong.

### Migration

Migration (gene flow) is a major evolutionary force operating on small, isolated highland populations. It is very easy to detect migration by analysing birthplaces of population samples (Baker, 1977). Parental immigration rate is relatively high in most peoples studied in America and Africa. In the Andes, immigration rates vary between 6 and 33% (Cruz-Coke *et al.*, 1966; Dutt, 1976). Data from Ethiopia by Harrison *et al.* (1969) showed that in some villages more than half of the population were migrants. In some regions there is an unidirectional migration from highlands to lowlands (Cruz-Coke *et al.*, 1966), but in other areas, migration is bidirectional so that there is also some migration from lowland up to highland areas (Harrison *et al.*, 1969; Dutt, 1976). Migration flow is generally not continuous but cycles according to environmental variations in the ecosystems such as drought, floods, earthquakes, etc.

Migration may explain some of the clines in genetic polymorphisms previously described. In the Andean range, the clines of Rhesus haplotypes *Rh₁* and *Rh₂* and the alleles *Diᵃ* and *Hp₁*, from north to south, may be an expression of the migration of Quechuas during the expansion of

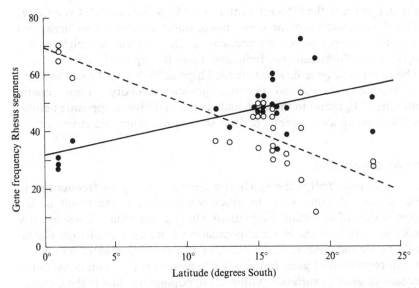

Fig. 3.1. Distribution of the gene frequencies of the Rhesus system segments $Rh_1$ (broken line and open circles) and $Rh_2$ (solid line and filled circles), from north to south along the Andes, from Ecuador (0°) to Northern Chile (25° S). (From Cruz-Coke, 1972.)

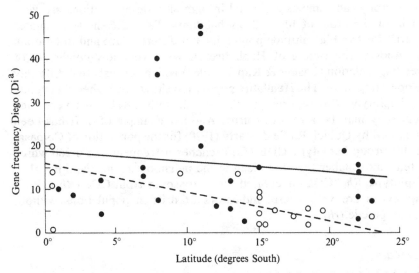

Fig. 3.2. Distribution of the gene frequencies of the Diego allele, $Di^a$, in the lowlands (solid line and filled circles) and highlands (broken line and open circles) of South America, from north to south. (From Cruz-Coke, 1972.)

59

## The biology of high-altitude peoples

the Inca Empire in the fifteenth century (see Figs. 3.1 and 3.2; Cruz-Coke, 1972). The distribution of the genotypes of Baltis peoples in the Karakoram range of Asia may also be a consequence of the migration of neighbouring peoples from Turkmanistan, India and Tibet (Clegg et al., 1961).

The process of gene flow has enabled high-altitude populations to share new genetic material and maintain genetic diversity within genetic continuity. Migration in the highlands is the main force opposing natural selection and random genetic drift in shaping its genotypic structure.

### Random genetic drift

Random genetic drift is the stochastic fluctuation of gene frequencies in a population of finite size. In other terms, drift is the result of the accumulation of sampling fluctuations along generations. Consequently genotypic distributions in small-population isolates may change significantly by simple chance. The possibility of detecting genetic drift lies in the differentiation of gene frequencies between small populations, in the decrease of genetic variation within small populations and in the increase of the frequency of homozygotes. The probability of drift is inversely related to the population size and positively related to the degree of isolation (Cruz-Coke et al., 1966).

Generally, all samples surveyed in high-altitude populations are very small, of the order of 100 to 200 inhabitants. Table 3.6 shows estimates of drift for two high-altitude populations of different size and isolation in the Andes. The isolate of Huallatire shows a very low coefficient of breeding isolation (Lasker & Kaplan, 1964) which suggests that drift may be operating there. The Huallatire population shows the highest frequency for haplotype $Rh_2$ reported in the world, with 0.847. This very high frequency may be a result of drift. Another example of drift has been suggested by Quilici, Ruffie & Marty (1970) for the population of Coipasa. In this group haplotype CDE $(Rh_z)$ reaches a frequency of 0.260 which is ten times higher than the mean in the neighbouring highland peoples. Haplotype $Rh_z$ (CDE) is created by a crossover from CDe $(Rh_1)$. Both haplotypes are very useful indicators of drift in populations without haplotype cde (rh).

### Inbreeding

Inbreeding is a process which increases the frequency of homozygotes, and, consequently, in small isolates contributes to increased random genetic drift. Theoretically, small isolates should have a high degree of inbreeding, because in a finite population all individuals are related to some extent. Unfortunately only a few studies on inbreeding have been performed in highland populations.

60

Genetic description

According to Freire-Maia (1971) the largest national inbreeding levels among Latin-American countries occurred in the Andean areas of Peru (mean coefficient 0.00270) and Ecuador (mean coefficient 0.00230). Nevertheless, Bolivia in the *altiplano* shows a mean coefficient of only 0.00030. In a detailed study in Belen, north of Chile, at 3,300 m, Cruz-Coke *et al.* (1966) found only one consanguineous marriage among forty-seven couples, with an inbreeding coefficient $F = 0.00008$. Exogamy is a cultural value among Aymarans. Garruto & Hoff (1976) also reported a trend to exogamy among a Quechua-speaking population. Harrison *et al.* (1969) reported a trend towards exogamy in the marriages occurring in both the highland and lowland villages of Ethiopia. On the other hand, the high inbreeding behaviour of the Inca dynasty of the Andes is well documented. Cruz-Coke (1965) gives a value for the inbreeding coefficient $F$ of the last true Inca, Huascar, as 0.44037. It appears the degree of inbreeding of high-altitude peoples remains an open question.

**Genetic adaptation to altitude**

Our review of the genetic description of high-altitude populations has not provided any evidence to support the idea of a specific human genetic adaptation to altitude.

The affinity of haemoglobin for oxygen is an essential variable in the process of physiological adaptation to high altitude and as previously noted not one variant of haemoglobin has been found specific to high-altitude populations. Consequently, at first it appears that the process of adaptation to high altitude is developmental and acclimatizational, not genetic.

Morpurgo *et al.* (1970) investigated whether the natives of Andean highlands had developed an enhanced Bohr effect, i.e. the decrease in affinity of haemoglobin for oxygen at lower tissue pH, as a possible mechanism of adaptation. Morpurgo and others found that Andean natives did have a higher Bohr effect than the acclimatized Europeans. As both groups have normal haemoglobins the different Bohr effect found in Peruvians must be ascribed to some factor which is capable of decreasing the oxygen affinity of haemoglobin at lower pH. A higher Bohr effect may be adaptive at the low oxygen pressures of high altitudes, because under some conditions it increases oxygen delivery to tissue. Perhaps this is an evolutionary adaptive phenomenon; however, it is difficult to judge the degree of genetic causality in the trait. In any case the physiological implications and potential genetic features of this trait deserve further study.

61

The biology of high-altitude peoples

**References**

Allison, A. C. (1964). Polymorphisms and natural selection in human populations. *Cold Spring Harbor Symposium on Quantitative Biology*, **xxiv**, 137–48.
Bada, J. L., Schroeder, R. A. & Carter, G. F. (1974). New evidence for the antiquity of man in North America deduced from aspartic acid racemization. *Science, Washington*, **184**, 791–2.
Baker, P. T. (1977). Migration and biological fitness: A case study in Southern Peru. *American Anthropologist* (in press).
Baker, P. T. & Weiner, J. S. (1966). *The Biology of Human Adaptability*. Oxford: Clarendon Press.
Bodmer, J. & Bodmer, W. F. (1973). Population genetics of the HLA system. *Israel Journal of medical Sciences*, **9**, 1257–69.
Bodmer, W. F. (1973). Population studies and the measurement of natural selection with special reference to the HLA system. *Israel Journal of medical Sciences*, **9**, 1503–32.
Cavalli-Sforza, L. L. & Bodmer, W. F. (1971). *The Genetics of Human Populations*. San Francisco, Calif.: W. H. Freeman & Co.
Cavalli-Sforza, L. L. & Edwards, A. W. F. (1967). Phylogenetic analysis: Models and estimation procedures. *American Journal of human Genetics*, **19**, 233–40.
Clegg, E. J., Harrison, G. A. & Baker, P. T. (1970). The impact of high altitudes on human populations. *Human Biology*, **42**, 486–518.
Clegg, E. J., Ikin, E. W. & Mourant, A. E. (1961). The blood groups of the Baltis. *Vox Sanguis*, **6**, 606–14.
Constans, J. & Quilici, J. C. (1973). Hémotypologie des indiens chipaya de Bolivie. *Cahiers d'Anthropologie et Ecologie Humaine*, **1**, 147–59.
Crow, J. F. (1958). Some possibilities for measuring selection intensities in man. *Human Biology*, **30**, 1–13.
Cruz-Coke, R. (1965). Consanguinidad parental en una poblacion hospitalizada. *Revista Medica de Chile*, **93**, 583–7.
Cruz-Coke, R. (1970). *Color Blindness: An Evolutionary Approach*. Springfield, Ill.: C. C. Thomas.
Cruz-Coke, R. (1972). Polimorfismos geneticos en poblaciones andinas. *V Congreso Argentino de Ciencias Biologicas*. Buenos Aires: Instituto de Fisiologia, Facultad de Medicina.
Cruz-Coke, R., Cristoffanini, A. P., Aspillaga, M. & Biancani, F. (1966). Evolutionary forces in human populations in an environmental gradient in Arica, Chile. *Human Biology*, **38**, 421–38.
Dutt, J. S. (1976). Population movement and gene flow. In *Man in the Andes: A Multidisciplinary Study of High Altitude Quechua*, ed. P. T. Baker & M. A. Little, pp. 115–27. Stroudsburg, Pa.: Dowden, Hutchinson & Ross.
Fisher, R. A. (1930). *The Genetic Theory of Natural Selection*. Oxford: Clarendon Press.
Freire-Maia, N. (1971). Consanguineous marriages and inbreeding load. In *The Ongoing Evolution of Latin American Populations*, ed. F. M. Salzano. Springfield, Ill.: C. C. Thomas.
Garruto, R. M. & Hoff, C. J. (1976). Genetic history and affinities. In *Man in the Andes: A Multidisciplinary Study of High Altitude Quechua*, ed. P. T. Baker & M. A. Little, pp. 98–114. Stroudsburg, Pa.: Dowden, Hutchinson & Ross.

Glasgow, B. G., Goodwin, M. J., Kopeć, A. C., Lehmann, H., Mourant, A. E. & Tills, D. (1965). The blood groups, serum groups, and haemoglobins of the inhabitants of Lunana and Thimbu. In *Report of I.B.P. Expedition to Bhutan, October–December 1965*, compiled by F. Jackson, T. Richard & M. Ward, pp. 52–63. London: I.B.P. Office.

Harrison, G. A., Küchemann, C. F., Moore, M. A. S., Boyce, A. J., Baju, T., Mourant, A. E., Godber, M. J., Glasgow, B. G., Kopeć, A. C., Tills, D. & Clegg, E. J. (1969). The effects of altitudinal variation in Ethiopian populations. *Philosophical Transactions of the Royal Society of London, Series B*, **256**, 147–82.

Ikin, E. W. & Mourant, A. E. (1962). A survey of some genetical characters in Ethiopian tribes. V. The blood groups of the Tigre, Billen, Amhara and other Ethiopian populations. *American Journal of Physical Anthropology*, **20**, 183–9.

Lalouel, J. M. & Morton, N. E. (1973). Bioassay of kinship in South American Indian populations. *American Journal of Human Genetics*, **25**, 62–73.

Lasker, G. W. & Kaplan, B. (1964). The coefficient of breeding isolation. *Human Biology*, **39**, 327–38.

Li, C. C. (1955). *Population Genetics*. Chicago, Ill.: University of Chicago Press.

MacNeish, R. S. (1971). Early man in the Andes. *Scientific American*, **224** (4), 36–41.

Modiano, G., Bernini, L. & Carter, N. D. (1972). A survey of several red cell and serum genetic markers in a Peruvian population. *American Journal of Human Genetics*, **24**, 111–23.

Morpurgo, G., Battaglia, P., Bernini, L., Paolucci, A. M. & Modiano, G. (1970). Higher Bohr effect in Indian natives of Peruvian highlands as compared with Europeans. *Nature, London*, **227**, 387–8.

Neel, J. V. (1973). Diversity within and between South American Indian tribes. *Israel Journal of Medical Sciences*, **9**, 1216–24.

Neel, J. V. & Salzano, F. M. (1964). A prospectus for genetic studies of the American Indians. *Cold Spring Harbor Symposium on Quantitative Biology*, **xxiv**, 85–98.

Post, R. M., Neel, J. V. & Schull, W. J. (1968). Tabulations of phenotypes and gene frequencies for 11 different genetic systems studied in the American Indian. In *Biomedical Challenges Presented by the American Indian*, Scientific Publication No. 165, pp. 141–85. Washington, D.C.: Pan American Health Organization.

Quilici, J. C. (1975). *Structure hémotypologique des populations indiennes en Amérique du Sud*. Toulouse, France: Centre National de la Recherche.

Quilici, J. C., Ruffie, J. & Marty, Y. (1970). Hémotypologie d'un groupe paléo-amérindien des Andes: Les Chpaya. *Nouvelle Revue Française d'Hématologie*, **10**, 727–31.

Rothhammer, F. & Spielman, R. S. (1972). Anthropometric variation in the Aymara: Genetic, geographic and topographic contributions. *American Journal of Human Genetics*, **24**, 371–80.

Schull, W. J. (1963). *Genetic Selection in Man*. Ann Arbor, Michigan: University of Michigan Press.

Tiwari, S. C. (1966). The blood groups of the Tibetans. In *Human Adaptability to the Environments and Physical Fitness*, ed. M. S. Malhotra, pp. 281–9. New Delhi, India: Defence Institute of Physiology and Allied Sciences, Ministry of Defence.

# 4. Fertility and early growth

E. J. CLEGG

In this chapter the earliest events of the life-cycle will be discussed. We begin with the formation of the germ cells, continue with fertilisation, placental development and maternal–foetal interrelationships, foetal growth, the events of birth and the biology of the neonate and end with growth during the first year of life.

From the point of view of the investigator, this is one of the more inaccessible periods of the life-cycle, especially in the relatively simple societies which constitute the majority of the world's high-altitude populations. For example, data on gametogenesis at high altitudes are very sparse and the sophistication of techniques required to study maternal–foetal interrelationships has resulted in their application being limited very often to populations with advanced systems of health care; such populations are infrequent in most of the high-altitude areas of the world.

The result is that our knowledge of the biology of reproduction in high-altitude man is patchy. Fortunately a considerable amount of animal data is available, but extrapolations to man should be made with caution, partly of course because of possible interspecific variation in the responses to the various environmental stresses imposed and partly because, as a rule, animal experiments have been concerned with only one factor in the high-altitude environment, that of hypoxia. As much of the work on the biology of human populations shows, there may be significant interactions between different environmental components, both at high and at low altitudes, which modify phenotypes, and these environmental components include not only physical factors, but biotic, socio-economic and cultural ones as well. Notwithstanding these *caveats*, the results of animal studies may be valuable in the present context, and they will be discussed when human data are lacking, or as otherwise thought appropriate.

In all societies demographic data are of considerable value in the investigation of problems of fertility and mortality at the population level. Either census data or individual records may be used, and the same problems are found as occur in biological investigations – the simpler the society, the less complete and reliable these vital statistics are. Indeed, in some countries with appreciable numbers of people living at high altitudes (Nepal, Tibet, Ethiopia) demographic data on a country-wide basis are almost completely lacking, and the results of small-scale *ad hoc* studies must be regarded as indicating no more than the situation in a particular place at a particular time. The demographic structure of high-

*The biology of high-altitude peoples*

altitude populations is discussed elsewhere in this book (see Baker, Chapter 11) and in this chapter only information relevant to fertility will be included.

As indicated above, important interactions may exist between the different components of the high-altitude environment, resulting in the production of very varied phenotypes in the human population. It would be naïve to explain all the unique characteristics of high-altitude populations in terms of the reduced partial pressure of oxygen. Other physical factors, such as cold, levels of sunlight and ionizing radiation or precipitation may be important, and many biotic components such as the vectors or intermediate hosts of infectious disease and standards of nutrition obviously themselves depend on physical factors. Given the addition to these non-human (albeit humanly modifiable) components of the environment, of specific socio-economic and cultural structures, it can be seen that the total network of ecological interrelationships is one of great complexity. Few attempts have been made to place man in the context of his ecological relationships with the high-altitude environment – the example of Thomas (1973) seems unique in this respect. Indeed, the difficulty in gaining any overall picture of the total ecology of high-altitude man points out the danger of ascribing particular phenotypic characteristics as the result of particular environmental stresses. While in some situations we may feel justified in doing this, we must always remember the complexity of the system and not be over-ready to accept only the most simple explanations.

Studies of the biology of man at high altitude have often been designed on a comparative basis: either long-established highland and lowland groups have been contrasted, or migrants from highland to lowland (or vice versa) have been compared with *sedentes* (see Harrison, 1966). The validity of these types of study depends upon the assumption of genetic homogeneity of the groups compared; this can only be accepted if there is gene exchange between the populations which is at such a level that differentiating factors such as differential selection and/or genetic drift can be ruled out. Harrison's (1966) model for highland–lowland comparisons would reduce to a minimum the possibility of genetic inhomogeneity, but as Haas (1973) and Abelson, Baker & Baker (1974) have pointed out, such a model has yet to be realised totally in a field situation and, as in the case of the Ethiopian studies (Harrison *et al.*, 1969; Clegg, Pawson, Ashton & Flinn, 1972) and many of the Andean investigations, what information can be extracted from less-than-ideal field situations must be used.

These somewhat pessimistic qualifications of the scientific merit of research on high-altitude populations should not blind us to the fact that a great deal of useful information has been gathered on human reproduction and growth. Much of it is still controversial and undoubtedly the

66

result of IBP, in this sphere as in others, has been to raise more questions than it answers. But that is by no means a bad thing.

**Historical survey**

In only one part of the world, South America, is there any significant historical record of the effects of life at high altitudes on reproduction and growth. Indeed, this is not surprising; many highland regions in Africa and Asia, for example Ethiopia, Nepal and Tibet, were for long either so sparsely populated or so remote from major centres that the peculiar effects of life in these regions were simply not described. In South America though, the existence of large tracts of comparatively level land at high altitude and the climatic hostility of the coastal lands in the west and the lowland forests in the east resulted in the settlement, perhaps 6,000–12,000 years ago, of these upland regions and the development eventually of the Inca civilisation, which was based on the highlands and which extended from what is now Peru both northwards into Ecuador and southwards into Chile. The rulers of this civilisation were well aware of the ecological differences between highlands and lowlands and indeed, to be transported from one altitudinal zone to another was regarded as a punishment (Cobo, 1653).

In the sixteenth and seventeenth centuries the coming to South America of the Spanish *conquistadores* opened this vast area to the study, by relatively sophisticated observers, of the effects of life in the very varied environments which they exploited. In particular one of the most clear-cut differences between the resident highland Amerindians and the immigrant lowland Spanish was the reproductive success of the former compared with the early failure and later only partial success of the latter. There are many references to the adverse effects of life in the Andean highlands on the reproductive capacity of man and animals transported there from the lowlands. Perhaps the two most interesting accounts are those of de la Calancha (1639) and Cobo (1653). The former relates the early history of the city of Potosi, situated in what is now Bolivia at an altitude of about 4,000 m. The initial population was about 20,000 Spanish and 100,000 Amerindians. Although the latter were able to reproduce successfully the former were not: all Spanish children born in the city died either at birth or within a fortnight and pregnant women descended to the lowlands to have their children, whom they would not bring up to the highlands until they were a year old. It was not until the city had been established for fifty-three years that a child was successfully reared from birth there. As Monge (1948) says, this points to 'a slow process of adaptation to life in the high altitude, for which more than one generation may be necessary'. The existence of a change over time in the adaptation of animals to high altitude is also seen in Cobo's (1653) account of the early capital of Peru, Jauja (approx. 3,500 m). At first it was regarded as a 'sterile place', where horses, pigs and fowls could not be raised. A century later it yielded abundant crops and was one of the principal pig- and poultry-farming areas, supplying Lima (200 m) on the coast with its produce.

Perhaps the most revealing of all the comments on the contrast between the reproductive performance of the lowland Spanish and the highland Indian again comes from Cobo (1653). He first points out that whereas Spanish children born in the highlands mostly died, Indian children, living under much more rigorous conditions, mostly survived. He then comments that socio-economic differences between the two groups cannot account for this difference, because the Spanish children, in order to survive, had to be protected from the cold environment to a much greater extent than the Indians. He then says that children hybrid between Spanish and Indian stocks were intermediate in their resistance to the adverse highland environment, and the greater the proportion of Indian admixture, the greater the resistance. Thus, as early as the seventeenth century, the possibility of genetic adaptation to the environmental stress of high altitudes is suggested, at least implicitly.

After the days of the *conquistadores* there are few reports about the reproductive process in high-altitude populations. Only in the third decade of the present century did systematic studies begin in South America, prompted by the obvious differences between the low-level Spanish or Mestizo inhabitants and the high-level, largely Indian population. While research was at first focussed on respiratory and cardiac physiology it soon became apparent that there were major differences in the reproductive performances of highlanders and lowlanders. The pioneer worker in this field was Carlos Monge whose book *Acclimatisation in the Andes*, published in 1948, has become the basis of much subsequent work.

### Environmental factors influencing reproduction

As indicated previously, various factors operating at high altitude interact with one another to give the total environment its unique characteristics. Although all high-altitude environments have in common a reduced barometric pressure, and hence reduced partial pressure of oxygen, they may in other ways show greater variation. While invariably cooler than adjacent lowlands (air temperature falls by 0.67 deg C for every 100 m increase in altitude) there is often considerable temperature variation, either seasonal as in temperate, or diurnal, as in tropical, mountains. This difference in temperature regimes makes it virtually impossible for temperate-zone high-altitude regions to be permanently settled. Summer snow lines are often at or near the 3,000 m level and for much of the autumn and spring land down to below 2,500 m may be covered, making permanent settlements impossible. In such regions a transhumant type of economy is often practised, temporary settlements at up to 3,500 m being occupied during the late spring and evacuated during early autumn.

Even in tropical high-altitude regions there is some seasonal variation in temperature, except on or very near to the equator. Hence in the Himalayas, Karakoram and the more southern Andes transhumance is practised, albeit at a more elevated level, the pastures temporarily occupied often lying at altitudes of 4,000–4,750 m. In the last analysis, cold rather than altitude appears to be the most important factor determining the uppermost limit of human settlement, except for equatorial high-altitude regions in the Andes, where places of work are situated in some instances at altitudes of 5,500 m. At this level, the low atmospheric oxygen tension makes permanent habitation impossible and workers reside some 500–1,000 m lower.

At high altitudes the lessened density and reduced thickness of the shell of air allows greater penetration to ground level of ultraviolet and ionising radiation, principally cosmic rays. Muller (1954) has calculated that the excess cosmic radiation in highland areas of South America and Asia amounts to about 5 R per generation. This might be expected to increase the mutational load on human populations living under such conditions by about 6%.

These various physical factors, as well as affecting man directly, influence the biological components of his environment. Amounts of insolation, precipitation (as either rain or snow) and humidity are important factors in this respect, as well as the more obvious ones of cold and hypoxia. In both tropical and temperate mountain regions well-defined vegetational zones succeed one another with changing altitude and food crops also show specificity for particular altitudinal zones. The same is true for the faunal component of the biotic environment, both wild and domesticated. With increasing altitude the number of well-adapted species diminishes, so that the economies of highland populations depend frequently on the exploitation of a small number of well-adapted plants and animals. Such relatively simple ecosystems are less stable than the potentially more complex ones at lower altitudes.

The severity of the physical environment at high altitudes has its beneficial side. The micro-organisms causing infectious disease and/or their animal vectors or intermediate hosts are often peculiarly susceptible either to cold or hypoxia (Buck, Sasaki & Anderson, 1968) and often these diseases, which cause considerable morbidity and mortality, especially in early childhood, are less common in the highlands.

Frequently, cultural and social factors within highland populations are related to the physical characteristics of the environment and may have important consequences for reproduction. Thus terrain influences the positioning and separation of settlements within a highland region and may largely determine the extent to which communications, and hence gene interchange, with neighbouring regions exist. The social and kinship structures of populations may foster or inhibit exogamy and outbreeding, while patterns of marriage may affect the chances of fertile individuals contributing their share of genes to the next generation.

In all these cultural factors which may affect reproduction, the populations in which we are interested show few, if any, features in common which may be related specifically to their residence at high altitude. However, within particular populations it may sometimes be possible to indicate the importance of certain social structures in the overall pattern of reproduction.

## Gametogenesis

Hypoxia exerts a severe inhibitory effect on cell populations undergoing active division (Kim & Han, 1969; Cheek, Graystone & Rowe, 1969). It might be expected, therefore, that the cell proliferation which occurs in gametogenesis might be especially susceptible to hypoxic damage. Surprisingly, the evidence that this is so is by no means clear-cut. Undoubtedly, extremely severe hypoxia, at levels unable to support human life for more than short periods, has been shown in experimental animals to cause damage to the spermatogenic process (Sundstroem & Michaels, 1942; Gordon, Tornetta, d'Angelo & Charipper, 1943; Dalton, Jones, Peters & Mitchell, 1945–6; Walton & Uruski, 1946; Altland, 1949a; Altland & Highman, 1952; Atland & Allen, 1952), but at atmospheric pressures comparable with or even slightly below those at which human high-altitude populations live, the effect on spermatogenesis has been reported as severe in only a few isolated investigations (Monge, 1942, 1945). In most studies damage to spermatogenesis seems to be absent (Walton & Uruski, 1946; Altland, 1949b) or if it does occur, is of a temporary nature (Altland & Allen, 1952; Chiodi, 1964; Johnson & Roofe, 1965; Monge & Monge, 1968; Monge, San Martin, Atkins & Castañon, 1945).

In a sense none of these reports throws a great deal of light on the situation *vis-à-vis* human high-altitude populations. Even if the level of hypoxia was comparable with the human situation, the animals studied were invariably exposed acutely and long-term acclimatisation was never observed over more than one generation. Even then the temporary nature of at least some of the spermatogenic damage suggests that in animals reared at normal barometric pressures, adaptive processes allow some degree of recovery.

70

Experiments more closely related to human problems have been conducted by Moore & Price (1948) and by Timiras (1964). The former authors exposed male and female rats and mice to altitudes varying from 2,290 m to 4,330 m for considerable periods but were unable to find evidence for any damage to either spermatogenesis or oogenesis. Timiras studied *inter alia* the reproductive organs of rats born at sea level and exposed to an altitude of 3,800 m, and those of their hypoxia-born offspring. She found no significant differences in testis weights in any of the various high-altitude groups in comparison with sea level controls.

Clegg (1968) has shown that in mice reared from weaning at an atmospheric pressure of 390 mm Hg, growth in testicular weight, when standardized for variation in body weight, is the same as in control animals. Thus it appears that under conditions of hypoxia comparable to those to which human high-altitude populations are subjected, the effects on spermatogenesis are not severe.

Comparatively little work has been done on oogenesis and what reports are available are concerned with the process of ovulation rather than the dynamics of oogenesis. Fernandez-Cano (1959) and Chang & Fernandez-Cano (1959) have reported on the effects of temperature and atmospheric pressure variation on ovulation in rats. They found that exposure to an atmospheric pressure of 410 mm Hg for 5 h per day on two successive days resulted in an increase in the interval between cohabitation and conception from $4.7\pm0.7$ days in controls to $12.7\pm1.3$ days in hypoxic animals – two to three times the length of the normal oestrous cycle. Similar results were obtained when the stress was heat or cold. However, ovulation when it occurred was normal, as judged by the similar numbers of corpora lutea in experimental and control animals.

These findings have been extended by Donayre (1966) in mice exposed acutely to an altitude of 4,300 m for varying periods and in which superovulation was stimulated by pregnant mares' serum (PMS) and human chorionic gonadotrophin (HCG). With increasing periods of exposure the numbers of ova shed decreased from about forty per animal at sea level to fewer than twenty after 9 days' exposure. The effect was greater if the animals were exposed concurrently to cold (see Fig. 4.1).

There appear to be no reports of the effect of long-term residence at reduced atmospheric pressure on oogenesis although indirect evidence from vaginal smears in the mouse (Krum, 1957; Donayre, 1966, 1969) suggests that there is an initial period of anoestrus (and failure of ovulation). Donayre found this period to be succeeded after about 30 days' exposure by periods of prolonged oestrus, alternating with normal oestrous cycles, but Krum reported normal cycles once the period of anoestrous, varying from nine to 30 days, was over.

This evidence, from studies in experimental animals, suggests that acute

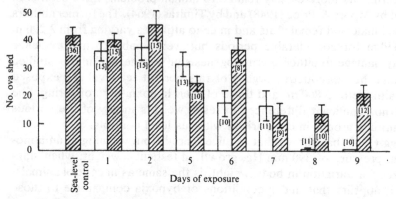

Fig. 4.1. Effects of high altitude on superovulation in mice. Animals were given 2 I.U. PMS and 1 I.U. HCG. Continued exposure leads to a fall in numbers of ova shed. The effects of altitude are greater at low temperatures. (Data from Donayre, 1966.) Open columns, cold room experiments; hatched columns, temperate room experiments. Numbers in brackets are the numbers of animals. Bars represent mean±S.E.

exposure to low atmospheric pressure, particularly when accompanied by cold, leads to some disturbance of gametogenesis, but rather less than might have been expected. More prolonged exposure results in some degree of recovery, but whether it is ever complete is open to question. It might be pointed out, though, that there appear to be no detailed quantitative histological analyses of gametogenesis in hypoxic animals of either sex, with or without cold stress, and it may be that such studies could reveal degrees of impairment of gametogenesis which while not of major extent, might appreciably reduce fertility.

Data on gametogenesis in resident or visiting man at high altitudes are also scanty and what reports there are come from workers in South America. Donayre, Guerra-Garcia, Moncloa & Sobrevilla (1968) studied the effects of 14 days' residence at an altitude of 4,300 m on sea-level dwellers. They found significant falls in sperm counts (from $216.2\times 10^6$ ml$^{-1}$ to $98.2\times10^6$ ml$^{-1}$) after 13 days' residence. Motile spermatozoa decreased from 70% to 55% and abnormal forms increased from under 5% to 15%. Live/dead staining showed only a small, non-significant change. Descent to sea level tended to reverse these adverse effects, but counts were still depressed after a week and abnormal forms rose to 39.3% after three days at sea level, falling to 20% after a week. This somewhat paradoxical change can be explained by a flushing out of abnormal forms due to an increase in the rate of sperm transport following return to normal atmospheric pressures and it suggests that investigations into the duration of the spermatogenic cycle and into epididymal transport, as well as into histological and/or cytological characteristics, might be rewarding.

The only report on the state of spermatogenesis in high-altitude residents is a brief reference by Sobrevilla *et al.* (1967) to the fact that eight healthy men at an altitude of 4,270 m had normal spermograms. There are no reports on the effect on gametogenesis of descent by high-altitude residents to lower levels.

Evidence of alteration in gametogenesis in the human female is less easy to obtain, but indirect evidence has been provided by Donayre (1966), who reported that seventeen out of twenty-nine women moving from low altitudes to an altitude of 4,300 m experienced dysmenorrhea, increased blood loss, some alteration in the regularity of the cycle and changes in the duration of flow. Broadly similar findings, also in Peruvian women, were reported by Sobrevilla (unpublished data). Harris, Shields & Hannon (1966) studied eight women who remained at an altitude of 4,300 m for ten weeks. Unlike the adverse effects reported by Donayre and Sobrevilla, there were few alterations in the menstrual cycle. Five of the subjects reported a lessened duration of flow and cycles tended to be shorter for two months after the end of the exposure. No increase in adverse symptoms occurred. Somewhat similar evidence of lack of change in menstrual pattern on moving from lowland to highland areas has been provided by Harrison *et al.* (1969) who compared residents and migrants in a lowland (1,500 m) and a highland (3,000 m) village in Ethiopia. Irrespective of altitude or of migrant status the frequency of menstrual irregularities was approximately the same.

It seems, therefore, that insofar as abnormalities in oogenesis are related to alterations in the menstrual cycle, the evidence is somewhat contradictory. Almost certainly environmental variables other than hypoxia play important parts in determining the regularity of the female reproductive cycle. Cold, for example, has been shown to cause anoestrus in unacclimatised hamsters (Grindeland & Folk, 1962) and in addition social and/or cultural factors associated with migration from lowlands to highlands may be important.

## Endocrinology of the reproductive organs at high altitudes

The events of gametogenesis are the final results of a complex series of neural and hormonal interactions. Environmental factors act locally on the various parts of the reproductive tract or they may act at a distance from any particular organ or process, influencing it not by direct action but by changing the neural or endocrine mechanisms which affect it. In addition, of course, the endocrine function of the reproductive organs subsumes many activities other than the purely gametogenic. Not surprisingly, the study of the reproductive endocrinology of high-altitude populations has been and is a rewarding one.

In the male an external indication of reproductive endocrine status is

## The biology of high-altitude peoples

libido, which is to some extent at least dependent upon the level of circulating male sex hormone. In experimental animals chronically exposed to an altitude of 4,330 m mating behaviour did not seem to be significantly affected (Moore & Price, 1948). However, acute exposure to atmospheric pressures of 380–400 mm Hg (approx. ≡ 4,800 m), while having little effect on mating behaviour for the first seven days, resulted in reduced libido at periods greater than that (Walton & Uruski, 1946). Observations on mice of both sexes (Baird & Cook, 1962) suggest that loss of libido in either sex is a temporary phenomenon, irrespective of the severity of the hypoxia.

Observations on libido in high-altitude man are scanty and of doubtful reliability, but what evidence there is does suggest that well-acclimatized populations show little reduction. Harrison *et al.* (1969) found the frequency of sexual intercourse to be approximately the same in both high- and low-altitude Ethiopian villages and Clegg (unpublished) has ascertained from well-adapted mountaineers at altitudes of over 6,000 m that libido seemed normal.

Changes in libido in the male, even if temporary, suggest *prima facie* that there is some reduction in circulating levels of male sex hormone. Studies on seven lowland subjects aged between 19 and 22 years have been reported by Guerra-Garcia, Donayre, Sobrevilla & Moncloa (1965). Urinary testosterone excretion over 24 h was measured at sea level and on the third, seventh and fourteenth day after transfer to an altitude of 4,300 m where living conditions excluded environmental differences other than in barometric pressure. On the third day there was a significant fall in testosterone excretion, from $99.8 \pm 0.43$ $\mu$g to $47.1 \pm 3.66$ $\mu$g. Levels had returned to normal by the seventh and fourteenth days. More recently Guerra-Garcia (1971) has expanded this report and increased the number of subjects to ten. As in the previous report, there was a marked decline in means of testosterone excretion by the third day of altitude exposure, but by the seventh and fourteenth days the mean levels, while not significantly different from the sea-level value, differed significantly from each of the preceding levels, as shown in Table 4.1.

This suggests that while the initial effect of hypoxia is to cause a considerable fall in testosterone excretion, sea-level values are recovered within the first week of exposure. Indeed the high levels after seven days suggest a 'rebound' effect and the general picture is reminiscent of a 'hunting' phenomenon, stability not being entirely achieved by the end of the second week of exposure.

Excretion levels of a hormone do not necessarily reflect accurately rates of production and Guerra-Garcia (1971) has shown that despite the reduction in testosterone excretion during the first three days at altitude, rates of production remained normal ($7.23 \pm 1.11$ mg.24 h$^{-1}$ at sea level;

74

Table 4.1. *Urinary testosterone excretion in ten coastal residents exposed for 2 weeks to an altitude of 4,300 m*

| | Urinary testosterone excretion (mg/day) | | | |
| --- | --- | --- | --- | --- |
| | | Days at altitude | | |
| Subject no. | Sea level | 3 | 7 | 14 |
| 1 | 52.0 | 24.1 | 236.0 | — |
| 2 | 96.0 | 76.4 | 98.2 | — |
| 3 | 72.0 | 18.1 | 164.0 | 128.0 |
| 4 | 80.0 | 31.8 | 124.0 | 68.4 |
| 5 | 107.6 | 68.5 | 130.0 | 51.2 |
| 6 | 176.0 | 37.2 | 90.0 | 95.0 |
| 7 | 53.2 | 32.0 | 68.1 | 55.1 |
| 8 | 113.6 | 66.2 | 241.0 | 111.0 |
| 9 | 177.0 | 19.3 | — | 24.0 |
| 10 | 71.0 | 16.8 | — | 30.7 |
| Mean±s.e. | 99.8±14.3 | 39.0[a]±7.1 | 144.0[a]±23.0 | 70.4[b]±13.3 |

Data from Guerra-Garcia (1971).
[a] $P < 0.001$ ⎫ In comparison of mean values after 3, 7 or 14 days' exposure with
[b] $0.02 < P < 0.05$ ⎭ control, 3- and 7-day values respectively.

$5.11 \pm 0.44$ mg . 24 $h^{-1}$ at altitude). In the same subjects urinary excretion of injected [$^{14}$C]testosterone was essentially similar at sea level and after 60 h at altitude, as were the excretion of 17-oxysteroids and creatinine.

Guerra-Garcia also studied, in four individuals, the response, in the shape of urinary excretion of androsterone and 17-oxysteroids, to injection of HCG at sea level and after transfer to altitude. Subjects were given injections of 2,500 IU of HCG on four successive days and urinary testosterone and 17-oxysteroids were estimated on each successive day and a day after the final injection. Comparison of daily levels of excretion of the two substances showed no significant altitudinal differences between means.

The principal discrepancy here, between the lowered rate of urinary excretion of testosterone and its apparently normal rate of production, may be explained by the error involved in estimating testosterone production when ACTH is being produced in excess amounts (Saez & Migeon, 1967), as happens during the first few days of residence at high altitude (Moncloa, Donayre, Sobrevilla & Guerra-Garcia, 1965). Guerra-Garcia gives the following explanation: increased ACTH stimulation of the adrenal cortex results in the production of large amounts of androstenedione, which is converted peripherally to testosterone, diminishing the specific activity of testosterone in the process of estimation and thus overestimating the rate of production. Since the response to HCG, as measured by the urinary excretion of testosterone, is normal, it may be concluded that acute exposure to altitude results in a temporary lowering of luteinizing hormone (LH) production leading to diminished production and excretion of testosterone. This finding is in accord with evidence from experimental animals. Sohval & Soffer (1951) and Faiman & Winter (1971) found a reciprocal relationship between levels of ACTH and LH, whether the latter was measured as plasma LH or as urinary gonadotrophin. Guerra-Garcia, Velasquez & Whittembury (1965), Guerra-Garcia, Velasquez & Coyotupa (1969) and Guerra-Garcia (1971) have also produced data on similar variables in Mestizo subjects native to an altitude of 4,300 m. They were compared with similar subjects residing at sea level. Urinary testosterone levels were similar in the two groups (99.8±14.3 $\mu$g. 24 h$^{-1}$ at sea level; 96.5±10.1 $\mu$g. 24 h$^{-1}$ at altitude). Elimination of [$^{14}$C]testosterone was equally effective and rates of testosterone production were virtually the same (4.28 mg. 24 h$^{-1}$ at sea level; 4.18 mg. h$^{-1}$ at altitude). However, the response to HCG, measured as daily urinary testosterone excretion, was very much less in the highlanders than in the lowlanders, despite essentially similar levels before HCG administration. From a mean of 97.6±20.5 $\mu$g. 24 h$^{-1}$ before treatment, sea-level residents showed a rise in testosterone excretion to 301.4±47.5 $\mu$g. 24 h$^{-1}$ on the third day; high-altitude natives, from 79.2±7.2 $\mu$g. 24 h$^{-1}$, rose only to 130.9±9.9 $\mu$g. 24 h$^{-1}$, again on the third day. Excretion of 17-oxysteroids rose in both groups after HCG administration, the maximum response being at sea level rather than at altitude (5.75±1.48 mg; 3.42±0.56 mg, respectively) but this difference, although suggestive, was not statistically significant.

These results obviously differ from those in individuals acutely exposed to high altitudes. The lack of differences in urinary testosterone excretion, and in testosterone production rates, coupled with the similarity in urinary gonadotrophins reported previously by Sobrevilla *et al.* (1967) suggest that the hypophysis–gonad relationship functions normally in high-altitude natives. Nevertheless the response of high-altitude natives to HCG is very significantly less than at low altitude and appears to be shorter lived. This response resembles that to ACTH reported by Subauste *et al.* (1958) and Moncloa & Pretell (1964) and to glucagon, as shown by Picón-Reátegui (1966). The difference in response to ACTH has been suggested by Moncloa (1966) to be due to more rapid catabolism of ACTH at high altitude. Guerra-Garcia (1971) suggests that the diminished response to gonadotrophic stimulation, which must ultimately be at the cellular level, may be due to lower cellular concentrations of cyclic adenosine mono-phosphate (cyclic AMP), which has been demonstrated to be important in determining the sensitivity of target organs to circulating hormones. In the comparison between acutely and chronically hypoxic men, Guerra-

76

Garcia suggests that the diminished response to HCG in the latter is due either to the length of time of exposure or to genetic adaptation to life at high altitudes. In view of the Mestizo origin of all his subjects, the former explanation seems the more likely. The lessened response is of course more like that in sexually immature individuals and comparative data on age changes in response to exogenous gonadotrophin in highland and lowland residents would be most interesting.

The reproductive endocrinology of the nonpregnant female at high altitude has not been studied to the same extent as in the male and data tend to be fragmentary. The only observation comparable with those of Guerra-Garcia (1971) in men is that of Moncloa (quoted by Donayre, 1966), who found that urinary excretion of pregnanediol is similar in high- and low-altitude resident women, irrespective of the phase of the menstrual cycle. Pregnanediol is a product of the catabolism of progesterone, so it might reasonably be inferred that there are no significant differences in the dynamics of progesterone metabolism between high-altitude and low-altitude women. However, data on rates of production of oestrogens and progesterone and on the response to exogenous gonadotrophins would be of considerable interest. Observations of Tajik women by Petranjuk (1967) (quoted by Mirrakhimov & Lebedeva, personal communication) confirmed Moncloa's findings in respect of pregnanediol excretion. In addition, this worker could demonstrate no significant changes in oestrogen or gonadotrophin excretion among highlanders.

In one field, that of the rate of phenotypic sexual maturation, data on high-altitude females are more numerous than on males. The reason is, of course, the relative ease with which menarche may be timed. Reports are available from the Andes, Ethiopia and the Himalayas, but they are by no means unanimous. Perhaps the best documentation is from the Andes where there seems general agreement that menarche is delayed in high-altitude girls (Cruz-Coke, unpublished data; Cruz-Coke, Cristoffani, Aspillaga & Biancani, 1966; Frisancho & Baker, 1970; Moncloa, quoted by Donayre, 1966; Sobrevilla, unpublished data). The average period of delay is about two years: Moncloa's data indicate that at the age of 13 years, menarche has been attained by 73% of lowland girls, but only 38% of highlanders. In the Himalayas most reports support these observations (Lang & Lang, 1971; Pawson, personal communication, among the Sherpas of Sola Khumbu; Ward, 1969, in Bhutan), mean ages of menarche varying from 17 to 19 years. However, Kawakita (1953) reports that in the village of Tsumje at 3,450 m in Western Nepal, menarche occurs as a rule between the ages of 13 and 14. One report from the Pamirs (Pulatova, Laptieva & Iskandarova, personal communication) states that mean menarcheal age at Khorog at 2,200 m is 15 years. In the lower valleys the mean age is 13.5 years. In Kirghizia at altitudes of 2,300–2,800 m menarche is at 15 years 2 months on average. At 700 m it is at 14 years 5 months (Miklashevskaya, Solovyeva & Godina, 1972). Rather greater delays in menarche (1.5–2.0 years) have been reported in Tajik women at moderate or high altitude (Toktorbaeva, 1966; Borzykh, 1971) (quoted by Mirrakhimov & Lebedeva, personal communication).

The data from Ethiopia (Harrison *et al.*, 1969) suggest that menarcheal ages in highlanders and lowlanders do not differ significantly, retrospective means in highland and lowland residents being 14.5 and 14.7 years respectively. Estimates based on recall are liable to considerable error in any one individual, but as Damon, Damon, Reed & Valadian (1969) have pointed out, the error is unbiased in direction and mean values obtained in this way are probably reasonably accurate. Some confirmatory evidence that in Ethiopia at least, altitude has little effect on rates of sexual maturation was provided by Clegg *et al.* (1972) who found no significant differences between highland and lowland children in the chronological or skeletal ages of entering the various stages of breast development.

There are also some data from Europe on the influence of altitude upon menarcheal age. Valšík, Štukovský & Bernátová (1963) suggest that in central Europe median menarcheal age increases by three months per 100 m altitude, a postulated rate of change very much greater than that found in Andean or Himalayan populations.

There are few data on age at puberty in boys at high altitude. Clegg *et al.* (1972) report that mean ages of attaining the various stages of genital development are not significantly different in Ethiopian boys at high or low altitude, but no data are available from the Andes or Himalayan regions where the hypoxic stress is more severe. However, the delay in commencement of the adolescent growth spurt which is known to occur in Andean highland populations (Frisancho & Baker, 1970) suggests that here, as with girls, puberty is delayed (see Frisancho, Chapter 5).

The delay in puberty found in most investigations on resident populations at high altitudes may be explained as a simple effect of oxygen-lack on all metabolic processes. The inhibitory effect of increased levels of circulating glucocorticoid hormones may be important in the reduction in growth rates which occurs on acute exposure to hypoxia, but it has been shown by several groups (Correa, Aliaga & Moncloa, 1956; Moncloa, Pretell & Correa, 1961; Moncloa & Pretell, 1964) that cortisol production levels are normal in high-altitude residents. The delay in menarche may also be explained by the delay in reaching the 'critical weight' of 47 kg as postulated by Frisch & Revelle (1971), but no reports are available which examine this particular hypothesis in high-altitude populations.

**Fertilization and implantation**

In this aspect of reproductive physiology the evidence is entirely derived from animal experiments. Some early experiments have been described by Fernandez-Cano (1959) and Chang & Fernandez-Cano (1959). Rats were exposed to an atmospheric pressure of 410 mm Hg for five hours per day on two consecutive days after mating, up to the eleventh day. Examination of the corpora lutea of ovaries and the number of normal or degenerated embryos on the fifteenth or eighteenth days enabled (1) the number of pre-implantation and (2) the number of post-implantation embryonic deaths to be calculated. Early exposure to hypoxia (days 1–4) caused a significant rise in the proportions of pre-implantation deaths (25.9% at days 3–4; control 2.4%) while hypoxia at day 6 or later caused an even larger rise in the proportion of post-implantation deaths (65.9% at days 10–11; control 0%). The explanation given for these changes is a hormonal one: adrenal glucocorticoids have been shown to cause foetal death in experimental animals, and further experiments (Fernandez-Cano, 1959) showed that adrenalectomy resulted in reductions in rates of foetal death to values not significantly different from controls. Baird & Cook (1962) exposed mice to various hypoxic regimes and found that almost all female animals became pregnant after cohabitation with males. They noted no significant effects on implantation, the development of the placenta, the length of pregnancy, or birth. However, the foetus at the 7-mm stage seemed to be at greater risk of resorption, the level of risk being positively related to the degree of hypoxic stress.

More recently, Blackburn & Clegg (1975) have studied the effect of various levels of reduced atmospheric pressure on the recovery of fertilized ova from previously mated mice. It was found that at the lower pressures the numbers of ova recovered 72 h after mating were reduced as compared with higher pressures. Thus at 390 mm Hg fifteen mice produced only thirty-two ova, all morphologically abnormal and of the fifteen, nine mice produced no ova at all. At 470 mm Hg, twenty mice produced forty-three ova, thirty-five (81.4%) of which were abnormal. Three of the mice produced no ova. At 530 mm Hg thirteen mice produced fifty-five ova, fourteen of which (25.5%) were abnormal. Three of the mice produced no ova. At a pressure as high as 630 mm Hg (corresponding to an altitude of about 1,500 m) eight mice produced sixty-nine ova of which ten (14.5%) were morphologically abnormal. This rate of abnormality was significantly higher than at sea-level pressures, where forty-eight mice produced 410 ova, twenty-six (6.3%) being abnormal. These results are shown in Fig. 4.2.

Thus it seems that acute hypoxia even of minor degree causes early destruction of ova, whether fertilised or unfertilised, and abnormal deve-

79

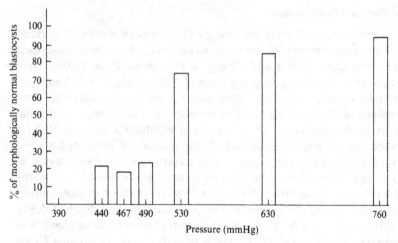

Fig. 4.2. Percentages of surviving blastocysts in mice exposed to varying barometric pressures from 7–86 h *post coitum*. At all pressures below normal atmospheric the reduction in the percentage of normal blastocysts is statistically significant. (Blackburn & Clegg, unpublished.)

lopment of a high proportion of those which survive. Since during this early stage of pregnancy implantation had not occurred, the most likely explanation for the dverse results of hypoxia is a simple diminution in the partial pressure of oxygen in the uterine secretions, possibly associated with a change in pH, although the hormonal explanation given by Fernandez-Cano (1959) may be valid towards the end of the pre-implantation period.

To what extent these findings are applicable to man, or how they are modifiable by prolonged residence in hypoxic conditions remains uncertain. Whether the explanation of the phenomena is due to hypoxia directly or to hormonal changes, the process of acclimatisation might be expected to reduce deviation from normality; further investigations are required here.

### The placenta

For the foetus *in utero* the placenta performs the exchange functions of the respiratory, gastro-intestinal and renal tracts in the free-living individual. Across the tissue barrier which separates the maternal and foetal circulations diffuse oxygen and carbon dioxide, the nutrients, minerals and water essential for growth and maturation and the waste products of foetal metabolism. As well as these functions the placenta is an endocrine gland in its own right, largely taking over the functions of the ovaries in the latter two-thirds of pregnancy.

In man the placenta is a haemo-chorial organ, that is, the trophoblast which forms the external, invasive, tissue of the implanting blastocyst burrows through the surface epithelium of the endometrium, the connective tissue ('stromal') cells and the walls of endometrial blood vessels, to form a thin placental barrier, consisting of syncytiotrophoblast, cytotrophoblast, trophoblastic connective tissue and blood-vessel wall. As pregnancy proceeds the placental sinusoids become lined by a thick layer of syncytiotrophoblast. Substances can pass from maternal to foetal circulation either by active transport or by diffusion, the rate of the latter depending on the concentration gradient between the two circulations and the thickness of the barrier as well as on the particular physical properties of the membrane (diffusion coefficient) and on the total surface area for exchange (Aherne & Dunnill, 1966). Generally speaking the diffusion coefficient remains fairly constant and diffusion is facilitated either by reducing the thickness of the placental barrier, by raising the maternal–foetal concentration gradient or by increasing the total area across which diffusion occurs. The first and last modifications may be achieved by morphological means, the second by changing the rate of blood flow through the placenta.

Experimental work on the effects of hypoxia on the development of the placenta is scanty and has the further disadvantage that the specific haemo-chorial relationship between maternal and foetal circulations is not seen in all experimental animals. In sheep living at an altitude of 4,300 m the syndesmo-chorial placenta appears to be able to maintain normal oxygen tensions in the developing foetus, the growth rate of which is also normal (Metcalfe *et al.*, 1962*b*). However, close inspection of their data suggests that between 80 and 120 days gestational age foetal growth may be retarded. Certainly birth weights in high-altitude and low-altitude sheep did not differ significantly despite the greatly reduced gradient in partial pressure of oxygen ($P_{O_2}$) between maternal and foetal circulations. In the sheep placenta the principal organs of exchange are the cotyledons, which consist of maternal syndesmal caruncles, overgrown by foetal trophoblast. The greater the number of cotyledons the greater the area for diffusion between maternal and foetal circulations. Surprisingly, Metcalfe *et al.* (1962*b*) found no increase in the numbers of cotyledons in high-altitude animals; indeed their total weight tended to be lower – possibly, as the authors suggest, because of the increased vascularisation, with replacement of the relatively heavy Wharton's jelly of the foetal side of the cotyledon by lighter vascular tissue. However, the authors did not report any histological observations, so the nature of any morphological changes in the high-altitude sheep placenta remains unknown.

81

Table 4.2. *Histopathological findings in the placenta of hypoxic and normal guinea-pigs*

|  | Hypoxic animals (N = 66) | | Control animals (N = 59) | |
|---|---|---|---|---|
|  | No. | % | No. | % |
| Acute, extensive infarcts | 12 | 38 | 2 | 7 |
| Small infarcts | 13 | | 2 | |
| Severe obstructive lesions in decidual arteries | 42 | 64 | 2 | 3.5 |
| Thrombosis of decidual veins or basal venous ring | 12 | 18 | 1 | 1.7 |
| Perilobular thrombi and/or fibrin deposit | | | | |
| Focal | 25 | 56 | 19 | 34 |
| Extensive | 12 | | 1 | |

Data from Delaquerriere-Richardson & Valdivia (1967).

In contrast with these obviously successful ovine adaptations, acute hypoxia in unacclimatised experimental animals may have severely adverse effects on the placenta. Delaquerriere-Richardson & Valdivia (1967) exposed pregnant guinea-pigs to reduced atmospheric pressures equivalent to 3,650–5,000 m and found massive placental infarcts and haemorrhages, their incidence significantly associated with the rate of foetal death. In addition microscopic infarcts were also frequent. These results are shown in Table 4.2. Thrombotic lesions of vessels were also seen, both in the mesometrium and decidua. Possible causes for such vascular changes will be considered later.

In comparing these results with those of Metcalfe *et al.* (1962*b*) not only the differences in the degree of adaptation of the pregnant ewes need to be taken into consideration, but also the structure of the maternal–foetal barrier. In the sheep, several structures, which include maternal vessel wall, endometrial stroma, foetal chorion and vessel wall contribute to the barrier. In the guinea-pig and other rodents the placenta is haemo–endothelial, the barrier consisting solely of the endothelium of the foetal blood vessels. It might be expected, therefore, that the effects of hypoxia on the placenta might well be different because of these morphological differences and independent of the state of adaptation of the animals. As will be discussed more fully later, one difference between the pregnant hypoxic sheep and the rat is that in the latter some degree of adaptation of the foetus occurs, and this may well be related to the differing placental structure.

Among animals specifically native to high-altitude regions only the llama

Fertility and early growth

appears to have been studied. Meschia *et al.* (1960) reported their findings on the placentae and on the oxygenation of the foetus in animals living at 4,500 m altitude. They found that, per unit foetal weight, the area for diffusion was much greater than for lowland animals with comparable epithelio–chorial placentae (e.g. horse and pig). The $P_{O_2}$ gradient across the maternal–foetal barrier was only about 15 mm Hg as compared with 40 mm Hg in sea-level sheep and about 20 mm Hg for altitude adapted sheep (Barron *et al.*, 1964). No data on lowland llama, for example in zoos, appear to be available for comparison.

While the animal material is thus small in amount, a good deal of information is available on the placenta in human high-altitude populations. Much of the work has been summarised and discussed by McClung (1969) in her admirable and comprehensive monograph.

Most authors report no significant differences between weights of placentae in high- and low-altitude Andean populations. Chabes *et al.* (1967) found mean placental weight in a highland population (4,000 m) to be 511 g; at sea level the mean was 541 g. McClung (1969) found means in highlanders (3,400 m) and lowlanders (200 m) to be 606±15 and 584±14 g for male births and 601±17 and 604±22 g for female births respectively. As usual, placentae from first pregnancies were lighter than from subsequent ones. Sobrevilla (1971) reports the results of two surveys in highland (4,300 m) and lowland (200 m) populations. Mean placental weights in highlands and lowlands respectively were 539±32 g and 474±15 g for the first study and 515±33 g and 522±17 g for the second. None of these differences is statistically significant and, from the data reported by all the authors, there seems to be no particular trend of placental weight change with increasing altitude. However, Krüger & Arias-Stella (1970) compared two populations at 4,570 m and at 200 m and found mean placental weight at the former (561 g) to be significantly greater than at the latter (500 g). Placental volumes did not differ significantly, indicating that the density of the placenta was greater at this very high altitude. The authors note the difference in their results from those of other authors, and attribute it to the greater degree of hypoxia in their highland population.

However, if the significance of differences in placental weights is somewhat controversial, there is no doubt that almost all reports show a notable increase in the placental ratio, the ratio between placental weight and birth weight. This occurs because birth weights are almost uniformly reported as depressed in high-altitude populations. Chabes *et al.* (1967) found no change in the ratio, but McClung (1969) gives data from her own investigation in Lima (200 m) and Cuzco (3,400 m), together with other reports comparing Lima and Cerro de Pasco (4,300 m) (see Table 4.3). These show clearly that despite the similarity of placental weights in highland and lowland, the placental ratio is significantly raised among

83

Table 4.3. *Comparison of measurements (means±S.E.) of placentae at Cuzco and Lima*

| Variable | Cuzco (3,400 m) | Lima (200 m) |
|---|---|---|
| Placental weight (g) | 603.6±15.1 | 592.4±12.5 |
| Placental ratio | 0.17±0.004[a] | 0.15±0.003 |
| Minimum placental diameter (cm) | 16.2±0.2 | 15.7±0.2 |
| Placental depth (cm) | 2.2±0.05[a] | 2.5±0.05 |
| Cord length (cm) | 43.9±1.3[a] | 49.7±1.0 |
| Number of infarcts per placenta | 0.9±0.1 | 0.4±0.1 |
| % placentae with infarcts | 53 | 31 |

Data from McClung (1969).
[a] $P > 0.001$.

highlanders. Similar results have been obtained by Krüger & Arias-Stella (1970) and Sobrevilla (1971). Two reports are available on the shape of the placenta. Chabes *et al.* (1968) reported the results of their studies at 3,800 m and at sea level. At high altitude the proportion of round placentae diminished significantly and variability in shape was greater. Variability was greater with male offspring than female. At high altitudes, round placentae were thicker than those of other shapes and variability in shape became greater with increasing age and parity. McClung's (1969) results showed slightly greater placental diameter and significantly smaller depth at high altitude (see Table 4.3). In view of the high correlation (0.88) between placental external surface area and the area across which diffusion can take place (Aherne & Dunnill, 1966) it looks as though high-altitude placentae do have a greater potential area for maternal–foetal exchange.

There are a few reports of studies on the microscopic anatomy of the placenta in high-altitude residents. Tominga & Page (1966) report some interesting observations on the histological consequences of perfusing fresh human placentae at varying atmospheric concentrations of oxygen. The most obvious changes were in the syncytiotrophoblast lining the inter-villous spaces. At 26% oxygen the thickness of this layer changed little after 48 h perfusion, but at 6% oxygen the syncytiotrophoblast formed aggregations ('knots') on one side of a villus, leaving only a very thin layer covering the connective tissue core elsewhere. These changes were quite obvious after 6 h perfusion and were reversible if oxygen tensions were returned to normal levels. They resulted in a thinning of the maternal–foetal barrier; the mean distance between syncytial surface and foetal capillary was reduced from 4.4 $\mu$m to 3.3 $\mu$m after six hours on 6% oxygen, potentially increasing the diffusion capacity by 25%. Similar results have been reported by Maclennan, Sharp & Shaw-Dunn (1972).

These experimental findings, obviously short-term responses to acute hypoxia, need not necessarily resemble long-term adaptations in high-altitude residents. Monroe-Rodriguez (1966, quoted by Krüger & Arias-Stella, 1970) reported that at high altitude the total inter-villous space is reduced in volume. Chabes *et al.* (1968) have presented, in abstract form, the results of morphometric studies on placentae from an altitude of 3,800 m and from sea level. At high altitude the numbers and areas of terminal villi were greater, together with the number of blood vessels and amount of trophoblast. As with Tominaga & Page's (1966) findings, the number of syncytiotrophoblastic 'knots' was increased. The authors suggest that the principal adaptive change is an increase in the surface area for exchange due to the increased number of villi and capillaries.

There appears to be no human evidence of changes in uterine blood flow at high altitude, but Metcalfe *et al.* (1962*a*) report that in altitude adapted pregnant ewes uterine blood flow is greater than at sea level, at least during the second half of pregnancy. Despite this, Metcalfe *et al.* (1962*a*) calculate that the effective oxygen tension, that constant tension required to produce the gas transfer observed, is only 41.3 mm Hg, compared to about 63 mm Hg in sea-level sheep.

Most of the changes so far described in the placenta may be regarded as adaptive in that they facilitate maternal–foetal transfer. However, as Chabes *et al.* (1968) demonstrate, the shape of the placenta in their high-altitude population was more variable and shapes other than round or oval were 25.7% of the total compared with 8.0% at sea level. There appear to be no data on the arrangement of placental vessels or on the insertion of the umbilical cord but McClung (1969) found cords to be shorter on average in her highland sample (Table 4.3).

The data of Delaquerrier-Richardson & Valdivia (1967) suggest that pathological vascular changes leading to infarction are common in guinea-pigs exposed acutely to hypoxia. Infarcts are seen in many human placentae and in McClung's (1969) series mean numbers of infarcts were significantly greater at high altitude (0.9) than at low altitude (0.4) (Table 4.3). Infarct frequency was negatively correlated with parity at altitude, but not in the lowlands, and significantly related to the race of the mother, women with a large European admixture having more. This last finding suggests that, as in guinea-pigs, relative lack of long-term population adaptation may lead to a greater risk of infarcts. The degree of infarction has been shown to be related to foetal growth and is presumably detrimental to placental function. McClung (1969) suggests that infarction may be the result of hypoxia-mediated constriction of placental vessels, or

85

increased viscosity of blood due to polcythaemia. However it has been shown that the response of uterine vessels to hypoxia is dilation (Panigel, 1962; Tominaga & Page, 1966). It may be that it is a response to the respiratory alkalosis which develops at altitudes.

A final placental abnormality which is worthy of consideration is placenta praevia. Obviously, the greater the surface area of the placenta the greater the chance that it will encroach on the lower uterine segments, possibly resulting in antepartum haemorrhage. McClung quotes obstetricians at Cuzco (3,400 m) as stating that in 27% of seventy-three deliveries placenta praevia was present as against less than 1% of eighty-eight deliveries at Lima. Why this should be so is uncertain. Other than excessive growth in area of the placenta, error of implantation of the blastocyst could account for the phenomenon. Many factors causing stress to an animal delay implantation (Daniel, 1970) and it seems at least plausible to suggest that at high altitude this stress could cause greater variability in the time taken to implant with greater risk of an abnormally low implantation. It would be particularly interesting to see if the frequency of the condition were any greater in European residents at high altitude, who presumably would be less well adapted than native highlanders.

## The foeto-placental unit

In recent years it has become apparent that the conjoined foetus and placenta constitute a single physiological entity, in many respects largely independent of the maternal environment which surrounds them. An illustration of this is the gradual transfer of the hormonal control of pregnancy from the ovaries to the placenta. The concept of the foeto-placental unit is to some extent an endocrine one since the control over the growth and maturation of the foetus, given adequate supplies of oxygen and nutrients, is largely by hormonal means and the relative independence of the unit from the hormonal environment of the mother means that whatever the stresses imposed upon her, foetal growth remains to a large extent unaffected.

In man the placenta has taken over most, if not all, of the endocrine functions of the ovary by the end of the first trimester of pregnancy. Yoshimi, Strott, Marshall & Lipsett (1969) have estimated the effective life-span of the human corpus luteum to be about 70 days, so that for the greater part of pregnancy the sole important source of progesterone is the syncytiotrophoblast of the placenta. However, the precursors necessary for the biosynthesis of this substance come from both mother and foetus (Hytten & Leitch, 1971, p. 181).

Likewise, oestrogens are also produced mainly by the placenta, but the part played by foetal tissues is probably more important than with

Table 4.4. *Birth and placental weights of newborns and endocrine status of newborns and mothers at altitudes of 4,300 m and sea level*

| | Mother | | | | Foetus | | |
|---|---|---|---|---|---|---|---|
| | | Urinary oestrogen excretion mg/24 h | | | | | |
| Altitude (m) | Oestriol in venous blood ($\mu$g %) | Oestrone | Oestradiol | Oestriol | Birth weight (g) | Placental weight (g) | Oestriol in cord blood ($\mu$g %) |
| 4,300 | 22.08±3.09 | 1.24±0.29 | 0.27[a]±0.06 | 10.66[b]±2.10 | 2995[b]±179 | 474[a]±15 | 128.13[a]±13.95 |
| 0 | 27.26±2.85 | 1.88±0.29 | 0.45±0.07 | 23.03±2.75 | 3603±154 | 539±32 | 161.72±10.48 |

Data from Sobrevilla (1971).
[a] $0.02 < P < 0.05$.        [b] $0.001 < P < 0.01$.

progesterone (Hytten & Leitch, 1971, p. 193). Both the liver and adrenal glands contain the enzyme systems necessary for the synthesis of various oestrogens and Simmer (1972) suggests that, at term, most of the precursors of placental oestriol and about half the oestrone and oestradiol are produced by the foetal adrenal cortex. However, Charles *et al.* (1970) suggest that the maternal adrenal glands may produce a greater proportion (up to 50%) of the precursors of placental oestriol.

Apart from these steroid hormones, the placenta produces large amounts of protein and polypeptide hormones, notably gonadotrophin (HCG) and a growth hormone and prolactin-like substance often called human chorionic somato-mammotrophin (HCS).

As yet most of the work on these placental hormones in high-altitude populations has been concentrated on the oestrogens. Sobrevilla, Romero, Krüger & Whittembury (1968) have shown oestrogen excretion to be decreased during pregnancy at high altitude and Rodriguez (1969, quoted by Sobrevilla, Romero & Krüger, 1971*b*) has shown that this decrease begins in the second trimester of pregnancy and continues to term. At term Sobrevilla (1971) and Sobrevilla *et al.* (1971*b*) found significantly reduced levels of oestriol in cord-blood of neonates at 4,300 m, compared with sea level. Levels in maternal venous blood and amniotic fluid were lower at altitude but not significantly so. As indicated above, urinary excretion of oestriol was significantly reduced in the high-altitude subjects. These results are shown in Table 4.4. There were significant correlations between cord–blood oestriol and birth weight and between cord and maternal venous blood oestriol levels, correlations which have been recognized in normal and abnormal births at low altitude (Coyle & Brown, 1963). These data point to a diminished production or decreased hydroxylation of oestrogen precursors, certainly by foetal and probably by maternal tissues.

87

The exact nature of the biochemical lesion, whether quantitative or qualitative, appears to have been the object of only one study (Sharp, Carty & Young, 1971). These authors found that in human placentae cultured in hypoxic conditions (6% $O_2$) there was within a short time a severe fall in hydroxysteroid dehydrogenase (HSD) activity. The trophoblast was more affected than the core of the villus and $3\beta$-HSD seemed more sensitive than $17\beta$-HSD. The activity of the latter enzyme was more affected when oestradiol-$17\beta$ was used as a substrate than when testosterone was used. While recognizing the artificiality of their system, the authors suggest that similar changes might take place in hypoxic conditions *in vivo*.

While the effects of high altitude on oestriol excretion are marked, Sobrevilla's (1971) study showed that amounts of oestrone and oestradiol are not affected to the same extent. Differences in the former were not statistically significant and in the latter barely significant. These differences from the response of oestriol seem to confirm Simmer's (1972) suggestion that the predominant source of these two steroids is not the same as of oestriol. Sobrevilla (1971) points out that the exceedingly low levels of urinary oestriol among pregnant highland women (mean 10.66 mg/day) are below the threshold level for normal foetal viability generally accepted in Europe and North America. In three of his cases excretion was less than 4 mg/day, a level which has usually been taken to indicate foetal death. Despite this, the newborns in his high-altitude group, although smaller in weight than newborns at low altitude, were otherwise normal.

The explanation of these discrepancies is probably related to the dynamics of oestrogen metabolism. Sobrevilla suggests that together with lessened foetal production of the hormone, the mother's kidney responds with a lesser rate of clearance, thus maintaining normal levels in her venous blood.

The problem of oxygen transport across the placenta has been briefly referred to above in experimental animals. For man there is a certain amount of data. Ever since the days of Barcroft's (1936) early observations, it has been realized that even at sea level the developing foetus is exposed to an appreciable level of hypoxia. However, there are major difficulties in the estimation of oxygen tension in the inter-villous space (Fuchs, Spackman & Assali, 1963). Uterine blood flow in women at term has been measured (Bartels, Moll & Metcalfe, 1962) at 500–750 ml. min$^{-1}$ which after allowing 25% for supply of the uterus alone, gives 375–560 ml. min$^{-1}$ as the flow through the inter-villous spaces.

Romney, Reid, Metcalfe & Burwell (1955) calculated uterine blood flow and umbilical arterio-venous oxygen differences in twelve sea-level women. These results have been re-assessed by Hytten & Leitch (1971, p. 255) who, on the basis of a 3.4 kg foetus, estimate oxygen requirements as follows: foetus 12 ml, placenta 3.7 ml, and uterus 3.3 ml, making a total of 19 ml. min$^{-1}$. At an arterio-venous oxygen difference of 4.5–5.0 ml. min$^{-1}$ this gives a blood flow to foetus and placenta of 310–360 ml. min$^{-1}$ and to the total uterus and its contents of 380–420 ml. min$^{-1}$

While these levels of flow are easily attained at sea level, it may be more difficult to do so at high altitude and adaptive changes may be necessary to maintain adequate oxygenation of the foetus. The degree of hyperventilation is greater in pregnant than in nonpregnant highland women (Hellegers *et al.*, 1961; Karavaeva & Svistov, personal communication), and cardiac enlargement is also greater than at sea level (Hellegers *et al.*, 1961). While this might suggest that the increase in cardiac output at altitude is greater than that in pregnant women at sea level, Yusupova & Svistov (personal communication) found no significant changes in arm–lung and arm–tongue circulation times in pregnant women as compared with non-pregnant women at an altitude of 2,200 m, a finding similar to that made in many surveys in sea-level populations (Hytten & Leitch, 1971, p. 89). Furthermore the changes in pulse rate found in Yusupova & Svistov's survey were in no way different from what might be expected at sea level. Admittedly the hypoxic stress at an altitude of 2,200 m is not severe and perhaps more pronounced differences might be found at higher altitudes. However, there appear to be no reports of such studies. It seems probable then that changes in cardiac output in pregnant women at high altitude are not very great. Nevertheless in view of the vasodilatory effect of hypoxia on placental blood vessels (Panigel, 1962; Tominaga & Page, 1966), rates of perfusion of the placenta may be increased. Mann (1970) has·shown that in the sheep acute maternal hypoxia leads to diminution in vascular resistance in the umbilical–placental circulation. Since it seems reasonably certain that the area for oxygen transfer is also increased, it might reasonably be expected that the efficiency of oxygen transport across the placental barrier is greater at high than at low altitudes.

One further point about oxygen transport might be noted. The mechanism generally accepted is diffusion by the Fick principle, depending on the area for exchange, the thickness of the barrier, the pressure gradient and the diffusion coefficient. However, there is some evidence that active-transport processes may be involved. Gurtner & Burns (1972) have demonstrated that oxygen transport by the placenta is approximately two orders of magnitude more efficient than that of carbon monoxide or of argon despite the relatively similar physical properties of the gases. Since in the liver oxygen transport is at least in part mediated by a microsomal cytochrome (P450) the same mechanism may exist in the placenta. The administration of various drugs which bind to this cytochrome does in fact reduce oxygen transport without affecting argon transport, so the idea has

## The biology of high-altitude peoples

some experimental support. Histochemical and/or ultrastructure investigations on microsomal cytochrome in animals exposed to hypoxia would be of considerable interest.

Lumley & Wood (1967) have suggested that glucose transport across the placenta is linked with oxygen transport. They found that exposing women to 10% oxygen for short periods during the first stage of labour resulted in glucose mobilisation (maternal blood-glucose rose significantly). However, there was little change in foetal blood–glucose levels. This result is contrary to that of Lebedeva (1973) (quoted by Mirrakhimov & Lebedeva, personal communication). She found cord-blood in high-altitude newborns to have a significantly higher glucose content than that in low-altitude newborns ($120.3 \pm 3.0$ mg % as compared with $103.1 \pm 1.8$ mg %). The explanation of this discrepancy is uncertain. Lebedeva states that glucose utilisation by the high-altitude foetus is greater than normal; possibly this reflects the utilisation of less efficient anaerobic pathways of glucose metabolism, so that for a given energy output (all foetal energy requirements are met from glucose), more glucose has to be burnt. The problem obviously requires further attention.

### The foetus

The normal development of the foetus at high altitude is to a very great extent dependent on the adequacy of the adaptive responses made by the mother. Nevertheless it is itself capable of adaptive response to reduced oxygen tensions.

It has been suggested that acute exposure to severe hypoxia in the experimental animal results in increased loss of fertilized ova at the pre-implantation and implantation stage. Almost certainly, such losses in man would not be recognised as terminations of a pregnancy, but would at most result in a few days' delay in menstruation, since implantation of the blastocyst occurs about the end of the first week after ovulation. It is at this time that the new individual is at greatest risk of death, even at sea level, so that it is not unreasonable to expect that at this early stage of pregnancy, before the development of the placental 'shield', the risk of loss of the zygote among high-altitude populations would be at its greatest. There is no direct evidence for high pre-implantation loss in high-altitude man, but a good deal of evidence from the Andes suggests that at high altitude, fertility, as measured by the total number of offspring per woman (total fertility), is reduced. Hoff (1968) and Hoff & Abelson (1976) give a mean total fertility at age 45 of 6.7 for women in Nuñoa, Peru (4,300 m) while for Huallatire, Chile (4,300 m), Cruz-Coke (unpublished data) gives a value of 7.3. For Cerro de Pasco (4,300 m) fertility averaged 7.7 (CISM, 1968). By contrast Abelson et al. (1974) found total

90

fertility of non-migrant lowland (100 m altitude) Quechua Indians (the same ethnic group as the Nuñoa highlanders) to be 8.3. While the numbers involved in these surveys were small, the altitudinal difference in fertility seems appreciable. In none of these communities are contraceptive measures practised to any significant extent and in these circumstances the highlanders would be counted as relatively infertile (Henry, 1961). Because of the cultural practise of serial monogamy (Baker, 1966), almost all women are exposed to risk of pregnancy and Hoff (1968) noted that of the 136 women in his sample only three were nonparous. It seems then, that fecundity is not seriously impaired, but fertility, in the sense of bearing living children, may be. Similar results have been reported among highland Tajik women. Mirrakhimov & Lebedeva (personal communication) report a total fertility of $4.6\pm0.09$ among highlanders and $5.0\pm0.08$ among lowlanders. Among Andean populations rates of abortion seem extremely low. Hoff (1968) and Hoff & Abelson (1976) report a rate of 0.6% in Nuñoa; Cruz-Coke *et al.* (1966) and Cruz-Coke (unpublished data) 1% at Huallatire as opposed to rates between 17% and 30% in the lowlands, and Buck *et al.* (1968) a zero rate in their Peruvian village at 3,500 m altitude. In their study of downward migrant Peruvian Indians, Abelson *et al.* (1974) found abortion rates of 2.2%. While the extremely low rates in the highlands must be suspect, their consistency does suggest that among these well-adapted populations pre-natal loss rates may be lower than the world average of about 15% (Penrose, 1959). It was this discrepancy which evoked the suggestion by Clegg, Harrison & Baker (1970) that very early abortions – within the first two weeks of conception – would not be recognized as such.

The data from Ethiopia (Harrison *et al.*, 1969) are rather different from those from the Andes. Here the abortion rate among highlanders was twenty-one out of 232 pregnancies (seventy-seven individuals), a rate of 9.1%. In lowlanders it was one out of 173 pregnancies (0.6%) among 51 individuals. In sixty-four highland *sedentes* there were sixteen miscarriages out of 189 pregnancies (8.5%). Among fourteen highland migrants to the lowlands there was one miscarriage in thirty-eight pregnancies (0.3%). The converse was true for lowlanders. In thirty-seven *sedentes* there were no miscarriages in 135 pregnancies, but among thirteen migrants to the highlands there were five miscarriages in forty-three pregnancies (11.6%). Increased rates of abortion have also been reported among highland Tajik women by Mirrakhimov & Lebedeva (personal communication).

These results suggest strongly that in Ethiopia there is a significantly increased risk of abortion in the highlands, a finding completely at variance with the Andean data. Clearly there are so many differences, both genetic and environmental, between the two ecological situations that strict comparisons cannot be made, but it seems reasonable to point out two pertinent facts. Firstly, the hypoxic stress in the Ethiopian highlands (at about 3,000 m altitude) is much less than in the Andes: it has been shown experimentally that the rate of pre-implantation loss is related to the degree of hypoxia, so that at lower altitudes a number of pregnancies, which at higher altitudes would be lost before implantation, may continue past the pre-implantation stage and be lost later as recognisable abortions. Secondly, while at relatively mild levels of hypoxic stress implantation losses in mice are greatly reduced, there is still a high incidence of abnormality (25%) (Blackburn & Clegg, 1975). Such abnormal foetuses would have a greater probability of being lost before term, either through resorption or abortion and the later the loss the greater the probability of it being recognised as such.

If the abortion frequencies from South American highland populations are to be believed, once the foetus has passed through the critical first few weeks of its life, it is not greatly at risk. However, there is very good evidence that in man its rate of growth is less than at low altitudes. This is a finding by no means universal in high-altitude animals. Metcalfe *et al.* (1962*b*) and Barron *et al.* (1964) found foetal growth rates in well-adapted Andean sheep to be essentially similar at high and low altitudes, although examination of their figures suggests that during early pregnancy foetal growth in both weight and crown–rump length may be retarded, with subsequent 'catch-up' so that, at term, foetal size is not different from sea-level values.

These findings may be contrasted with those in other experimental animals. Johnson & Roofe (1965) exposed rats to hypoxia (5,500 m equivalent) for two weeks before mating. At term, mean total litter size in the hypoxic animals was $6.9 \pm 1.71$ compared with $12.0 \pm 2.17$ at sea-level pressure, and mean body weight $4.7 \pm 0.14$ g compared with $5.5 \pm 0.23$ g. Stillbirths per litter were not significantly more frequent in hypoxic conditions, but of course formed a greater proportion of total litter size.

However this was a severe degree of hypoxia. Other workers (Moore & Price, 1948; Timiras, 1964) have found that milder degrees of hypoxia have no significant effect on birth weight in rats, although litters tended to be smaller.

The findings of Baird & Cook (1962) in the mouse have been described previously. Unlike the rat, the mouse foetus seems relatively unaffected by hypoxia except that the risk of resorption around the 7-mm stage is increased.

Delaquerriere-Richardson, Forbes & Valdivia (1965) exposed guinea-pigs to hypoxia ($\equiv 3,750$ m or $\equiv 5,000$ m) during the latter half of pregnancy. At the greater degree of hypoxia pregnancy was shortened significantly and in both hypoxic groups birth weights were reduced, in proportion to the severity of the hypoxia.

Thus it seems that there is a good deal of interspecific variation in the influence that exposure to hypoxia has on the foetal growth rates of experimental animals. Man falls unequivocally into that group of animals which shows a reduced birth weight in such conditions. This is a pheno-menon which has frequently been documented, in the Andes (Alzamora, 1958; Chavez, 1964; Mazess, 1965; McClung, 1969; Sobrevilla, unpub-lished data, 1971; Sobrevilla *et al.*, 1968, 1971*b*; Frisancho, 1970; Krüger & Arias-Stella, 1970; Haas, 1973, 1976) and in the United States (Lichty, Ting, Bruns & Dyar, 1957; Howard, Lichty & Bruns, 1957*b*; Grahn & Kratchman, 1963). That the effect is an altitudinal one is indicated by the finding of Howard *et al.* (1957*b*) that in the United States, mothers who had previously born children at low altitude had offspring on average 380 g lighter at an altitude of 3,100 m. Since for mothers aged older than 20–24 yr increased birth weight is related to both increasing age and parity (up to five) (Selvin & Janerisch, 1971) it might have been expected that in the study by Howard *et al.* the birth-weight difference would be in the reverse direction to that found. While there seems no doubt that, in man, birth weight is reduced at high altitude, this reduction cannot be ascribed to a reduction in foetal growth rates until the effect of the highland environment on the duration of pregnancy is evaluated. Neither experimental evidence, nor the results of surveys on human populations, suggest at all strongly that pregnancy is shortened in high-altitude conditions. McClung (1969) concluded from data on neonatal length and head circumference that her high-altitude (3,400 m) infants fell within the range of London infants born during the fortieth week of gestation. Other maturational indexes, including head circumference and chest circumference (Usher, McLean & Scott, 1966) and relative crown–rump length (Dawkins & Macgregor, 1965) gave values which, although indicating less maturity than in the sea-level neonates, could not be rated as immature. Sobrevilla's (1971) data suggest that the maturity score of high-altitude neonates is similar to that of lowlanders. Finally, McClung's data reveal that 70% and 63%, respectively, of highland and lowland mothers said their pregnancies had been of normal length; 22% and 20% said they had been shorter than normal, and 8% and 17% said they had been longer than normal. Sobre-villa's (1971) data also suggest that pregnancies in his high- and low-altitude groups were of the same mean lengths. Thus the evidence from both these studies is that the length of pregnancy is not significantly decreased among highlanders. Other than the reduction in birth weight, which in this context is not a sufficient indicator of true prematurity, the only evidence from the United States which bears on the problem is that of Howard *et al.* (1957*b*) who found that body length and head circumference were reduced in neonates at 3,100 m.

Thus, while the evidence is slightly contradictory, its weight suggests that the reduced birth weight at altitude is to only a small extent, if at

93

all, due to reduction in the duration of pregnancy and it seems likely that the main factor responsible is a slowing of foetal growth. It seems probable that this slowing is only manifested during the last trimester of gestation (Grahn & Kratchman, 1963), an unsurprising occurrence since it is at that time that the human placenta, even at sea level, ceases to be able adequately to nourish the growing foetus. This situation may be compared with that of the sheep, where any retardation in growth seems to be during the first half of pregnancy and is anyway compensated for in the second half; the placenta in this animal continues to grow until term.

The cause of this retardation in growth is uncertain. Nutritional impairment has a surprisingly small effect on birth weight (Ounsted & Ounsted, 1973) and in any case, there is no evidence for undernutrition or of malnutrition in the Andean or US populations studied. Furthermore, hypoxia appears to have no effect on the placental transfer of amino acids (Hopkins, Reynolds & Young, 1970). It seems likely, then, that the effect is due mainly to hypoxia but, as with the lamb, human foetal oxygen tensions at high altitude seem to be similar to those of lowlanders (Lichty *et al.*, 1957; Sobrevilla, 1971; Sobrevilla, Casinelli, Carcelen & Malaga, 1971*a*). However, in Sobrevilla's study $P_{O_2}$ in high-altitude foetuses was very low ($29.88 \pm 3.43$ as compared with $45.82 \pm 1.68$ mm Hg in lowlanders). Plasma bicarbonate was likewise significantly lowered and the base excsss was $-7.41 \pm 1.64$ in comparison with $-1.78 \pm 1.18$ meq/l in lowland foetuses. To what extent these differences, found during labour, reflect intra-uterine conditions at earlier stages of pregnancy is not known, and their possible effects on foetal growth remain highly speculative.

Other evidence reinforces the idea that the human foetus at high altitude is not hypoxic. At birth the haematocrit is not dissimilar from that at lower levels (Howard, Bruns & Lichty, 1957*a*) and Reynafarje (1959) found no evidence of increased erythropoiesis in neonatal bone marrow. Neither are there any of the characteristic changes in the pulmonary trunk or the right ventricle which are found in hypoxic conditions (Saldaña & Arias-Stella, 1963; Recavarren & Arias-Stella, 1962). While foetal mice, rats, rabbits and dogs on exposure to hypoxia show an increased production of foetal haemoglobin (Hb F), sheep and human foetuses do not appear to do so (Barker, 1964), although in pathological hypoxic conditions (hypertension, pre-eclamptic toxaemia), increased Hb F production may be observed (Barker, 1964; Haworth, Dilling & Younoszai, 1967). Barker suggests that this difference may be due to more efficient placental adaptations at high altitude which reduce the need for increased Hb F production.

The estimates given earlier of foetal oxygen requirements, plus the cardiorespiratory changes seen in the mother at high altitude, suggest that there is plenty of reserve in the maternal circulation for the provision of adequate amounts of oxygen for aerobic metabolism. However, acute maternal hypoxia may result in a switchover in the foetus from aerobic to anaerobic metabolism. Huckabee, Metcalfe, Prystowsky & Barron (1962) and Mann (1970) have shown that in acute maternal hypoxia in goats or sheep, the foetus produces increased amounts of lactic acid. This excess acid is arrested in the placenta and is there reoxidised. The extent to which this mechanism operates in man is uncertain. In normal pregnancy, especially at high altitude, it would be most likely at birth, but as far as can be ascertained oxygenation of the foetus at this time is normal (Sobrevilla, 1971).

In conclusion, the main effect of life at high altitude on the human foetus appears to be a slowing in growth rate, most marked during the last trimester of pregnancy. The proximate cause appears to be the reduced oxygen tension in the mother's circulation, yet as far as can be ascertained foetal oxygenation is normal. Undoubtedly there are also hormonal changes, particularly a reduced adrenal cortical activity and also perhaps changes in acid–base balance. All these differences may be related to the reduced growth rate, but whether causally or otherwise remains uncertain.

Yet in another context these 'small-for-dates' (SFD) foetuses need not cause surprise. Birth weight is said to be correlated with maternal size, particularly stature (Cawley, McKeown & Record, 1954; Baird, 1964) and in the high Andes at least, mean stature is small (Frisancho & Baker, 1970). However, Ounsted & Ounsted (1973) have found an excess of underweight women among SFD mothers but no departure from control values for stature, the latter finding confirming a previous report by Scott & Usher (1966). Thus the relationship between maternal size and birth weight is uncertain, but whether low maternal stature or low body weight are the factors associated with low birth weight, high-altitude populations in the Andes have both characteristics.

Presumably these possible maternal constraints on foetal growth rate do not operate in North America. In particular, the finding of reduced birth weight of offspring born at high altitude to mothers who had previously had children at low altitude suggests that here at least any effect of maternal size is swamped by external environmental effects.

Lichty *et al.* (1957) show the distribution of birth weight in their populations at Leadville (3,100 m) and Cripple Creek (2,750 m), together with similar data for Denver (1,550 m) and the United States as a whole. While lowland birth-weight distributions are approximately symmetrical, those for Leadville and Cripple Creek are negatively skewed, particularly the former. In other words there is a pronounced deficiency of high birth-weight newborns.

There appear to be no comparable distributions of birth weight in well-adapted Andean populations, but McClung (1969) states that, altitude for altitude, the proportion of low birth-weight children is less than in North America, suggesting perhaps a more symmetrical distribution. What this difference means, if anything, is uncertain. It may reflect genetic adaptation on the part of Andean populations, whose colonization of their highland habitat dates back perhaps 10,000 years. Under any circumstances the loss of large-for-dates (LFD) or post-mature infants would be advantageous, since they are more liable to hypoxia both before and during birth (Hytten & Leitch, 1971, p. 251). If directional natural selection removes from the population those genes which determine large body size, acting either directly in the foetus, or indirectly through the mother, then the mean size of offspring will be reduced and the shape of the distribution curve will be altered, while the directional selection pressure continues. Ultimately it might be that a symmetrical distribution of birth weight would be re-established with a shift of the mean to the left. However, the selection pressures at high altitude are undoubtedly complex; while there is probably an advantage in the foetus being of small size, after birth there may be some advantages in being larger (greater maturity, better cold tolerance), so that there may well be a balance of selective pressures which may prevent the re-establishment of a symmetrical distribution of birth weights. Unfortunately, apart from the figures given by Lichty *et al.* (1957) there appear to be no published data on distributions of birth weights in high-altitude populations. It is interesting in this connection to note that Frisancho, Sanchez, Pallardel & Yanez (1973) have reported greater reproductive success for smaller women in a high-altitude population. The reason for this advantage is uncertain.

Clearly evolutionary factors would be less important in North American white populations since the period of colonization of the highlands is extremely short. The relatively greater effect compared with that in Amerindians in the Andes, together with the positive relationship between amount of Amerindian admixture in Andean Mestizos and birth weight at altitude (McClung, 1969), does suggest some degree of genetic adaptation on the part of the Andean Indian – possibly, as McClung (1969) suggests, because of greater efficiency of the placenta. That the reduced rate of foetal growth, however adaptive at altitude, is not advantageous after birth, is illustrated in the next section.

## Birth and the neonate

For both mother and child, parturition is a period of risk. For the mother there is the possibility of haemorrhage due to premature separation of the placenta or to placenta praevia and, particularly during the second stage of labour, oxygenation of the tissues may be less than adequate. For the foetus there is a risk of birth injury, and, as with the mother, hypoxia may occur. Perhaps most importantly though, the changeover from a situation of partial dependence upon the homeostatic efficiency of the mother to one of almost complete independence is a severe test of physiological flexibility and the extent to which adaptive mechanisms have matured.

There is some evidence that women at high altitude have difficult labour less frequently than women at low altitude. Harrison *et al.* (1969) found that among high- and low-altitude residents in Ethiopia twenty out of forty-two of the former (48%) had difficulty compared with nineteen out of twenty-six (73%) of the latter, a statistically significant difference. The nature of the difficulties was not ascertained, but in view of the finding of reduced birth weights in all high-altitude regions where the problem has been studied, one would expect labour to be easier among highlanders. The observations of McClung (1969) on the increased frequency of placenta praevia at Cuzco (3,400 m) suggests that this cause of difficult labour may be important at altitude although the same author's data show that placenta praevia has no effect upon the size of the newborn. Sobrevilla (1971) reports on arterial $P_{O_2}$ in mothers during the second stage of labour at Cerro de Pasco (4,300 m). He found a mean value of $60.8\pm2$ mm Hg, which is significantly greater than that of adult natives at rest at the same place ($47.6\pm4$ mm Hg) (Sørenson & Severinghaus, 1968). Conversely, $P_{CO_2}$ in arterial blood was significantly reduced ($24.5\pm1.5$ compared with $32.4\pm2.1$ mm Hg). These material values are very similar to those of Hellegers *et al.* (1961) in pregnant women before term, suggesting that the second stage of labour is not associated with any change in oxygenation either of the mother or, as Sobrevilla (1971) has demonstrated, of the foetus.

The adequate oxygenation of the foetus is also demonstrated, albeit indirectly, by the post-mortem findings of Naeye (1965). Comparison of children born and dying at Leadville, Colorado (3,100 m) and children of similar ages born and dying at sea level shows that while at birth the relative thicknesses of the wall of the pulmonary vessels are not different in the two groups, at 3,100 m they remain thick, while at sea level the thickness gradually decreases. The thickness of the wall in high-altitude neonates is due at least in part, to increased numbers of medial muscle cells.

This suggests that the child born at an altitude of about 3,000 m enters an environment approximately as hypoxic as it encounters in the uterus. Hence the normal post-natal changes in the cardiopulmonary system fail to occur. Surprisingly, there is no evidence during early life of any adaptive changes in the erythropoietic system in the shape of increased erythropoiesis. Normally there is a reduction in haematocrit after birth as the infant adjusts to a more oxygen-rich environment. Whether this occurs at high

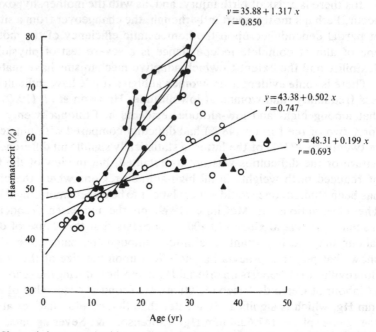

Fig. 4.3. Change in haematocrit with age at Puno, 3,800 m (triangles); Cerro de Pasco, 4,300 m (open circles) and Morococha, 4,500 m (solid circles). Regression lines meet about the age of 15 yr. The lines joining points represent 'longitudinal' data of change in haematocrit with age in particular individuals at Morococha. (Data from Whittembury & Monge, 1972.)

altitude is unknown. Howard *et al.* (1957*a*) found no differences in haematocrit during the first few days of life between children at 3,100 m and 1,550 m. Whittembury & Monge (1972) studied age changes in haematocrit in males between four and forty-six years at three different places; Puno (3,800 m), Cerro de Pasco (4,300 m) and Morococha (4,500 m). In all three places the haematocrit rose with age, and the slope of the line of increase correlated well with altitude and hence with degree of hypoxic stress (Fig. 4.3). However the regression lines tended to meet about the age of fifteen years, suggesting that below that age the haematocrit is less dependent upon altitude. Inspection of their figures suggests that a significant rise in the haematocrit does not occur much before the end of the first decade of life. Other recent work by Garruto (1973) supports this suggestion. At Macusani, Peru (4,400 m) mean haematocrits varied between 42.0 and 44.2 over the age range six to fourteen years. At fifteen years the mean was 47.3 and for the remainder of the second decade of life it varied between 49.1 and 56.8. Similar results were obtained at Nuñoa (4,000 m) and to a rather less extent at Ollachea (2,680 m). At lower altitudes haematocrits rose gradually through childhood and adolescence, with no sudden in-

98

Table 4.5. *Comparison of measurements (means±s.e.) of newborns at Cuzco and Lima*

| Variable | Cuzco (3,400 m) | Lima (200 m) |
|---|---|---|
| Birth weight (g) | 3092.8±52.6 | 3311.5±51.1 |
| Crown–heel length (cm) | 49.6±0.2[a] | 48.9±0.2 |
| Crown–rump length (cm) | 32.1±0.2[b] | 33.2±0.2 |
| Chest circumference (cm) | 32.6±0.2[c] | 33.2±0.2 |
| Arm length (cm) | 19.5±0.1[b] | 20.2±0.1 |
| Mean skinfold (eight sites) (cm) | 4.4±0.1[c] | 4.7±0.1 |

Data from McClung (1969).
[a] $0.001 < P < 0.01$.     [b] $P < 0.001$.     [c] $0.02 < < 0.05$.

creases. Why haematocrits in high-altitude dwellers should show the sudden rise during the second decade of life is uncertain, but since cardiac output is proportional to the square of linear dimensions and body weight to the cube (Smith, 1968, p. 6) the cardiac output per unit body weight per unit time, given constant heart rate, will be greater in small individuals who therefore, under hypoxic conditions, will have the least necessity to increase the oxygen-carrying capacity of the blood.

To return to the problem of the small size of the highland neonate: studies by McClung (1969) and Haas (1973, 1976) have involved fairly extensive anthropometric examinations of newborns at high altitude. In McClung's study seventy-three children born at Cuzco (3,400 m) and eighty-three born in Lima (200 m) were compared. The results are shown in Table 4.5. Head circumference showed no significant difference – a finding dissimilar from that of Howard *et al.* (1957b) at Leadville. Surprisingly, the data reveal that high-altitude children are longer than their low-altitude counterparts, although trunk length is less. Insofar as relative sitting height decreases with increasing age this could be regarded as indicative of increased maturity of the highland sample.

A much more detailed study was made by Haas (1973, 1976) utilizing data collected by himself and by T. Baker, A. Little & P. Baker (unpublished data). In all, six samples were studied: high-altitude rural Indian, high-altitude urban Indian, high-altitude Mestizo; low-altitude rural Indian, low-altitude urban Indian; low-altitude urban Mestizo. Such subgrouping enabled the interactions of altitude effects with rural or urban residence and ethnic/socio-economic group affiliations to be made. group affiliations to be made.

Generally, lowland newborns were heavier than highland newborns, although the difference was statistically significant only for males. The greatest difference was in the comparison between Mestizos.

Total length was again significantly greater in the total low-altitude sample, but in none of the subsample comparisons were differences statistically significant. There were no significant differences in sub-scapular skinfold.

These results are in general comparable with those of McClung (1969); they emphasize that the greater altitudinal effect appears to be on Mestizos rather than on pure-bred Indians and suggests a genetic component to the advantage of the latter since, in general, socio-economic differences at both high and low altitude should make pre-natal care better for the Mestizo than for the Indian mothers and the greater difference between Mestizo means is contrary to what might be expected.

In Haas' data there were no significant between-sex differences at either altitude in any of the measurements, but whereas at low altitude males were bigger and longer than females, the reverse was the case at high altitude. Surprisingly this reversal of normal sexual dimorphism was also found for subscapular skinfold; the increased thickness in low-altitude females was reversed in high-altitude females.

This result differs from that of McClung (1969). At Cusco (3,400 m) mean birth weight for males was $3,130\pm76$ and for females $3,053\pm71$ g. At Lima (200 m) the male mean was $3,290\pm68$ and the female mean $3,342\pm81$ g. The difference between first- and later-born means was 310 g at Cusco, compared with 150 g at Lima and McClung suggests that altitude effects may be more critical during a first than a later pregnancy.

But perhaps the most interesting finding is the relationship between sex and the probability of being of low birth weight. Chavez (1964) found 70% of males among his low-birth-weight ($< 2,500$ g) infants at Huaron (4,750 m) and McClung gives proportions of 53% for upper class and 78% for lower class mothers in Cuzco, the latter percentage statistically significant compared with 36% in Lima. Among most lowland populations males are heavier than females at birth and in the United States the proportion of males among infants of low birth weight is about 45% (McClung, 1969).

Generally speaking, males have a higher mortality than females at all ages and since low birth weight is significantly associated with increased mortality (Karn & Penrose, 1951) it might be expected that at high altitudes there should be an excess of male neonatal deaths. This is borne out by Spector's (1971) study at Nuñoa (4,300 m), where hebdomadal (first week) and neonatal deaths were slightly higher in males than in females (hebdomadal, 24.6/1,000 for males, 23.6/1,000 for females; neonatal, 70.5/1,000 for males, 66.8/1,000 for females). These results are comparable with those from the United States (Shapiro & Unger, 1965), where in neonates of below 2,500 g the male:female mortality ratio in the Mountain States was 235:131 and 214:139 in the country as a whole. Thus it appears

100

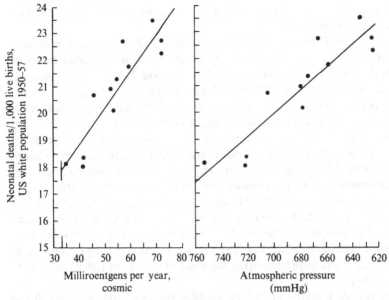

Fig. 4.4. Regressions of age and income-adjusted neonatal death rate on annual cosmic radiation dose and atmospheric pressure in 12 States of the United States. (Data from Grahn & Kratchman, 1963.)

that males are at greater risk in the high-altitude environment, perhaps because they are less effective at 'buffering' environmental insult.

Irrespective of the influence of sex, neonatal mortality is significantly increased among all high-altitude populations in which the data have been examined. Grahn & Kratchman (1963) found significant increases with decreasing atmospheric pressure for the white population of 12 States and conclude that the principal cause is hypoxia, although increased levels of cosmic radiation cannot be excluded as a contributory factor (Fig. 4.4). Similar results are shown by Shapiro & Unger's (1965) survey, where for the Mountain States the neonatal mortality rate was 24/1,000 compared with 20/1,000 for the United States as a whole. In South America increased perinatal or neonatal mortality has been reported by Mazess (1965), Cruz-Coke *et al.* (1966), Cruz-Coke (unpublished data), James (1966) and Heer (1967) and in Ethiopia by Harrison *et al.* (1969). Mazess' data suggest almost a doubling of mortality in highland as compared with lowland departments.

This increased mortality is to be expected on the basis of Karn & Penrose's (1951) data. Yerushalamay (1967) has recorded an overall neo-natal mortality rate 53% higher in SFD children than for children of

*The biology of high-altitude peoples*

average birth weight, some at least of the excess being due to respiratory insufficiency. The seriousness of respiratory disease in high-altitude neonates needs no emphasis.

Frisancho & Cossman (1970) have shown that neonatal mortality declined faster in the Mountain States of the United States between 1957 and 1967 than it did in the low-altitude States. They suggest that this shows that the differential cannot be due to hypoxia alone. Neither is the secular trend due to greater birth weights, since there was, over this period, no decrease in the frequency of SFD newborns. Perhaps the adverse effects of hypoxia and cold on the newborn are compounded by less adequate medical care. Undoubtedly the recognition by physicians and nurses of the increased risk to the newborn at altitude will have the effect of improving this particular aspect of child care, irrespective of any change in overall standards. However, Sobrevilla (1971) believes the effects of hypoxia to be important and regards the lowered functional reserve of oestrogen production in high-altitude neonates, coupled with metabolic abnormalities due to maternal hyperventilation, hypocapnia and metabolic acidosis as predisposing the child to increased risk of morbidity and mortality.

There is little information which enables this diversity of viewpoint to be resolved. In experimental animals hypoxia results in increased neonatal mortality (Moore & Price, 1948; Philips, Saxon & Quimby, 1950; Chiodi & Sammartino, 1952; Krum, 1957; Weihe, 1964; Kelley & Pace, 1968) and it has been shown that environmental temperature (Weihe, 1964) and humidity (Kelley & Pace, 1968) are important variables which affect survival. Lactation has also been shown to be important (Moore & Price, 1948; Chiodi & Sammartino, 1952; Krum, 1957; Weihe, 1964), but Kelley & Pace (1968) found no impairment of the mothers' milk production in rats suckling at 3,800 m. Indeed the stomachs of neonates at altitude contained more milk than at sea level, although this may be explained at least partially by reduced rates of stomach emptying (Van Liere & Stickney, 1963, p. 212). An important finding in several experimental investigations has been the low level of liver glycogen (Timiras, Hill, Krum & Lis, 1958; Kelley & Pace, 1968). The importance of this store for energy requirements immediately after birth has been emphasised by Shelley (1961) and Dawkins (1966) and clearly a failure to lay down this reserve, which is utilised during the first two days of life before lactation has been fully established, could have serious consequences for the neonate.

Almost certainly the causes of the increased neonatal mortality at altitude are complex. Environmental factors such as cold, lack of humidity and low standards of care are important, but it would be imprudent to exclude endogenous causes, manifested as specific physiological or biochemical deficiencies.

102

Several studies have revealed an interesting variation in the sex ratio at birth among high-altitude populations. In particular Hoff (1968) and Hoff & Abelson (1976) found a male preponderance of 120:100 at Nuñoa (4,300 m) and similar findings are reported by Abelson *et al.* (1974) in highlanders both before and after migrating to the lowlands. Baker & Dutt (1972) report a ratio of 111:100 for all births at Nuñoa between 1940 and 1964, a level higher than the 105–106:100 reported for most lowland areas of the world.

It is uncertain why these ratios are so high. It has been shown (Weir, 1953, 1955) that when male mice are selected for increased blood alkalinity (the situation at high altitudes) the frequency of male offspring increases, and Foote & Quevedo (1971) have shown that in cattle inseminated with semen previously subjected to reduced atmospheric pressure there is a preponderance of male offspring. It seems unlikely that post-zygotic selection acting against females is responsible, as females seem in general to be able to withstand environmental stress better than males, so that meiotic drive and/or pre-zygotic selection against X-bearing sperm are most likely explanations.

Finally in this section, we shall consider the frequency of congenital malformations. Grahn & Kratchman (1963) found no evidence in the United States for a relationship between altitude and frequency of malformations but studies in the Andes suggest that at very high altitudes malformations of the cardiovascular system are more common (Alzamora *et al.*, 1953; Chavez, Espino-Vela, Limon & Dorbecker, 1953). It might be suggested that cardiac defects, such as patent ductus arteriosus – approximately 15 times as common at Cerro de Pasco (4,300 m) as at Lima (Hellriegel, 1963) – might be due to the persistence of increased pulmonary arterial pressure after birth (Arias-Stella & Castillo, 1966) or the reduced stimulus, because of extra-uterine hypoxia, to physiological closure of the ductus.

Experimentally hypoxia is a good teratogenic agent. As stated above, Blackburn & Clegg (1975) found increased numbers of abnormal blastocysts in mice exposed to hypoxia after fertilisation. Ingalls, Curley & Prindle (1950, 1952), Curley & Ingalls (1957) and Haring (1965) have shown that hypoxia produces an increased incidence of congenital defects in the mouse. More recently Ingalls & Yamamoto (1972) found that hypoxia resulted in polyploidy in hamster embryos exposed to severe hypoxia during the pre-implantation stage. Once this stage was passed there was an increased frequency of mosaic embryos.

These severe effects occur of course on exposure of unacclimatised mothers to hypoxia and in resident human populations at high altitude these effects, if present, would presumably be less marked. However, as suggested previously, abnormal embryos probably stand a greater chance

103

of being lost either through resorption or abortion than do normal embryos and the reduced fertility and/or greater abortion rate found among some high-altitude populations may indicate an increased incidence of foetal abnormality.

## Growth during the first year of life

As seen in the last section, the newborn infant at high altitude starts life with the handicap of being of small size. During the neonatal period this handicap is manifested as an increased rate of mortality, and it continues during the whole of the period of growth. While there have been many reports on child growth at high altitude (see Frisancho, Chapter 5) the period of infancy – the first year of life – has not received a great deal of attention, and the only systematic study appears to be that of Haas (1973, 1976) utilising his own data and that collected by T. Baker, A. Little & P. Baker (unpublished data).

As noted before, the subjects studied by Haas were subdivided into groups by altitude (high or low), residence (rural or urban), and social/ethnic class (Indian or Mestizo). Variables measured included body weight, crown–heel (CH) and crown–rump (CR) lengths, leg length (CH–CR), head circumference and triceps and subscapular skinfolds. In addition radiographs were taken of the arm (for tissue component analysis) and of the wrist (for skeletal development). The number of erupted teeth was recorded and the Bayley scales of infant development (Bayley, 1969) were used to estimate motor development. Socio-economic, fertility and nutritional data were obtained from each child's mother.

In terms of body size (weight, CH length and head circumference) growth at high altitudes during the first year was less than at low altitudes. This difference held for both rural and urban Indians, and for urban Mestizos. While at both altitudes males were bigger than females, sexual dimorphism was significant only among lowlanders. Altitudinal differences in growth tended to increase with increasing age. While there was a significant difference in weight between rural and urban infants at high altitude, the latter being the larger, there was virtually none at low altitude. Similarly, Indian–Mestizo weight differences were significant at high altitude only.

As with most populations studied, skinfold thicknesses increased with age, the increment being greater in females, particularly in the trunk (subscapular skinfold). Tissue component analyses showed that trunk and limb fat was thicker in the low-altitude sample. Increments of muscle in the arm were greater in low-altitude infants for the first six months, but thereafter the situation was reversed and high-altitude children gained more so that there was an apparent 'catch-up' Rural–urban differences in trunk fat were significant only at high altitude, as were Indian–Mestizo differences. Surprisingly, between-altitude Mestizo differences were not statistically significant, whereas differences between Indian groups were.

In the radiographs of the humerus, measurements of cortical and medullary bone areas showed that, increasingly, total bone areas and proportions of cortical bone in high-altitude children fell below those at low altitude. The effect was more marked in Mestizos, possibly because of the significantly larger marrow cavity at birth. Wrist radiographs showed that low-altitude children had more ossific centres at the end of the first year, an advancement which failed to persist beyond the end of the second year of life. Low-altitude rural Indians and urban middle-class Mestizos had more erupted teeth than similar subjects in the highlands.

Generally speaking then, low-altitude infants had significantly more trunk and limb fat, cortical bone and wrist ossific centres than high-altitude infants.

The results of psychomotor testing suggest that in general Peruvian children develop at first at approximately the same rate as US children. However, in the later-developed skills (finger prehension, walking) they seem significantly behind US children. Generally speaking, low-altitude infants were more advanced than highlanders, but differences were small. These differences were small compared with sex (females quicker in walking and in acquiring finger prehension), Indian–Mestizo (Indians more advanced in early infancy, but more retarded later on) and rural–urban differences (urban dwellers generally more advanced).

The differences observed between highland and lowland children may be due to essentially non-specific factors inherent in the different ecological situations, or to more specific causes, such as nutritional differences, differences in the incidence of infectious disease, etc. Haas presents data on comparative feeding practices. Generally speaking almost all children were breast-fed. For the Indian and the high-altitude samples demand feeding was the rule, whereas among the Mestizo and low-altitude samples regular intervals between feeds were more often seen. Weaning from the breast occurred earlier among Mestizos and lowlanders and low-altitude Mestizos, in particular, were more aware of the value of milk substitutes if breast milk was insufficient.

The incidence of infectious disease (colds, coughs, fevers, tonsillitis, colic, diarrhoea, vomiting and constipation) was greatest among the low-altitude sample, particularly during the first six months of life. This contrasts with the very much inferior sanitary practices among high-altitude dwellers, particularly Indians.

Possibly the overall advantage of the lowlanders over the highlanders might be explained on a nutritional basis. Certainly it seems that, in rural areas in particular, highland diets although generally adequate are poorer than lowland, but the greater duration of breast feeding at high altitude would tend to minimize any effects on the infant. However, there is little or no evidence of the effects of high altitude on milk yield and quality in man. Also, when only well-cared-for infants were compared, the altitudinal differences were still there, more among the Mestizos than the Indians and this finding suggests that in the latter, long residence in the highlands has resulted in some degree of genetic adaptation.

The generally reduced sexual dimorphism seen among highland infants suggests that life at altitude imposes some environmental stress (Stini, 1969, 1972). In those variables where males are normally bigger than females, dimorphism was reduced, but where the female is normally bigger or more advanced than the male, as in skinfold thickness and number of wrist ossific centres, the dimorphism was marked.

Haas found that altitudinal differences in infant mortality were small compared with Indian–Mestizo differences although more general surveys by Mazess (1965) and Spector (1971) showed that the highland areas have consistently higher infant mortality rates than for Peru as a whole. Possibly this difference may be as much an ethnic as an altitudinal one.

The discrepancy between the general retardation in physical growth in highland infants and the close similarity in the development of psychomotor skills suggests that despite the adverse environment, brain development is not greatly affected. As with many other studies, it appeared that ethnic socio-economic and cultural differences were more important than those relating to the physical environment in determining the pattern and rate of psychomotor development.

**Conclusions**

The results of the various investigations described in this survey, lead unmistakably to the conclusion that life at high altitudes impairs reproductive efficiency. While the production of germ cells is probably not seriously impaired it seems at least likely that the embryo during the pre-implantation and early post-implantation stages is at increased risk. Foetal growth is slow, and the newborn is smaller than at low altitude with perhaps some immaturity in the capacity to make adaptive responses. As a result neonatal and infantile mortality are increased and growth during the first year of life is generally retarded.

While *prima facie*, these altitudinal differences in reproduction and early growth may be regarded as due to such physical environmental factors as hypoxia and cold, differences in culture, nutrition, standards of hygiene, etc. need to be taken into account. In general, however, when comparisons are made with altitude as the only varying factor, differences are still seen.

The overall conclusion which one may draw is that at high altitudes high birth-rates and a rapid rate of population increase (see Baker, Chapter 12) are achieved by maximisation of reproductive potential. Contraceptive methods are little used, and the patterns of mate selection and marriage (Baker, 1966) ensure that all women are exposed to risk of pregnancy. Thus no potentially fertile woman is wasted. Insofar as socio-economic factors may play some part in causing some of the reduced fertility at high altitude, programmes of social amelioration may need to be accompanied by instruction in contraceptive techniques if even greater rates of population increase are to be avoided. Since some at least of the reproductive losses occur during the perinatal, neonatal and infantile periods, programmes designed to mitigate such losses would be comparatively easily instituted. The necessity of information and instruction on birth-control measures thus become doubly important.

**References**

Abelson, A. E., Baker, T. S. & Baker, P. T. (1974). Altitude, migration and fertility in the Andes. *Social Biology*, **21**, 12–27.

Aherne, W. & Dunnill, M. S. (1966). Morphometry of the human placenta. *British medical Bulletin*, **22**, 5–8.

Altland, P. D. (1949a). Effect of discontinuous exposure to 28,000 feet simulated altitude on growth and reproduction of the albino rat. *Journal of experimental Zoology*, **110**, 1–17.

Altland, P. D. (1949b). Breeding performance of rats exposed repeatedly to 18,000 feet simulated altitude. *Physiological Zoology*, **22**, 235–46.

Altland, P. D. & Allen, E. (1952). Studies on degenerating sex cells in immature animals. III. *Journal of Morphology*, **91**, 541–53.

Altland, P. D. & Highman, B. (1952). Acclimatisation response of rats to discontinuous exposures to simulated high altitudes. *American Journal of Physiology*, **167**, 261–7.

Alzamora, O. (1958). Algunas observaciones sobre alteraciones de la placenta en la altura. *Revista de la Asociacíon medica de Yauli*, **3**, 75–81.

Alzamora, V., Rotta, A., Battilana, G., Abugattas, R., Rubio, C., Bouroncle, J., Zapata, C., Santa Maria, E., Binder, T., Subira, R., Paredes, D., Pando, B. & Graham, G. (1953). On possible influence of great altitudes on determination of certain cardio-vascular anomalies; Preliminary report. *Pediatrics*, **12**, 259–62.

Arias-Stella, J. & Castillo, Y. (1966). The muscular pulmonary arterial branches in stillborn natives of high altitude. *Laboratory Investigation*, **15**, 1951–9.

Baird, B. & Cook, S. F. (1962). Hypoxia and reproduction in Swiss mice. *American Journal of Physiology*, **202**, 611–15.

Baird, D. (1964). The epidemiology of prematurity. *Journal of Pediatrics*, **65**, 909–24.

Baker, P. T. & Dutt, J. S. (1972). Demographic variables as measures of biological adaptation: a case study of high altitude human populations. In *The Structure of Human Populations*, ed. G. A. Harrison & A. J. Boyce, pp. 352–78. Oxford: The Clarendon Press.

Baker, T. S. (1966). Quechua marriage: Some ecological determinants of marriage patterns in the southern highlands of Peru. Master of Science Thesis. The Pennsylvania State University.

Barcroft, J. (1936). Fetal circulation and respiration. *Physiological Reviews*, **16**, 103–28.

Barker, J. N. (1964). Adaptations of pregnant ewes and their fetuses. In *The Physiological Effects of High Altitude*, ed. W. H. Weihe, pp. 125–8. Oxford: Pergamon Press.

Barron, D. H., Metcalfe, J., Meschia, G., Huckabee, W., Hellegers, A. & Prystowsky, H. (1964). Adaptations of pregnant ewes and their fetuses to high altitude. In *The Physiological Effects of High Altitude*, ed. W. H. Weihe, pp. 115–25. Oxford: Pergamon Press.

Bartels, H., Moll, W. & Metcalfe, J. (1962). Physiology of gas exchange in the human placenta. *American Journal of Obstetrics and Gynecology*, **84**, 1714–30.

Bayley, N. (1969). *Manual for the Bayley Scales of Infant Development*. New York: The Psychological Corporation.

Blackburn, R. O. & Clegg, E. J. (1976). Blastocyst survival in the hypoxic mouse. *Journal of Anatomy*, **119**, 406.

Buck, A. A., Sasaki, T. T. & Anderson, R. I. (1968). *Health and Disease in Four Peruvian Villages*. Baltimore: The Johns Hopkins Press.

Calancha, A. de la (1639). *Crónica moralizada de la Orden de San Agustin*, Vol. 1. Barcelona: Imp. Pedro Lacaballeria.

Cawley, R. H., McKeown, T. & Record, R. G. (1954). Parental stature and birth weight. *American Journal of human Genetics*, **6**, 448–56.

Chabes, A., Pareda, J., Hyams, L., Barrientos, N., Perez, J., Campos, L., Monroe, A. & Mayorga, A. (1968). Comparative morphometry of the human placenta at high altitude and at sea level: the shape of the placenta. *Obstetrics and Gynecology*, **31**, 178–85.

Chabes, A., Pareda, J., Perez, J., Barrientos, N. & Campos, L. (1967). Morphometry of human placenta at high altitude. *American Journal of Pathology*, **50**, 14a–15a.

Chang, M. C. & Fernandez-Cano, L. (1959). Effects of short changes of environmental temperature and low atmospheric pressure on the ovulation of rats. *American Journal of Physiology*, **196**, 653–5.

Charles, D., Harkness, R. A., Kenny, F. M., Menini, E., Ismail, A. A. A., Durkin, J. W. & Loraine, J. A. (1970). Steroid excretion patterns in an adrenalectomised woman during three successive pregnancies. *American Journal of Obstetrics and Gynecology*, **106**, 66–74.

Chavez, I., Espino-Vela, R., Limon, R. & Dorbecker, N. (1953). La persistencia del conducto arteriel: Estudio de 200 casos. *Archivo del Instituto cardiologica Mexicana*, **23**, 687–755.

Chavez, M. H. A. (1964). Algunas aspectos del nino premaduro en las alturas. Thesis, Facultad de Médicina, Lima.

Cheek, D. B., Graystone, J. E. & Rowe, R. D. (1969). Hypoxia and malnutrition in newborn rats: effects on RNA, DNA and protein in tissues. *American Journal of Physiology*, **217**, 642–5.

Chiodi, H. (1964). Action of high altitude chronic hypoxia on newborn animals. In *The Physiological Effects of High Altitude*, ed. W. H. Weihe, pp. 97–112. Oxford: Pergamon Press.

Chiodi, H. & Sammartino, R. (1952). Fatty degeneration of the liver caused by high altitude hypoxia in young albino rats. *Acta physiologica latinoamerica*, **2**, 228–37.

Fertility and early growth

CISM (1968). *Encuesta de Fecundidad en la Ciudad de Cerro de Pasco.* Lima: Centro de Investigaciones Sociales por Muestro y Ministerio de Trabajo y Communidades del Peru.

Clegg, E. J. (1968). Some effects of reduced atmospheric pressure on secondary sexual characteristics of male mice. *Journal of Reproduction and Fertility,* 16, 233–42.

Clegg, E. J., Harrison, G. A. & Baker, P. T. (1970). The impact of high altitudes on human populations. *Human Biology,* 42, 486–518.

Clegg, E. J., Pawson, I. G., Ashton, E. H. & Flinn, R. M. (1972). The growth of children at different altitudes in Ethiopia. *Philosophical Transactions of the Royal Society of London,* 264B, 403–37.

Cobo, B. (1653). *Historia del Nuevo Mundo.* Madrid: Biblioteca de Autores Españoles 91, 92.

Correa, J., Aliaga, R. & Moncloa, F. (1956). *Study of the Adrenal Function at High-Altitudes with the Intravenous ACTH Test,* Report 56–101. Randolph Field: USAF School of Aviation Medicine.

Coyle, M. G. & Brown, J. B. (1963). Urinary excretion of oestriol during pregnancy. II. Results in normal and abnormal pregnancy. *Journal of Obstetrics and Gynaecology of the British Commonwealth,* 70, 225–31.

Cruz-Coke, R., Cristoffani, A. P., Aspillaga, M. & Biancani, F. (1966). Evolutionary forces in an environmental gradient in Arica, Chile. *Human Biology,* 38, 421–38.

Curley, F. J. & Ingalls, T. H. (1957). Hypoxia at normal atmospheric pressure as a cause of congenital malformations in mice. *Proceedings of the Society for experimental Biology and Medicine,* 44, 87–8.

Dalton, A. J., Jones, B. F., Peters, V. B & Mitchell, E. R. (1945–6). Organ changes in rats exposed repeatedly to lowered oxygen tension with reduced barometric pressure. *Journal of the National Cancer Institute,* 6, 161–85.

Damon, A., Damon, S. T., Reed, R. B. & Valadian, I. (1969). Age of menarche of mothers and daughters, with a note on recall. *Human Biology,* 41, 161–75.

Daniel, J. C. (1970). Dormant embryos of mammals. *Bioscience,* 20, 411–15.

Dawkins, M. J. R. (1966). Biochemical aspects of developing function in newborn mammalian liver. *British medical Bulletin,* 22, 27–33.

Dawkins, M. J. R. & MacGregor, W. G. (1965). The small-for-dates baby. In *Gestational Size, Age and Maturity,* ed. M. J. R. Dawkins & W. G. MacGregor, *Clinics in Developmental Medicine,* Vol. 19, p. 33. London: The Spastics Society with Heinemann.

Delaquerriere-Richardson, L., Forbes, S. & Valdivia, E. (1965). Effects of simulated high altitude on the growth rate of albino guinea pigs. *Journal of applied Physiology,* 20, 1022–5.

Delaquerriere-Richardson, L. & Valdivia, E. (1967). Effects of simulated high altitude on pregnancy. *Archives of Pathology,* 84, 405–17.

Donayre, J. (1966). Population growth and fertility at high altitude. In *Life at High Altitudes,* ed. A. Hurtado, pp. 74–9. Scientific Publication No. 140. Washington, DC: Pan American Health Organization.

Donayre, J. (1969). The oestrus cycle of rats at high altitude. *Journal of Reproduction and Fertility,* 18, 29–32.

Donayre, J., Guerra-Garcia, R., Moncloa, F. & Sobrevilla, L. A. (1968). Endocrine studies at altitude. IV. Changes in the semen of men. *Journal of Reproduction and Fertility,* 16, 55–8.

Faiman, C. & Winter, J. S. D. (1971). Diurnal cycles in plasma FSH, testosterone

109

and cortisol in man. *Journal of clinical Endocrinology and Metabolism*, **22**, 186–92.

Fernandez-Cano, L. (1959). The effects of increase or decrease of body temperature or of hypoxia on ovulation and pregnancy in the rat. In *Recent Progress in the Endocrinology of Reproduction*, ed. C. W. Lloyd, pp. 97–106. New York & London: Academic Press.

Foote, W. D. & Quevedo, M. M. (1971). Sex ratio following subjection of semen to reduced atmospheric pressure. *Journal of Animal Science*, Supplement: *Sex Ratio at Birth – Prospects for Control*, 55–8.

Frisancho, A. R. (1970). Developmental responses to high altitude hypoxia. *American Journal of physical Anthropology*, **32**, 401–8.

Frisancho, A. R. & Baker, P. T. (1970). Altitude and growth: a study of the patterns of growth in a high altitude Peruvian Quechua population. *American Journal of physical Anthropology*, **32**, 279–92.

Frisancho, A. R. & Cossman, J. (1970). Secular trend in neonatal mortality in the mountain states. *American Journal of physical Anthropology*, **33**, 103–6.

Frisancho, A. R., Sanchez, J., Pallardel, D. & Yanez, L. (1973). Adaptive significance of small body size under poor socioeconomic conditions in southern Peru. *American Journal of physical Anthropology*, **39**, 255–62.

Frisch, R. E. & Revelle, R. (1971). Height and weight at menarche and a hypothesis of menarche. *Archives of Disease in Childhood*, **46**, 695–701.

Fuchs, F., Spackman, T. & Assali, N. S. (1963). Complexity and non-homogeneity of the intervillous space. *American Journal of Obstetrics and Gynecology*, **86**, 226–33.

Garruto, R. M. (1973). Polycythemia as an adaptive response to chronic hypoxic stress. Doctoral Dissertation in Anthropology, The Pennsylvania State University.

Gordon, A. S., Tornetta, F. J., D'Angelo, S. A. & Charipper, H. A. (1943). Effects of low atmospheric pressures on the activity of the thyroid, reproductive system and anterior lobe of the pituitary in the rat. *Endocrinology*, **33**, 366–83.

Grahn, D. & Kratchman, J. (1963). Variation in neonatal death rate and birth weight in the United States and possible relations to environmental radiation, geology and altitude. *American Journal of human Genetics*, **15**, 329–52.

Grindeland, R. E. & Folk, G. E. (1962). Effects of cold exposure on the oestrus cycle of the golden hamster. *Journal of Reproduction and Fertility*, **4**, 1–6.

Guerra-Garcia, R. (1971). Dinamica de la androgenesis en las grandes alturas. Thesis, Universidad Peruana 'Cayetana Heredia', Instituto de Investigaciones de la Altura, Lima.

Guerra-Garcia, R., Donayre, J., Sobrevilla, L. A. & Moncloa, F. (1965). Cambios en la testosterone urinaria en sujetos expuestos a la altura. *Proceedings of the VIth Pan American Congress of Endocrinology*, p. 130.

Guerra-Garcia, R., Velasquez, A. & Coyotupa, J. (1969). A test of endocrine gonadal function in men: urinary testosterone after the injection of HCG. II. A different response of the high altitude natives. *Journal of clinical Endocrinology and Metabolism*, **29**, 179–82.

Guerra-Garcia, R., Velasquez, A. & Whittembury, J. (1965). Urinary testosterone in high altitude natives. *Steroids*, **6**, 351–5.

Gurtner, G. H. & Burns, B. (1972). Possible facilitated transport of oxygen across the placenta. *Nature, London*, **246**, 473–5.

Haas, J. D. (1973). *Altitudinal variation and infant growth and development in*

*Peru.* Doctoral Dissertation in Anthropology, The Pennsylvania State University.

Haas, J. D. (1976). Infant growth and development. In *Man in the Andes: A Multidisciplinary Study of High Altitude Quechua*, ed. P. T. Baker & M. A. Little. Stroudsburg, Pa.: Dowden, Hutchinson & Ross.

Haring, O. M. (1965). Effects of prenatal hypoxia on the cardiovascular system in the rat. *Archives of Pathology*, **80**, 351–6.

Harris, W., Shields, J. L. & Hannon, J. P. (1966). Acute altitude sickness in females. *Aerospace Medicine*, **37**, 1163–7.

Harrison, G. A. (1966). Human adaptability with reference to the IBP proposals for high altitude research. In *The Biology of Human Adaptability*, ed. P. T. Baker & J. S. Weiner, pp. 509–19. Oxford: Clarendon Press.

Harrison, G. A., Küchemann, C. F., Moore, M. A. S., Boyce, A. J., Baju, T., Mourant, A. E., Godber, M. J., Glasgow, B. G., Kopeć, A. C., Tills, D. & Clegg, E. J. (1969). The effects of altitudinal variation in Ethiopian populations. *Philosophical Transactions of the Royal Society of London, Series B*, **256**, 147–82.

Haworth, J. C., Dilling, L. & Younoszai, M. K. (1967). Relation of blood glucose to haematocrit, birth weight and other body measurements in normal and growth-retarded newborn infants. *Lancet*, **2**, 901–5.

Heer, D. M. (1967). Fertility differences between Indian and Spanish speaking parts of Andean countries. *Population Studies*, **18**, 71–84.

Hellegers, A., Metcalfe, J., Huckabee, W. E., Prystowsky, H., Meschia, G. & Barron, D. H. (1961). Alveolar $pCO_2$ and $pO_2$ in pregnant and nonpregnant women at high altitude. *American Journal of Obstetrics and Gynecology*, **82**, 241–5.

Hellriegel, K. O. (1963). El ductus arterioso persistente: Observaciones hechos en las grandes alturas. *Revista de la Asociacíon médica de Yauli*, **8**, 20–31.

Henry, L. (1961). Some data on natural fertility. *Eugenics Quarterly*, **8**, 81–91.

Hoff, C. J. (1968). Reproduction and viability in a high altitude Peruvian Indian population. *Occasional Papers in Anthropology*, **1**, 85–160. The Pennsylvania State University.

Hoff, C. J. & Abelson, A. E. (1976). Fertility. In *Man in the Andes: A Multidisciplinary Study of High Altitude Quechua*, ed. P. T. Baker & M. A. Little, pp. 128–46. Stroudsburg, Pa.: Dowden, Hutchinson & Ross.

Hopkins, L., Reynolds, M. L. & Young, M. (1970). The influence of blood flow and hypoxia on free $\alpha$-amino nitrogen transfer across the placental membrane. *Journal of Physiology*, **207**, 13–15.

Howard, R. C., Bruns, P. D. & Lichty, J. A. (1957a). Studies of babies born at altitude. III. Arterial oxygen saturation and haematocrit values at birth. *American medical Association Journal of Diseases of Children*, **93**, 674–7.

Howard, R. C., Lichty, J. A. & Bruns, P. D. (1957b). Studies of babies born at high altitude. II. Measurement of birth weight, body length and head size. *American medical Association Journal of Diseases of Children*, **93**, 670–4.

Huckabee, W. E., Metcalfe, J., Prystowsky, H. & Barron, D. H. (1962). Insufficiency of $O_2$ supply to pregnant uterus. *American Journal of Physiology*, **202**, 198–204.

Hytten, F. E. & Leitch, I. (1971). *The Physiology of Human Pregnancy*, 2nd edition. Oxford: Blackwell Scientific Publications.

Ingalls, T. H., Curley, F. J. & Prindle, R. A. (1950). Anoxia as a cause of fetal

The biology of high-altitude peoples

death and congenital defect in the mouse. *American medical Association Journal of Diseases of Children*, **80**, 34–45.

Ingalls, T. H., Curley, F. J. & Prindle, R. A. (1952). Experimental production of congenital anomalies. *New England Journal of Medicine*, **247**, 758–68.

Ingalls, T. H. & Yamamoto, M. (1972). Hypoxia as a chromosomal mutagen. *Archives of Environmental Health*, **24**, 305–15.

James, W. H. (1966). The effect of high altitude on fertility in Andean countries. *Population Studies*, **20**, 87–101.

Johnson, D. & Roofe, P. D. (1965). Blood constituents of normal newborn rats and those exposed to low oxygen tensions during gestation: Weight of newborn and litter size also considered. *Anatomical Record*, **153**, 303–9.

Karn, M. N. & Penrose, L. S. (1951). Birth weight and gestation time in relation to maternal age, parity and survival. *Annals of Eugenics*, **16**, 147–64.

Kawakita, J. (1953). Ethno-geographical observations on the Nepal Himalaya. In *Peoples of Nepal Himalaya*, ed. H. Kihara, pp. 1–362. Kyoto: Fauna and Flora Research Society.

Kelley, F. C. & Pace, N. (1968). Etiological considerations in neonatal mortality among rats at moderate high altitude (3,800 m). *American Journal of Physiology*, **214**, 1168–75.

Kim, J. H. & Han, S. (1969). Studies on hypoxia. V. Effects of anoxia on developing connective tissue cells in rats. *Anatomical Record*, **165**, 531–41.

Krüger, H. & Arias-Stella, J. (1970). The placenta and the newborn infant at high altitude. *American Journal of Obstetrics and Gynecology*, **106**, 586–91.

Krum, A. A. (1957). Reproduction and growth of laboratory rats and mice at altitude. Thesis. University of California.

Lang, S. D. R. & Lang, A. (1971). The Kunde Hospital and a demographic survey of the upper Kumbu, Nepal. *New Zealand medical Journal*, **74**, 1–8.

Lichty, J. A., Ting, R. Y., Bruns, P. D. & Dyar, E. (1957). Studies of babies born at high altitude. I. Relation of altitude to birth weight. *American medical Association Journal of Diseases of Children*, **93**, 666–9.

Lumley, J. M. & Wood, C. (1967). Influence of hypoxia on glucose transport across the human placenta. *Nature, London*, **216**, 403–4.

McClung, J. (1969). *Effects of High Altitude on Human Birth*. Cambridge, Mass.: Harvard University Press.

Maclennan, A. H., Sharp, F. & Shaw-Dunn, J. (1972). The ultrastructure of human trophoblast in spontaneous and induced hypoxia using a system of organ culture. *Journal of Obstetrics and Gynaecology of the British Commonwealth*, **72**, 113–21.

Mann, L. I. (1970). Effects of hypoxia on umbilical circulation and fetal metabolism. *American Journal of Physiology*, **218**, 1453–8.

Mazess, R. B. (1965). Neonatal mortality and altitude in Peru. *American Journal of physical Anthropology*, **23**, 209–14.

Meschia, G., Prystowsky, H., Hellegers, A., Huckabee, W., Metcalfe, J. & Barron, D. H. (1960). Observations on the oxygen supply to the fetal llama. *Quarterly Journal of experimental Physiology*, **45**, 284–91.

Metcalfe, J., Meschia, G., Hellegers, A., Prystowsky, H., Huckabee, W. & Barron, D. H. (1962a). Observations on the placental exchange of respiratory gases in pregnant ewes at high altitude. *Quarterly Journal of experimental Physiology*, **47**, 74–92.

Metcalfe, J., Meschia, G., Hellegers, A., Prystowsky, H., Huckabee, W. & Barron, D. H. (1962b). Observations on the growth rates and organ weights

112

of fetal sheep at altitude and sea level. *Quarterly Journal of experimental Physiology*, **47**, 305–13.

Miklashevskaya, N. N., Solovyeva, V. S. & Godina, E. Z. (1972). Growth and development in high altitude regions of southern Kirghizia, U.S.S.R. *Vopros Anthropologii*, **40**, 71–91.

Moncloa, F. (1966). Natural acclimatisation to high altitudes: Endocrine factors. In *Life at High Altitudes*, ed. A. Hurtado, pp. 36–9. Scientific Publication No. 140. Washington, DC: Pan American Health Organization.

Moncloa, F., Donayre, J., Sobrevilla, L. A. & Guerra-Garcia, R. (1965). Endocrine studies at high altitude. II. Adrenal cortical function in sea level natives exposed to high altitudes (4,300 m) for two weeks. *Journal of clinical Endocrinology and Metabolism*, **25**, 1640–2.

Moncloa, F. & Pretell, E. (1964). Cortisol secretion rate, ACTH and methopyrapone tests in high altitude native residents. *Journal of clinical Endocrinology and Metabolism*, **24**, 915–18.

Moncloa, F., Pretell, E. & Correa, J. (1961). Studies on urinary steroids of men born and living at high altitude. *Proceedings of the Society for experimental Biology and Medicine*, **108**, 336–7.

Monge, C. (1942). Life in the Andes and chronic mountain sickness. *Science, Washington*, **95**, 79–84.

Monge, C. (1945). Aclimatacíon en los Andes – confirmaciones históricas sobre la 'agrésion climatica' en el desenvolvimiento las sociedades de América. *Anales de la Facultad de Médicina de Lima*, **28**, 307–82.

Monge, C. (1948). *Acclimatization in the Andes*. Baltimore, Md.: The Johns Hopkins Press. (Republished by Blaine Ethridge–Books, Detroit, 1973.)

Monge, C. & Monge, C. (1968). Adaptacíon de los animales domesticos: Adaptacíon a las grandes alturas. *Archivos del Instituto de Biologia Andina*, **2**, 276–91.

Monge, C., San Martín, M., Atkins, J. & Castañon, J. (1945). Aclimatacíon del ganado ovino en las grandes alturas: fertilidad y infertilidad reversible en la fasa adaptativa. *Anales de la Facultad de Médicina de Lima*, **28**, 15–31.

Moore, C. R. & Price, D. (1948). A study at high and low altitude of reproduction, growth, sexual maturity and organ weights. *Journal of experimental Zoology*, **108**, 171–216.

Muller, H. J. (1954). Manner of dependence of 'permissible dose' of radiation on amount of genetic damage. *Acta Radiologica*, **41**, 5–20.

Naeye, R. L. (1965). Children at high altitude: pulmonary and renal abnormalities. *Circulation Research*, **16**, 33–8.

Ounsted, M. & Ounsted, C. (1973). *On Fetal Growth Rate. Clinics in Developmental Medicine, No. 46*, p. 204. Spastics International Publications. London: William Heinemann Medical Books Ltd.

Panigel, M. (1962). Placental perfusion experiments. *American Journal of Obstetrics and Gynecology*, **84**, 1664–83.

Penrose, L. S. (1959). Natural selection in man: some basic problems. In *Natural Selection in human Populations*, ed. D. F. Roberts & G. A. Harrison, pp. 1–10. *Symposium of the Society for the Study of Human Biology*, **2**. Oxford: Pergamon Press.

Phillips, N., Saxon, E. P. & Quimby, F. H. (1950). Effect of humidity and temperature on the survival of albino mice exposed to low atmospheric pressure. *American Journal of Physiology*, **161**, 307–11.

Picón-Reátegui, E. (1966). Efecto de la exposicíon crónica a la altura sobre el metabolismo de los hidratos de carbono. *Archivos del Instituto de Biologia Andina*, **1**, 255–85.

113

The biology of high-altitude peoples

Recavarren, S. & Arias-Stella, J. (1962). Topography of right ventricular hypertrophy in children native to high altitude. *American Journal of Pathology*, **41**, 467–75.

Reynafarje, C. (1959). Bone marrow studies in the newborn infant at high altitudes. *Journal of Pediatrics*, **54**, 152–61.

Romney, S. L., Reid, D. E., Metcalfe, J. & Burwell, C. S. (1955). Oxygen utilization by the human fetus in utero. *American Journal of Obstetrics and Gynecology*, **70**, 791–9.

Saez, J. M. & Migeon, C. J. (1967). Problems related to the determination of the secretion and interconversion of androgens by 'urinary' methods. *Steroids*, **10**, 441–56.

Saldaña, M. & Arias-Stella, J. (1963). Studies on the structure of the pulmonary trunk. II. The evolution of the elastic configuration of the pulmonary trunk in people native to high altitudes. *Circulation*, **27**, 1094–1100.

Scott, K. E. & Usher, R. (1966). Fetal malnutrition, its causes and effects. *American Journal of Obstetrics and Gynecology*, **94**, 951–63.

Selvin, S. & Janerisch, D. T. (1971). Four factors influencing birthweight. *British Journal of social and preventive Medicine*, **25**, 12–16.

Shapiro, S. & Unger, J. (1965). *Weight at Birth and its Effect on Survival of the Newborn in the United States, Early 1950*. National Center for Health Statistics, Publication Series 21, No. 3. Washington, DC: US Department of Health, Education, and Welfare.

Sharp, F., Carty, J. & Young, H. (1971). The effects of hypoxia on hydroxysteroid dehydrogenase activity in placental villi maintained in organ culture. *Journal of Obstetrics and Gynaecology of the British Commonwealth*, **79**, 44–9.

Shelley, H. J. (1961). Glycogen reserves and their changes at birth. *British medical Bulletin*, **17**, 137–43.

Simmer, H. H. (1972). Disorders of placental endocrine functions. In *Pathophysiology of Gestation*, **2**, ed. N. S. Assali & C. R. Brinkman, pp. 78–155. New York & London: Academic Press.

Smith, J. Maynard. (1968). *Mathematical Ideas in Biology*. London: Cambridge University Press.

Sobrevilla, L. A. (1971). Nacer en los Andes: Estudios fisiologicas sobre el embarazo y parto en la altura. Thesis. Universidad Peruana 'Cayetana Heredia', Instituto de Investigaciones de la Altura, Lima.

Sobrevilla, L. A., Casinelli, M. T., Carcelen, A. & Malaga, J. M. (1971a). Human fetal and maternal oxygen tension and acid–base status during delivery at high altitude. *American Journal of Obstetrics and Gynecology*, **111**, 1111–18.

Sobrevilla, L. A., Romero, I. & Kruger, F. (1971b). Estriol levels of cord blood, maternal venous blood and amniotic fluid at delivery at high altitude. *American Journal of Obstetrics and Gynecology*, **110**, 596–7.

Sobrevilla, L. A., Romero, I., Kruger, F. & Whittembury, J. (1968). Low estrogen excretion during pregnancy at high altitude. *American Journal of Obstetrics and Gynecology*, **102**, 828–33.

Sobrevilla, L. A., Romero, I., Moncloa, F., Donayre, J. & Guerra-Garcia, R. (1967). Endocrine studies of high altitude. III. Urinary gonadotrophins in subjects native to and living at 14,000 feet and during acute exposure of men living at sea level to high altitude. *Acta Endocrinologica*, **56**, 369–75.

Sohval, A. R. & Soffer, L. J. (1951). The influence of cortisone and adrenocorticotrophin on urinary gonadotrophin excretion. *Journal of clinical Endocrinology and Metabolism*, **11**, 677–87.

Sørensen, A. & Severinghaus, J. W. (1968). Respiratory sensitivity to acute

114

hypoxia in man born at sea level living at high altitude. *Journal of applied Physiology*, **25**, 211–16.

Spector, R. M. (1971). *Mortality characteristics of a high altitude Peruvian population*. Master's Thesis in Anthropology. The Pennsylvania State University.

Stini, W. A. (1969). Nutritional stress and growth: sex difference in adaptive response. *American Journal of physical Anthropology*, **31**, 417–26.

Stini, W. A. (1972). Reduced sexual dimorphism in upper arm muscle circumference associated with protein-deficient diet in a South American population. *American Journal of physical Anthropology*, **36**, 341–52.

Subauste, C., Aliaga, R., Silva, C., Correa, J. & Moncloa, F. (1958). *Comparative Study of adrenal Function at Sea Level and at High Altitude*. Report 58–95. Randolph Field: USAF School of Aviation Medicine.

Sundstroem, E. S. & Michaels, G. (1942). *The Adrenal Cortex in Adaptation to Altitude, Climate and Cancer*. Berkeley & Los Angeles: University of California Press.

Thomas, R. B. (1973). Human Adaptation to a High Andean Energy Flow System. *Occasional Papers in Anthropology*, **7**. The Pennsylvania State University.

Timiras, P. S. (1964). Comparison of growth and development of the rat at high altitude and at sea level. In *The Physiological Effects of High Altitude*, ed. W. H. Weihe, pp. 21–30. Oxford: Pergamon Press.

Timiras, P. S., Hill, R., Krum, A. A. & Lis, A, W. (1958). Carbohyrate metabolism in fed and fasted rats exposed to an altitude of 12,470 feet. *American Journal of Physiology*, **193**, 415–24.

Tominaga, T. & Page, E. W. (1966). Accommodation of the human placenta to hypoxia. *American Journal of Obstetrics and Gynecology*, **94**, 679–91.

Usher, R., McLean, F. & Scott, K. E. (1966). Judgement of fetal age: clinical significance of gestational age and an objective method for its assessment. *Pediatric Clinics of North America*, **13**, 835–48.

Valsík, J. A., Štukovský, R. & Bernátová, Ĺ. (1963). Quelque facteurs géographiques et sociaux ayant une influence sur l'age de la puberté. *Biotypologie*, **24**, 109–23.

Van Liere, E. J. & Stickney, J. C. (1963). *Hypoxia*. Chicago: The University of Chicago Press.

Walton, A. & Uruski, W. (1946). The effects of low atmospheric pressure on the fertility of male rabbits. *Journal of experimental Biology*, **23**, 71–6.

Ward, M. P. (1969). *Report of IBP Expedition to Bhutan*. London: The Royal Society.

Weihe, W. H. (1964). Some examples of endocrine and metabolic functions in rats during acclimatisation to high altitude. In *The Physiological Effects of High Altitude*, ed. W. H. Weihe, pp. 33–44. Oxford: Pergamon Press.

Weir, J. A. (1953). Association of blood pH with sex ratio in mice. *Journal of Hereditary*, **44**, 133–8.

Weir, J. A. (1955). Male influence on sex ratio of offspring in high and low blood pH strains of mice. *Journal of Heredity*, **46**, 277–83.

Whittembury, J. & Monge, C. (1972). High altitude, haematocrit and age. *Nature, Lond.* **238**, 278–9.

Yerushalamy, J. (1967). The classification of newborn infants by birth weight and gestational age. *Journal of Pediatrics*, **71**, 164–72.

Yoshimi, T., Strott, C. A., Marshall, J. R. & Lipsett, M. B. (1969). Corpus luteum function in early pregnancy. *Journal of clinical Endocrinology and Metabolism*, **29**, 225–30.

115

# 5. Human growth and development among high-altitude populations

A. R. FRISANCHO

From the time of conception the growth and development of an individual depends on the interaction of genetic potentialities and internal and external environmental conditions. At all stages of life environmental factors are constantly conditioning and modifying the expression of inherited potentials. It is a basic principle that the environment provides the external factors that make development possible and permit genetic potentials to find expression. These factors include nutrition, temperature, humidity, drugs, infection, radiation and atmospheric gases. Experimental research on animals has indicated that, independent of nutritional factors, hypoxic stress resulting from low barometric pressure reduces the rate of growth and adult size. Thus, the possibility that growth and development of high-altitude populations may also be affected by hypoxic stress becomes evident. In order to test this proposition, and with the purpose of understanding the mechanism whereby the non-adult organism adapts to low oxygen availability, studies on growth and development of children in the South American Andes, Ethiopian highlands, Himalayas, and Tien Shan mountains of the USSR were conducted. Since in each study there is a wide divergence in the sample size, age structure and method of analysis, in the present chapter the growth of Andean, Ethiopian, Tien Shan and Himalayan populations are described separately. Following the description of the population-specific patterns of growth the overall trends are discussed within an adaptive context. The objective is not to compare the growth or morphology of populations but to derive an overview of the possible influences of high altitude on growth and development.

**Growth in the Andes**
*Source of data*

The main data for the Andean region come from studies conducted in the Southern and Central Highlands of Peru. These populations live at altitudes that range from 3,000 to 5,500 m. The climatic conditions are characterized by low temperatures and low humidity. The economic focus of the indigenous populations inhabiting areas around 3,000–3,500 m is the cultivation of corn, wheat and potatoes, and the herding of sheep and cattle. In contrast, the subsistence pattern of the indigenous populations over 4,000 m is based primarily upon the herding of sheep, llamas and

117

Table 5.1. *Distribution of major samples included in the evaluation of growth of Andean populations*

| Location and type of sample | Altitude range (m) | Sex | Non-adult | | Adult | | Reference |
|---|---|---|---|---|---|---|---|
| | | | Age range (yr) | N | Age range (yr) | N | |
| Nuñoa | 4,000–5,500 | M | 2–22 | 702 | 25–35 | 52 | (Frisancho & |
| Quechua | | F | 2–22 | 400 | 25–35 | 50 | Baker, 1970) |
| Nuñoa | 4,000–5,500 | M | 12–19 | 253 | 25–35 | 123 | (Boyce *et al.*, 1974) |
| Quechua | | | | | | | |
| Mollendo | Sea level | M | 12–19 | 183 | 25–35 | 137 | (Boyce *et al.*, 1974) |
| Quechua | | | | | | | |
| Nuñoa-Macusani | 4,000–5,500 | M | 11–17 | 119 | 18–30 | 62 | (Hoff, 1974) |
| Quechua | | | | | | | |
| San Juan del Oro | 1,340 | M | 11–17 | 59 | 18–30 | 21 | (Hoff, 1974) |
| Quechua | | | | | | | |
| Ondores | 4,500 | M | 6–17 | 157 | 18–40 | 62 | (Frisancho, Borkan |
| Quechua | | F | 6–17 | 156 | 18–40 | 58 | & Klayman, 1975) |
| Pamashto | 980 | M | 6–17 | 145 | 20–40 | 57 | |
| Quechua | | F | 6–17 | 129 | 20–40 | 60 | |
| Cerro Pasco | 4,200 | M | 7–16 | 96 | 20–40 | 50 | (Llerena, 1973) |
| Mestizo | | F | 7–16 | 44 | 20–40 | 20 | |
| Lima | Sea level | M | 7–16 | 107 | 20–40 | 70 | |
| Mestizo | | F | 7–16 | 89 | 20–40 | 15 | |
| San Mateo | 3,280 | M | 6–16 | 330 | — | — | (Peñaloza, 1971) |
| Huancayo Mestizo | | F | 9–17 | 270 | — | — | |
| Lima | Sea level | F | 6–16 | 120 | — | — | |
| Mestizo | | | | | | | |
| La Paz (Bolivia) | 3,850 | M | 14–16 | 31 | — | — | (Bouloux, 1968) |
| Aymara | | F | 13–15 | 33 | — | — | |
| La Paz (Bolivia) | 3,850 | M | 14–16 | 24 | — | — | (Bouloux, 1968) |
| Aymara | | | | | | | |

alpacas and the cultivation of potatoes and varieties of native *Chenopodium* (quinua and cinihua). However, in the central highlands of Peru and La Paz (Bolivia) the economy is based upon both herding and mining.

The distribution by sex and age range of the Andean samples is presented in Table 5.1. Data on Nuñoa Quechua samples were obtained between 1964 and 1966 and taken from the District of Nuñoa which is situated at a mean altitude of 4,250 m. The data include both cross-sectional and 3-year-longitudinal samples. Detailed characteristics of these samples are already given elsewhere (Baker, Frisancho & Thomas, 1966; Frisancho, 1966; Frisancho, 1969*a, b*; Frisancho & Baker, 1970; Boyce *et al.*, 1974; Frisancho, 1976). The ages of these samples were obtained either directly from birth records or from school records (which in turn were derived from birth records). Data on the lowland Quechua samples were obtained from the coastal town of Mollendo (Boyce *et al.*, 1974) and from the lowland village of San Juan del Oro (Hoff, 1974). The parents of these subjects were Quechuas who migrated from the highlands, but their children were born at low altitudes. As indicated by Boyce *et al.* (1974), only people who identified themselves as Quechua and who on visual assessment by the observers had clearly Indian characteristics were included in the study. The highland and low-altitude Mestizo samples were derived from populations living in the central highlands and in the city of Lima. As indicated by Peñaloza (1971) and Llerena (1973), both highland and lowland samples were derived from the same socio-economic strata. The ages of the subjects were derived from school records and calculated in years and months. The Aymara samples from Bolivia were obtained from a school located on the outskirts of the city of La Paz and an orphanage school. The ages, to the nearest month, were obtained from school records.

### Body size and composition

#### Height, weight, and chest size

As shown in Fig. 5.1, growth in stature and weight appears to continue until the twenty-second year. On the other hand, growth in chest width and depth stops at about the age of 19 years. Sexual dimorphism in stature, weight and chest size is not well defined until about the age of 16 years.

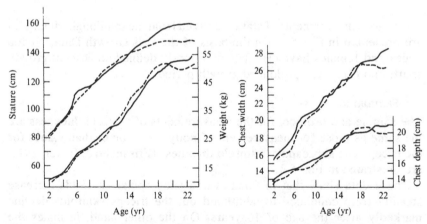

Fig. 5.1. Development of stature, weight, chest width and chest depth of Nuñoa indigenes. The pattern of growth in body size is characterized by a slow growth and late sexual dimorphism. Unbroken lines, males; broken lines, females. Adapted from Frisancho & Baker (1970).

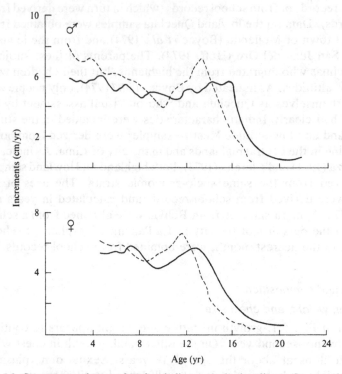

Fig. 5.2. Stature growth rate of Nuñoa indigenes (unbroken lines). Compared to the USA Fels Standards (broken lines), the Nuñoa males and females (unbroken lines) have a late and poorly defined spurt. Adapted from Frisancho & Baker (1970).

The growth increments of stature derived from the semilongitudinal data are presented in Fig. 5.2. On the basis of the Fels Growth Data, Nuñoa males and females have (1) a late and poorly defined adolescent growth spurt, and (2) a very prolonged growth period.

### Skinfold thickness

The sample and protocol for the measurements of skinfold thickness are the same as those for measurement of body size. The median values for the triceps and subscapular skinfold thickness (Frisancho & Baker, 1970) are illustrated in Fig. 5.3.

These data show that in females and subscapular skinfolds increase steadily from childhood to adulthood but the triceps skinfolds decline markedly after the age of 16 years. On the other hand, in males the increase in the subscapular skinfold thickness is very slow, while the triceps skinfolds decline drastically after the age of 10 years and through adulthood. In other words, the distribution of subcutaneous fat during

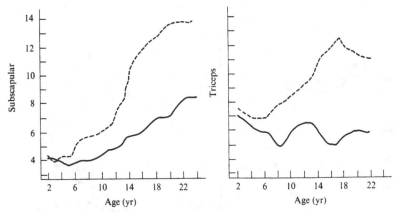

Fig. 5.3. Skinfold thickness of Nuñoa indigenes. Subcutaneous fat deposition from eight years on shows a well-defined sex difference. Adapted from Frisancho & Baker (1970). Unbroken lines, males; broken lines, females.

growth follows a centripetal pattern. This means that while trunk fat increases steadily during the middle teens and early twenties, limb fat decreases in females but not in males. It would follow that 'fat gain' and 'fat loss' in the body may take place simultaneously in different locations over the same time span. Thus, as in Western populations (Garn & Rohmann, 1966), the adult pattern of subcutaneous fat among Nuñoans is the result of a developmental rearrangement of adipose tissue.

### Brachial tissue areas

The data were drawn from both cross-sectional and semilongitudinal samples (Baker *et al.*, 1966; Frisancho, 1969*b*). The cross-sectional sample consisted of 300 subjects, aged 2 to 35 years. The semilongitudinal sample consisted of 110 subjects, aged 2 to 22 years, who were remeasured after 27 months.

Standardized lateromedial radiographs of the right brachium were taken at a focal distance of 86 cm with a portable X-ray machine of 15 mA. The protocol for the measurement of radiographic shadows of fat, muscle, bone and marrow cavity was that of Baker *et al.* (1958). The tissue diameters were converted into areas of cross-section according to the formula:

$$A = \frac{\pi}{4}d^2$$

where $A$ represents the area of cross-section in mm² and $d$ is any of the brachial diameters listed above.

Fig. 5.4 illustrates the development of the brachial tissues in Nuñoa

121

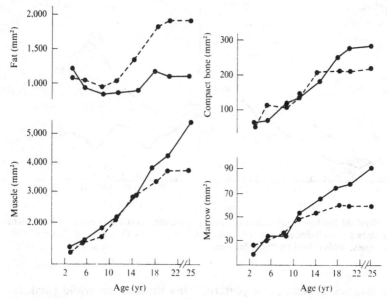

Fig. 5.4. Development of brachial tissue areas. Compared to USA data the Nuñoa sample show a late sexual dimorphism in muscle and fat. Unbroken lines, males; broken lines, females. Adapted from Baker, Frisancho & Thomas (1966) and Frisancho (1969*b*).

using the method developed for US samples (Baker *et al.*, 1958). These data show that in girls the deposition of fat proceeds at a rapid and steady rate until the age of 14 years, and between 15 and 18 years of age has a marked spurt. In this manner, by adulthood women have about twice as much fat as at an age of 2 years, while in men the trend of fat deposition proceeds at a slow and irregular pace so that by adulthood they have nearly the same amount of fat as during childhood. In Nuñoa, a clear sexual difference in fat deposition occurs after the age of 14 years, while in the US samples it occurs by the age of 8 years.

The development of muscle proceeds at a steady and regular pace in both sexes until about the age of 18 years. In boys, the adolescent spurt in muscle size ocurs between the ages of 15 and 19 years, while in girls it occurs between 14 and 17 years of age. In Nuñoa, sexual dimorphism is well defined after the age of 18 years, while in the US it appears after the age of 14 years.

The growth in compact bone proceeds at a steady and regular pace until the age of 12 years in both sexes. The adolescent spurt occurs between the age of 15 and 18 years in boys and between 13 and 15 years in girls. In Nuñoa the sexual dimorphism is well defined after the age of 18 years.

The development of marrow cavity proceeds at the same rate until the

122

age of 12 years in both males and females. Unlike that of the brachial tissues, the development of the marrow cavity does not show an adolescent spurt. In Nuñoa the sexual dimorphism is well defined after the age of 14 years, while in the US boys exceed girls at the age of 12 years.

During post-natal life the red blood cells are produced exclusively by the marrow of all bones. Although after about the age of 20 years the marrow of the long bones diminishes in hematopoietic function, its continuing activity and growth depends entirely on the organism's requirements for oxygen (Wintrobe, 1967). Therefore, the growth or hyperplasia of the bone marrow is regulated by the demands for increased functional activity. X-ray measurements of marrow cavity at high altitude indicate a continued growth through the age of 22 years in men, and 18 years in women, as compared to 10 and 16 years at sea level (Baker *et al.*, 1958; Johnston & Malina, 1966; Frisancho *et al.*, 1970).

In view of the importance of red blood cells in oxygen transport at high altitude, the differences in the pattern of growth in bone marrow and in marrow space are probably related to the increased demand for functional activity. Through experimental studies, Hunt & Schraer (1965) demonstrated that under hypoxic conditions the bone marrow of rats is increased by 20%. Studies on man (Merino & Reynafarje, 1949) indicate that in the bone marrow the ratio of nucleated red-cell elements to granulocytes in high-altitude natives is 1:1, while at sea level the ratio is 1:3. These indications suggest that at high altitude the increased requirements for oxygen result in increased hematopoietic activity of the bone marrow.

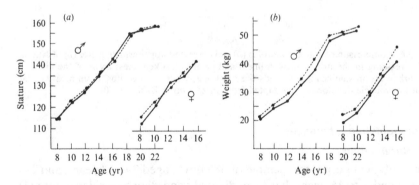

Fig. 5.5. Comparison of stature (*a*) and weight (*b*) of Nuñoans residing in the urban (broken lines) and rural (unbroken lines) areas of the district. Age-for-age, both rural and urban groups attain nearly the same stature, while the urban are slightly heavier than their rural counterparts. Adapted from Frisancho & Baker (1970).

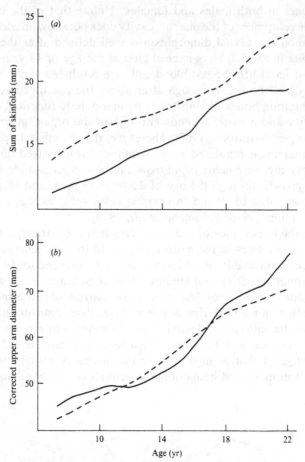

Fig. 5.6. Comparison of sum of skinfolds (*a*) and corrected upper arm diameter (*b*) of Nuñoa males residing in the urban (broken lines) and rural (unbroken lines) areas of the district. The urban group significantly exceeds the rural one, while both groups attain nearly equal values of muscle development. Adapted from Frisancho & Baker (1970).

## Intrapopulation differences

### Rural–urban

For analytical purposes a sample of 388 boys, aged 7 to 22 years, and 216 girls, aged 7 to 16 years, has been divided according to residence into (1) urban, comprising those who live in the central town capital of the district of Nuñoa and attend the school, and (2) rural, comprising those who live in the native communities and privately owned ranches (haciendas) and who do not attend the school in the central town.

Figs. 5.5 and 5.6 illustrate the development of stature, weight, sum of skinfolds, and upper-arm muscle diameter of the rural and urban male samples. The urban group is heavier than the rural sample, but age-for-age, both groups attain nearly the same stature. The urban group is also considerably fatter (about 20% more than the rural sample), while they both appear to have nearly the same upper-arm muscle diameter.

### Altitude

Since the population of Nuñoa lives at altitudes which range from 4,000 to 5,500 m, the sample was divided according to the altitude of residence into a lower-altitude group, those who live at 4,000 m, and a higher-altitude group, those who live at altitudes above 4,500 m.

Both groups attained nearly the same values for body size and composition in stature, weight, sitting height, biacromial diameter, sum of skinfolds and upper-arm diameter. On the other hand the ratio of chest circumference to height (the relative chest circumference) in the higher-altitude group, especially between the ages of 14 and 18, was significantly larger ($P < 0.01$) than those residing at lower altitudes (Fig. 5.7).

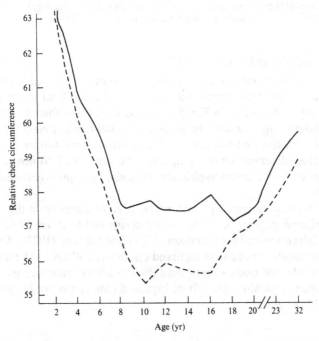

Fig. 5.7. Comparison of relative chest circumference at maximum inspiration of Nuñoa males residing at around 4,000 m (broken lines) and above 4,500 m (unbroken lines). The higher altitude group attains a significantly greater chest circumference than the group which lives at around 4,000 m. Adapted from Frisancho & Baker (1970).

125

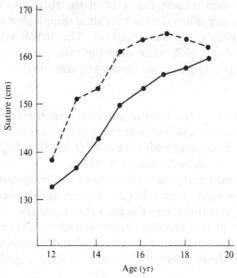

Fig. 5.8. Comparison of Absolute and Relative growth in height of Nuñoa Highland (unbroken line) and sea-level Mollendo Quechua (broken line) samples. The Nuñoa highland Quechuas show a slow development in stature. Adapted from Boyce *et al.* (1974).

### Interpopulation differences

The development of stature of Nuñoa cross-sectional samples given by Boyce *et al.* (1974) was compared to that in the lowland sample from Mollendo and is illustrated in Fig. 5.8. These data show that the highland Quechua children age-for-age are significantly shorter in height than their low-altitude Quechua counterparts. These differences follow the same pattern as that observed when comparing the growth of Nuñoa children with sea-level and moderate-altitude Peruvian samples (Frisancho & Baker, 1970).

In Fig. 5.9 the patterns of growth of Mestizo children from the central highlands (Cerro de Pasco, 4,200 m) are compared to those of sea-level Mestizo children studied by Peñaloza (1971) and Llerena (1973). As in the southern highlands, the central highland children at all ages have a slower growth in height and body weight than their sea-level counterparts. Thus it would appear that slow growth of highland children is not limited only to the southern region.

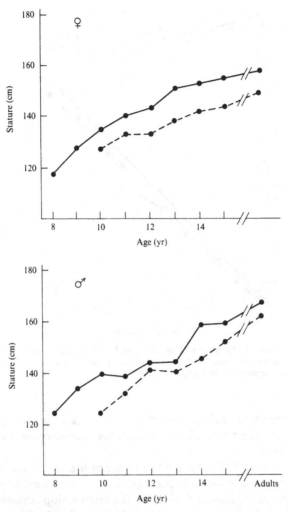

Fig. 5.9. Comparison of growth in stature of Mestizos from the Central Highlands (broken lines) and sea-level Lima (unbroken lines). As in Fig. 5.8, the highland samples exhibit a slower growth than those at sea level. Adapted from Peñaloza (1971) and Llerena (1973).

### Indices of maturation

#### Skeletal age

As before, both cross-sectional and semilongitudinal samples were studied. The cross-sectional sample consisted of 270 subjects aged 1 month to 24 years. The semilongitudinal sample comprised 110 subjects aged 2 to 22 years who were remeasured after 27 months.

127

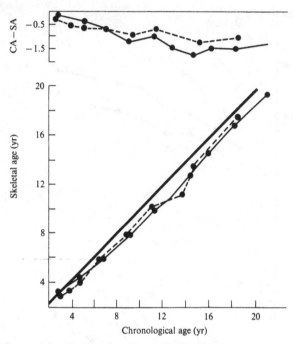

Fig. 5.10. Comparison of chronological age (CA) and skeletal age (SA) of Nuñoa children. In both males (unbroken lines) and females (broken lines) the skeletal age is behind chronological age. Adapted from Frisancho (1966b).

For assessments of skeletal age, the Greulich & Pyle atlas (1959) was the standard for reference. These evaluations excluded the round bones of the carpal area.

As shown in Fig. 5.10, the Nuñoa children are markedly delayed in skeletal maturation for their chronological ages, in comparison with US norms. Regression equations describing the relationship between skeletal age ($y$) and chronological age ($x$) yield $y$ values of $-0.165+0.910x$ for boys, and $-0.168+0.881x$ for girls, indicating that the rate of maturation in boys and girls tends to be about equal. Furthermore, the negative value for the $y$-intercept suggests that the average Nuñoa child is 0.168 years (i.e. 2 months) behind US children in maturation age at birth.

### Adolescent maturation

As illustrated in Fig. 5.11 the studies of Donayre (1966) indicated that menarche among high-altitude girls occurred at a later age than among sea-level girls. Recently, Peñaloza (1971) pointed out that the mean age at menarche among highland girls from Cerro de Pasco is 13.58 years, while

128

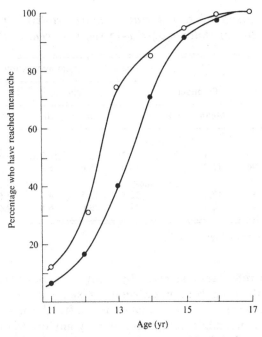

Fig. 5.11. Comparison of age at menarche among Mestizos from sea level and Central Highlands. The mean age at menarche of the highland girls (●) is higher than that of girls at sea level (○). Adapted from Peñaloza (1971).

at sea level it is 11.58 years. Bouloux (1968) also reported a mean age of 13.58 years for highland Aymara girls, which is comparable to that of Nuñoa highland girls.*

Table 5.2 summarizes the ratings of secondary sexual maturation among high-altitude and sea-level samples derived from the data of Llerena (1973). These data show that within the age range of 7 to 15 years, about 76% of the boys at high altitudes were in the prepubertal stage (Stage I), while at sea level only 39% were prepubertal. Consequently, the proportion of boys in the second, third, fourth, and fifth pubertal stages was lower at high altitude than at sea level. Indeed, none of the boys at high altitude were in the fourth and fifth stages. In the same manner, at sea level about 60% of the girls were already in the fourth pubertal stage, while at high altitude, there were only 30% in this stage. In other words, in both males and females the age at which the secondary sexual characteristics are attained at high altitude is markedly delayed.

* These values are averages only of those girls who reported having had menarche and therefore the values are lower than those derived from the status quo method (which includes those who have not menstruated).

Table 5.2. *Comparison of sexual stage among sea-level (Lima) and high-altitude (4,200 m) children aged 7–15 years, in central Peru*

| | | | | Secondary sexual maturation stage (%) | | | | |
| | | Prepubertal | | Pubertal | | | | |
| | $N$ | Mean | S.D. | I | II | III | IV | V |
|---|---|---|---|---|---|---|---|---|
| | | Males | | | | | | |
| Sea level | 105 | 11.1 | 2.2 | 39.3 | 23.4 | 18.7 | 16.8 | 1.9 |
| High altitude | 96 | 11.2 | 2.4 | 76.0 | 15.6 | 8.4 | | |
| | | Females | | | | | | |
| Sea level | 89 | 12.2 | 2.0 | 22.5 | 13.5 | 4.5 | 59.6 | |
| High altitude | 44 | 12.7 | 1.7 | 34.1 | 29.6 | 6.8 | 29.6 | |

Adapted from Figs. 10–13 of Llerena (1973).

Studies on skin reflectance as given by Conway & Baker (1972) also demonstrate that the prepubescent darkening in skin color which occurs at sea level before the age of 10 years, takes place after the age of 13 years among Nuñoans. Thus, adolescent maturation by any criteria is delayed among Nuñoans. Furthermore, measurements of luteinizing hormones indicate that the adult values in girls are attained by the age of 11 years at sea level and 12 years at high altitude (Llerena, 1973). It is also important to note that according to the studies of Guerra-Garcia (1971), the excretion of urinary testosterone in adults is lower at high altitude than at sea level.

*Pulmonary function*

*Data*

The subjects were a cross-sectional sample of 150 Quechua boys aged 11 to 20 years, part of the student population of the school of the central town of Nuñoa. Standard anthropometric measurements were obtained on all subjects as described above.

Forced expiratory volumes were measured according to standard procedures (Comroe *et al.*, 1955; Consolazio, Johnson & Peccora, 1963; Cotes, 1965) with a dry spirometer (Jones–Pulmonor). Each subject was carefully instructed in Spanish and Quechua, the maneuvres were practiced several times, and recordings were made of several attempts to insure precision. The readings used were the highest values. All measurements were obtained in the standing position. In order to get records of the best quality, the subjects were stimulated to compete with each other and were rewarded with a photograph of themselves.

130

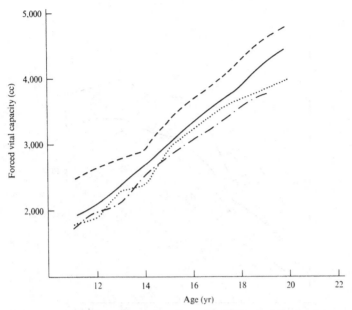

Fig. 5.12. Development in Forced Vital Capacity (FVC) of Nuñoa boys (– – – –) compared to US Norms (——) and Peruvian boys from Ica (sea level) (– · – · –) and Huanuco (2,300 m) (......). As in Fig. 5.1, despite their smaller stature, the Nuñoa boys have a systematically higher forced vital capacity than the US and the Peruvian samples. Adapted from Frisancho (1969c).

### Age trends

Fig. 5.12 illustrates the development of forced vital capacity (FVC) in Nuñoa boys compared to that for other Peruvian samples (Preto & Calderon, 1947) and US norms (Bjure, 1963). The Nuñoa boys show an accelerated growth in FVC, so that by the age of 20 years their values of FVC are nearly twice those attained at 11 years. The Nuñoa boys, age-for-age, attain a significantly higher value than the other Peruvians and US sea-level norms. In other words, the growth in FVC follows the same rapid developmental pattern as that of the chest (Frisancho, 1969a; Frisancho & Baker, 1970). These findings confirm the earlier studies of Hurtado (1932).

Fig. 5.13 illustrates the results of the studies on highland (Nuñoa) and lowland (Mollendo) Quechua samples studied with the McDermott dry spirometer by Rimmer and recently analysed by Boyce *et al.* (1974). These data show that FVC and forces expiratory volume (FEV), adjusted for height and weight, of the highland Quechuas are greater than those of their lowland Quechua counterparts.

131

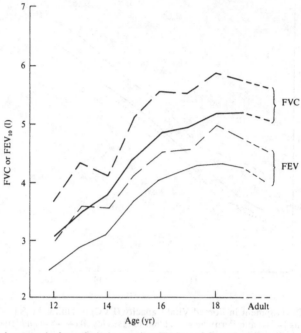

Fig. 5.13. Development in forced vital capacity (FVC) and forced expiratory volumes in 1 min ($FEV_{1.0}$) of highland Quechua natives from Nuñoa (broken lines) and lowland Quechua natives from Mollendo (unbroken lines). The highlanders for their stature have a significantly greater forced expiratory capacity. Adapted from Boyce *et al.* (1974).

### Interrelationship

As shown by the multiple regression analyses presented in Table 5.3, of all the variables used to characterize body size (height, weight, surface area) and chest size (chest depth, chest width, chest circumference), the chest circumference at maximum inspiration appears to be the best predictor of FVC, thereby explaining about 90% of the variance. As indicated by the standard partial regression, the next important variable that predicts FVC includes chest volume and sitting height.

In terms of pulmonary function, the Nuñoa children showed rapid and accelerated development in FVC. In view of the fact that at sea level, during childhood, growth in lung volume is associated with a proliferation of alveolar units and alveolar surface area (Dunnill, 1962), the rapid growth in FVC seen among high-altitude children is probably also associated with an increase in alveolar units and alveolar surface area. Since there is a direct relationship between alveolar quantity and diffusion capacity (Cotes, 1965), the accelerated development in lung volume almost surely has adaptive significance in facilitating the diffusion of oxygen. Several

Table 5.3. *Correlation coefficients and regression equations of prime predictor of forced vital capacity (FVC), as determined by up-rank, multiple regression of 150 high-altitude Peruvian Quechua boys, 11–20 years of age*

| Forced vital capacity versus | r | Equations | Standard and partial regression coefficients |
|---|---|---|---|
| Chest circumference at maximum inspiration | 0.90 | $y = -6113.10 + 117.56x_1$ | $b_1 = 0.90$[b] |
| Chest circumference at maximum expiration and sitting height | 0.91[a] | $y = -6801.10 + 920.80x_1 + 358.00x_2$ | $b_1 = 0.70; b_2 = 0.50$[c] |
| Chest circumference at maximum expiration, chest volume | 0.91[a] | $y = -5412.90 + 729.86x_1 + 291.26x_2 + 859.92x_3$ | $b_1 = 0.55; b_2 = 0.18$[c]; $b_3 = 0.20$ |

[a] Multiple regression coefficient.    [b] Standard regression coefficient.
[c] Partial regression coefficient.
From Frisancho (1969*b*).

studies, after adjustment for variations in body size, have demonstrated that adult highland natives have an enlarged lung volume (Hurtado *et al.*, 1956; Hurtado, 1964; Frisancho, 1969*b*; Garruto, 1969; Frisancho *et al.*, 1973*a*). Through this rapid pattern of growth the high-altitude native attains his enlarged lung volume.

In a recent investigation designed to determine the mechanisms of functional adaptation to high-altitude hypoxia (Frisancho *et al.*, 1973*a, c*), we measured the FVC of sea-level and high-altitude natives (Table 5.4). The results of this study, after adjustment for variations in body size, demonstrated that sea-level subjects who were acclimatized to high altitude during growth attained the same values of FVC as the high-altitude natives. In contrast, sea-level subjects (Peruvian and US whites) acclimatized as adults had significantly lower vital capacity than high-altitude natives. On the basis of these findings, we have postulated that the enlarged lung volume of the high-altitude natives is the result of adaptations acquired during the developmental period.

The hypothesis that the enlarged lung volume at high altitude is probably developmental in nature is supported by experimental studies on animals. The studies of Burri & Weibel (1971*a*), Bartlett & Remmers (1971) and Bartlett (1972) demonstrate that with prolonged exposure to high-altitude hypoxia (at 3,450 m) young rats show an accelerated development in the proliferation of alveolar units and alveolar surface area, and in the growth in lung volumes. In contrast, adult rats with prolonged exposure to high-altitude hypoxia did not show changes in alveolar quantity and lung

Table 5.4. *Covariance adjustment of forced vital capacity (FVC) among subjects tested at high altitude*

| Test altitude (m) | N | Group | FVC, adjusted for age, weight, and height (ml) | | F-ratio |
|---|---|---|---|---|---|
| | | | Mean | S.E. | |
| 3,840 | 40 | High-altitude natives | 4830.3 | 69.9 | |
| 3,840 | 13 | Sea-level subjects acclimatized as adults | 4504.6 | 122.1 | |
| | | | | | 5.19* |
| 3,400 | 20 | High-altitude natives | 4990.3 | 128.6 | |
| 3,400 | 21 | Sea-level subjects acclimatized as children | 5055.0 | 121.5 | |
| | | | | | 0.36 |
| 3,400 | 10 | US whites acclimatized as adults[a] | 4573.9 | 231.6 | |
| | | | | | 5.53* |

\* $P < 0.02$.
[a] Compared with high-altitude natives at both altitudes.
From Frisancho *et al.* (1973).

volume (Burri & Weibel, 1971*b*). These findings together suggest that in experimental animals and man the development of an enlarged lung volume at high altitude is probably mediated by developmental factors. Recent work (Beall *et al.*, 1977) has shown that low-altitude Peruvian natives have equal or more rapid development of chest size than Nuñoa natives. Research on the same sample suggests that altitude-related differences in FVC are small when expressed relative to measures of chest size rather than the more conventional measure of stature (P. Baker, personal communication).

### Nutrition and growth

Based upon the dietary surveys of Mazess & Baker (1964), Gursky (1969) and Thomas (1972), Picón-Reátegui (1976) concludes that the dietary intakes in Nuñoa meet US dietary recommended allowances. Another way to test these findings can be derived by evaluating the relationship of dietary intakes and growth. In Table 5.5 are presented partial correlation coefficients (adjusted for age) between dietary intakes and height of Nuñoa subjects. These data show that the reported caloric and protein intakes of both boys and girls during childhood and adolescence are not significantly related to height. The lack of relationship between dietary intakes and growth in Nuñoa is probably due to the fact that evaluations of

Table 5.5. *Correlation coefficients between calorie intake and height, and protein intake and height, partialling out the effect of age among highland Quechua subjects from Nuñoa*

| | Age group (yr) | | | | | |
| | 1–12 | | 13–19 | | 20–80 | |
| Variables | N | r | N | r | N | r |
|---|---|---|---|---|---|---|
| | | Males | | | | |
| Calories–height | 25 | 0.29 | 15 | −0.03 | 15 | 0.32 |
| Protein–height | 25 | 0.28 | 15 | 0.15 | 15 | 0.35 |
| | | Females | | | | |
| Calories–height | 28 | −0.19 | 7 | 0.26 | 24 | 0.30 |
| Protein–height | 28 | −0.26 | 7 | 0.61 | 24 | 0.02 |

Calculated from individual values given by Gursky (1969).

dietary intakes in any population, because of technical and field difficulties as well as wide daily, monthly and seasonal variations, do not give a true picture of actual food consumption. Indeed, as pointed out in Chapter 8 by Picón-Reátegui, the various dietary surveys show great variability in both quantity and quality of dietary intakes. For these reasons, and in view of the fact that the height of a child reflects the cumulative effect of his developmental history, it is not surprising that in Nuñoa the dietary intakes show little relationship to growth.

Bouloux (1968) conducted a study among two selected Aymara samples from La Paz, Bolivia. One sample of adolescents was derived from a population living in the outskirts of La Paz (Comunidad de la Garita de Lima), and the second sample was derived from an orphanage of La Paz (Orfelinato de Mindes Arcos). According to the evaluations of Bouloux (1968), the orphanage provided the children with conditions of nutrition (not quantified), health, hygiene and housing that were better than those of the community. In Table 5.6 data from that study are summarized. These data show that the adolescents living in the orphanage are similar in weight, height, chest circumference, chest depth, arm circumference and leg circumference and skeletal age to those living in the community.

The interpopulation comparisons in Nuñoa indicate that the urbanized groups are slightly heavier and considerably fatter than their rural counterparts. These differences are probably related to differences in activity and dietary intake. In any event, whatever the factors responsible for the greater fat deposition of the urbanized group, they did not appear to be reflected in taller statures. In other words, it appears that in Nuñoa

135

*The biology of high-altitude peoples*

Table 5.6. *Maturation and body size of two groups of high-altitude Aymara adolescent Indian boys studied in La Paz, Bolivia*

| | Community (N = 31) | | Orphanage (N = 24) | |
|---|---|---|---|---|
| | Mean | s.d. | Mean | s.d. |
| Age (yr) | 15.9 | 0.3 | 16.0 | 0.3 |
| Height (cm) | 157.1 | 6.1 | 157.6 | 5.0 |
| Weight (kg) | 49.1 | 5.7 | 51.5 | 5.2 |
| Chest circumference (cm) | 81.3 | 4.0 | 82.5 | 1.5 |
| Chest depth (cm) | 20.2 | 1.9 | 21.0 | 1.1 |
| Arm circumference (cm) | 21.6 | 1.8 | 23.0 | 1.7 |
| Leg circumference (cm) | 31.5 | 1.8 | 32.8 | 1.7 |
| Skeletal age (yr) | 15.0 | 0.6 | 15.5 | 0.3 |

Adapted from Bouloux (1968).

greater calorie reserve as reflected by the increased fatness and weight is not associated with increased dimensional growth. This finding is contrary to those found among sea-level populations where increased fatness and weight are associated with advanced maturity and growth (Reynolds, 1950; Garn & Haskell, 1960; Lloyd, Wolff & Whelen, 1961).

*Parent–offspring correlations in height*

Table 5.7 gives the parent–offspring correlations in height of Nuñoa samples compared with similar (unpublished) data derived from a sample of impoverished Quechua Indians from Cuzco (Frisancho *et al.*, 1973*a*). In Nuñoa the parents' heights are significantly correlated with the height of the offspring. In contrast, among the impoverished Quechua from Cuzco, the parents' heights are *not* correlated with the height of the offspring, except during earlier childhood. These findings suggest that the environmental conditions for both parents and offspring are more uniform in Nuñoa than in the squatter settlement or *barriada* of Cuzco.

The parent–offspring correlations in Nuñoa range from 0.08 to 0.52, while among Western populations they range from 0.26 to 0.60 (Tanner, 1962). Thus, it would appear that genetic contributions to phenotypic variation in growth in height in Nuñoa are comparable to those of Western populations. However, it is important to note that the investigations of Rothhammer & Spielman (1972) among Aymara Indians living at various altitudes in Chile demonstrate that altitude difference makes the largest contribution to anthropometric variation while geographic and genetic distance contribute considerably less.

136

Table 5.7. *Parent–offspring correlations in height among two high-altitude samples: Quechua from Nuñoa and impoverished Quechua from Cuzco*[a]

| Age group (yr) | Father | | | | Mother | | | |
|---|---|---|---|---|---|---|---|---|
| | Son | | Daughter | | Son | | Daughter | |
| | N | r | N | r | N | r | N | r |
| | | | Quechuas from Nuñoa | | | | | |
| 1–4.9 | 24 | 0.12 | 25 | 0.08 | 20 | 0.18 | 21 | 0.10 |
| 5–10.9 | 115 | 0.31** | 98 | 0.29** | 96 | 0.36** | 86 | 0.32** |
| 11–16.9 | 80 | 0.40** | 60 | 0.42** | 60 | 0.48** | 50 | 0.52** |
| | | | Impoverished Quechuas from Cuzco | | | | | |
| 1–4.9 | 54 | −0.23 | 52 | −0.28* | 80 | −0.18 | 57 | −0.46* |
| 5–10.0 | 85 | 0.10 | 71 | 0.08 | 99 | 0.30 | 78 | 0.21 |
| 11–12.0 | 16 | −0.05 | 19 | 0.40* | 12 | 0.50 | 20 | 0.24 |

[a] Weighted mean $r$ derived from Z-transformed age-specific values of $r$. A given age-specific correlation coefficient is converted to $Z$ values and, in turn, the derived $Z$ values are transformed back into $r$ coefficients.

\* $P < 0.05$.            \*\* $P < 0.01$.

## Conclusions

Evaluations of growth and development of Andean populations both in the Southern and Central Highlands indicate the following:

(*a*) Growth in body size, skeletal maturation, and adolescent maturation in both Quechua and Mestizo samples is unquestionably slow.

(*b*) Variations in subcutaneous fat, weight and muscle area do not parallel differences in growth in height in the southern Quechua highlanders.

(*c*) Dietary intakes of selected samples from Nuñoa highland Quechua do not show any meaningful correlations with measurements of linear growth.

(*d*) As inferred from parent–offspring correlations, in Nuñoa the genetic contribution to phenotypic variation in growth in height is comparable to that of Western populations.

(*e*) Contrary to the slow growth in stature and delayed maturation, the development of lung size and chest morphology in the Central and Southern Highlands is as rapid as it is in sea-level Indian populations.

(*f*) Since there is a direct relationship between alveolar area and diffusion capacity, the rapid growth in forced vital capacity probably has an adaptive capacity in facilitating the diffusion of oxygen.

(*g*) As derived from analytical studies on sea-level populations at high-altitude, the enlarged forced vital capacity of the high-altitude native is probably developmental in nature.

(*h*) Since along with the rapid growth in lung volume high-altitude populations also have accelerated growth in heart size (Peñaloza *et al.*, 1960), the rapid development in chest size of Nuñoa subjects probably forms an integrated part of this system. The implication is that the enlarged thorax of high-altitude natives is not adaptive *per se*, but is rather a by-product of their enlarged lung and heart volumes.

(*i*) The continued growth in marrow space at altitude is probably related to the demand for increased hematopoietic activity at high altitudes.

### Growth in the Ethiopian mountains

*Source of data*

Populations inhabiting the areas of the Simien mountains live at altitudes that range from 1,200 m to 3,900 m. Between 1965 and 1968 two communities were studied by Harrison *et al.* (1969) and Clegg *et al.* (1972). The population selected for the study corresponded to the Amhara tribal group, and consisted of two villages: Adi-Arkai, located at an altitude of 1,524 m, and Debarech, situated at an altitude of 3,024 m. These two communities are separated by a distance of about 35 miles. Communication between the two villages is very good, thus facilitating migration. The barometric pressure (BP) corresponding to the altitudes is 524 mm Hg for the highland village of Debarech and 634 mm Hg for the lowland community of Adi-Arkai. Compared to the sea-level value (BP = 760 mm Hg), the barometric pressure of the lowland village is reduced by 17%, while that of the highland Debarech is lowered by 31%. According to the reports of Harrison *et al.* (1969) and Clegg *et al.* (1972), the climatic conditions in the low-altitude village of Adi-Arkai are characterized by high temperatures and high humidity. On the other hand, in the highland community of Debarech the climate is more temperate. Adi-Arkai and Debarech each have about 5,000 inhabitants.

Although information on socio-economic and nutritional data are lacking, as shown in Table 5.8 the incidence of communicable infectious diseases such as malaria and measles in the low-altitude village of Adi-Arkai is three times greater than in the highland Debarech (Harrison *et al.*, 1969). In both villages the subsistence pattern is based upon agriculture. The major staple crop is a native grain called ' *teff*' (*Eragrostis abyssinica*). Other crops include cereals like wheat and barley. Domestic animals include Longhorn cattle, oxen and chickens. In general, as stated by Harrison *et al.* (1969), Clegg *et al.* (1972) and Pawson (1971), the environmental conditions are worse in the low-altitude village of Adi-Arkai than in Debarech.

138

Table 5.8. *Frequency of communicable disease among lowland and highland Ethiopian tribal groups*

| Disease | Adi-Arkai (1,524 m) (N = 188) | | | Debarech (3,048 m) (N = 145) | | |
|---|---|---|---|---|---|---|
| | Number affected | Number not affected | % Affected | Number affected | Number not affected | % Affected |
| Measles | 90 | 98 | 91.8 | 33 | 112 | 29.5 |
| Malaria | 85 | 103 | 89.5 | 21 | 124 | 16.9 |
| Dysentery | 13 | 175 | 7.4 | 1 | 144 | 0.7 |
| Scabies | 9 | 179 | 5.0 | 17 | 128 | 13.3 |
| Syphilis | 47 | 141 | 33.3 | 31 | 114 | 27.2 |

Adapted from Harrison *et al.* (1969).

Table 5.9. *Distribution by age and sex of samples included in the evaluation of Ethiopian growth*

| Age group (yr) | Low altitude (Adi-Arkai, 1,524 m) | | | | Highland (Debarech, 3,048 m) | | | |
|---|---|---|---|---|---|---|---|---|
| | Males | | Females | | Males | | Females | |
| | N | Mean | N | Mean | N | Mean | N | Mean |
| < 2 | — | — | — | — | — | — | 1 | — |
| 2.01–4.00 | 9 | 3.10 | 5 | 3.47 | 7 | 3.01 | 6 | 3.35 |
| 4.01–6.00 | 11 | 4.93 | 3 | 5.37 | 2 | 5.21 | 4 | 5.21 |
| 6.01–8.00 | 9 | 6.60 | 14 | 6.66 | 24 | 6.88 | 18 | 6.78 |
| 8.01–10.00 | 12 | 8.67 | 15 | 8.99 | 2 | 9.97 | 2 | 8.40 |
| 10.01–12.00 | 14 | 10.84 | 10 | 11.07 | 14 | 10.69 | 16 | 10.96 |
| 12.01–14.00 | 13 | 12.90 | 8 | 12.56 | 26 | 13.21 | 24 | 13.32 |
| 14.01–16.00 | 17 | 15.07 | 5 | 15.25 | 7 | 15.47 | 5 | 14.95 |
| 16.01–18.00 | 7 | 17.30 | — | — | 8 | 16.26 | — | — |
| 18.01–19.00 | 3 | 18.65 | — | — | 3 | 18.18 | — | — |
| Adult | 71 | 29.7 | 36 | 27.1 | 82 | 34.5 | 58 | 29.9 |

Subjects under 19 yr are adapted from Clegg *et al.* (1972) and Pawson (1973). Adult subjects (aged 18–50 yr) as given in Harrison *et al.* (1969). No number indicates insufficient data.

A total of 155 children were studied in the village of Adi-Arkai and 169 children were studied from the highland community of Debarech. The age range was from 2 to 19 years. These children were attending the schools of the two villages; ages to the nearest year were obtained from the parents. In addition, a total of 107 native adults from the village of Adi-Arkai and 140 from the highland village of Debarech were studied.

A distribution by sex, age and altitude is given in Table 5.9. From these data it is evident that the size of the samples is too small for evaluations of mean values. For this reason, Clegg *et al.* (1972) analyzed the data as follows:

(1) Between-group comparisons of individual physical parameters were made by regression analyses, the independent variable being either chronological age or skeletal age. Differences in the elevations of regression lines (or adjusted means) were tested by covariance analysis.

(2) In the case of stature and weight, the individual Ethiopian values were plotted on percentile standards for British children (Tanner, Whitehouse & Takaishi, 1966) and between-group comparisons were made on the basis of the frequencies with which individuals fell into different British percentile ranges (Clegg *et al.*, 1972).

(3) For each group the regression of each variable (or its logarithm where appropriate) on chronological age was calculated and covariance adjustments of the mean values of each dependent variable were made to eliminate differences due to the varying age ranges of the four groups.

### Body size and composition

#### Height, weight and chest size

In Table 5.10 it can be seen that with reference to the British standing, the lowland males (but not the females) are more often in the lower

Table 5.10. *Distribution of Ethiopian children of the different groups in the British percentile ranges of stature and weight*

|  | Height percentile ranges | | | | |
|---|---|---|---|---|---|
|  | < 3 | 3–10 | 10–25 | 25–50 | 50+ |
| Highland male | 27 | 17 | 23 | 18 | 8 |
| Lowland male | 47 | 21 | 17 | — | — |
| $\chi^2 = 13.817, P < 0.001$ | | | | | |
| Highland female | 20 | 17 | 15 | 17 | 7 |
| Lowland female | 12 | 10 | 19 | 12 | 7 |
| $\chi^2 = 3.312$ NS | | | | | |
|  | Weight percentile ranges | | | | |
|  | 3 | 3–10 | 10–25 | 25+ | |
| Highland male | 34 | 16 | 26 | 17 | |
| Lowland male | 59 | 18 | 11 | 7 | |
| $\chi^2 = 16.995, P < 0.0001$ | | | | | |
| Highland female | 26 | 15 | 14 | 21 | |
| Lowland female | 32 | 12 | 10 | 6 | |
| $\chi^2 = 8.186, P < 0.05$ | | | | | |

Adapted from Clegg *et al.* (1972).

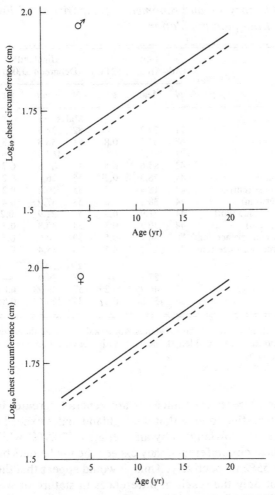

Fig. 5.14. Regression of age to chest circumference ($\log_{10}$) among highland (unbroken lines) and lowland (broken lines) Ethiopian samples. At a given age the highlanders show a greater chest size, for their stature, than do the lowlanders. Adapted from Pawson (1971).

percentile ranges for height and weight than the highlanders. The frequency distribution as tested by $\chi^2$ is significantly different at 0.001 level. However, Clegg *et al.* (1972) indicate through covariance analysis that the highlander males, when adjusted for differences in age, are unequivocally and significantly ($P < 0.05$) taller but not heavier than their lowland counterparts, and that the regressions of individual measurements on chronological age are steeper, although in females the overall differences in body size are not well defined. While the adjusted means are greater in the

141

Table 5.11. *Summary of anthropometric comparisons of lowland and highland adult Ethiopian populations*

| | Lowland (Adi-Arkai, 1,524 m) | | | Highland[a] (Debarech, 3,000 m) | | | |
|---|---|---|---|---|---|---|---|
| | N | Mean | S.E. | N | Mean | S.E. | t-test |
| | | | | Males | | | |
| Age (yr) | 71 | 29.7 | — | 82 | 34.5 | — | NS |
| Weight (kg) | 68 | 53.7 | 0.8 | 81 | 56.8 | 0.8 | P < 0.05 |
| Height (cm) | 71 | 168.8 | 0.9 | 81 | 167.3 | 0.6 | NS |
| Sitting height (cm) | 42 | 85.6 | 0.5 | 17 | 84.5 | 0.7 | NS |
| Transverse chest (cm) | 44 | 26.3 | 0.2 | 55 | 26.9 | 0.2 | P < 0.05 |
| Antero-posterior chest (cm) | 44 | 18.4 | 0.2 | 55 | 19.2 | 0.2 | P < 0.05 |
| Biacromial diameter (cm) | 44 | 38.1 | 0.2 | 53 | 37.9 | 0.4 | NS |
| Upper-arm circumference (cm) | 44 | 21.9 | 0.3 | 54 | 22.9 | 0.2 | P < 0.05 |
| Calf circumference (cm) | 44 | 30.3 | 0.3 | 54 | 30.8 | 0.3 | NS |
| Maximum chest circumference (cm) | 70 | 86.7 | 0.5 | 80 | 90.0 | 0.5 | P < 0.05 |
| Minimum chest circumference (cm) | 70 | 82.1 | 0.5 | 80 | 85.4 | 0.5 | P < 0.05 |
| | | | | Females | | | |
| Age | 36 | 27.1 | — | 58 | 29.9 | — | NS |
| Weight (kg) | 35 | 46.96 | 0.29 | 37 | 49.98 | 0.88 | P < 0.05 |
| Height (cm) | 34 | 152.64 | 0.83 | 37 | 156.64 | 0.86 | P < 0.05 |

[a] Highland Debarech also exceed the Lowlands Adi-Arkai in bi-iliac diameter, bicondylar femur, neck circumference, head length, head breadth, bigonial diameter. No figure indicates insufficient data.
Adapted from Harrison *et al.* (1969).

highlanders, the regression coefficients are generally greater in the low-landers. It is interesting to note that the highland males exceed in stature, adjusted for age, the lowlanders by an average of 2.89%, while in chest diameter and chest circumference they exceed the lowlanders by averages of 4.67% and 3.55% respectively. Thus, it would appear that the highland chest size is not only the result of differences in stature or weight. This can be seen in Fig. 5.14. These data, which are based upon the regression equations given by Pawson (1971), show that at a given age the highland boys and girls have greater chest circumferences ($Log_{10}$) than their low-altitude counterparts.

Harrison *et al.* (1969) have summarized the studies on adult Ethiopian populations. In Table 5.11 it can be seen that the highland Debarech and lowland Adi-Arkai males have similar stature, sitting height, and biacromial diameter. However, despite these similarities, the highland Debarech have significantly ($P < 0.05$) greater measurements of transverse chest diameter, antero-posterior chest diameter, maximum chest circumference and minimum chest circumference than the lowland Adi-Arkai. Highland females are both heavier and taller than their lowland counterparts. Table 5.12

Table 5.12. *Comparison of chest size between highland and lowland migrants and highland and lowland natives in Ethiopia*

| | Migrant | | | | | |
|---|---|---|---|---|---|---|
| | Highlanders living in the lowlands at Adi-Arkai | | | Lowlanders living in the highland at Debarech | | |
| | N | Mean | S.E. | N | Mean | S.E. |
| Weight (kg) | 30 | 55.9* | 1.2 | 19 | 55.7* | 1.6 |
| Height (cm) | 40 | 168.3 | 1.0 | 19 | 166.0 | 1.8 |
| Maximum chest circumference (cm) | 41 | 89.3* | 1.0 | 21 | 88.5* | 1.1 |
| Minimum chest circumference (cm) | 41 | 84.9* | 1.0 | 21 | 85.1* | 1.2 |
| | Natives | | | | | |
| | Highland Debarech | | | Lowland Adi-Arkai | | |
| | N | Mean | S.E. | N | Mean | S.E. |
| Weight (kg) | 81 | 56.8* | 0.8 | 68 | 53.7 | 0.8 |
| Height (cm) | 81 | 167.3 | 0.6 | 71 | 168.8 | 0.9 |
| Maximum chest circumference (cm) | 80 | 90.0* | 0.5 | 70 | 86.7 | 0.5 |
| Minimum chest circumference (cm) | 80 | 85.4* | 0.5 | 70 | 82.1 | 0.5 |

* $P < 0.05$ when compared with the lowland natives from Adi-Arkai.
Adapted from Harrison *et al.* (1969).

compares weight, stature and measurements of chest size of highlanders who migrated to the lowlands, lowlanders who migrated to the highlands, and lowland and highland natives. From these data it is evident that the lowlanders who migrated to the highlands had a greater chest size than the lowland natives, and their chest size was equal to that of the highland natives. The highlanders who migrated to the low altitude also appeared to retain their enlarged chest size. Thus, it would appear that the enlarged chest size of the highland natives can be attained by the lowland subjects living in the highlands.

### Skinfolds

Table 5.13 compares the distribution of the triceps skinfold thickness with reference to the British percentile ranges. Although there are no major differences between altitudes for the whole age range, if the samples are divided into those above and below the age of 8 years, more differences become evident. These data show that in the below-8-year age-group the low-altitude boys show a significant preponderance of thicker skinfolds compared with the high-altitude boys. In the over-8-year age-group, the

*The biology of high-altitude peoples*

Table 5.13. *Distribution of Ethiopian children of the different groups in British percentile ranges of triceps skinfold thickness*

| | Triceps percentile ranges | | | | |
|---|---|---|---|---|---|
| | < 3 | 3–10 | 10–25 | 25–50 | 50+ |
| | All age groups | | | | |
| Highland males | 17 | 28 | 18 | 14 | 6 |
| Lowland males | 22 | 16 | 18 | 14 | 15 |
| $\chi^2 = 7.75$, NS | | | | | |
| Highland females | 23 | 19 | 11 | 14 | 19 |
| Lowland females | 17 | 15 | 9 | 11 | 8 |
| $\chi^2 = 4.22$, NS | | | | | |
| | Under eight years of age | | | | |
| Highland males | 6 | 10 | 7 | 6 | 4 |
| Lowland males | 3 | 1 | 3 | 8 | 14 |
| $\chi^2 = 12.83, P < 0.001$ | | | | | |
| Highland females | 6 | 7 | 6 | 7 | 3 |
| Lowland females | 7 | 5 | 1 | 6 | 3 |
| $\chi^2 = 0.22$, NS | | | | | |
| | Over eight years of age | | | | |
| Highland males | 11 | 18 | 11 | 8 | 2 |
| Lowland males | 19 | 16 | 15 | 6 | 1 |
| $\chi^2 = 2.86$, NS | | | | | |
| Highland females | 7 | 12 | 5 | 7 | 16 |
| Lowland females | 10 | 10 | 8 | 5 | 5 |
| $\chi^2 = 5.64, P < 0.02$ | | | | | |

Adapted from Clegg *et al.* (1972).

high-altitude girls have a greater frequency of thicker skinfolds than their lowland counterparts.

## Indices of maturation

### Skeletal development

As shown in Fig. 5.15, both highland and lowland children exhibit a delayed skeletal age between chronological ages 2 and 10 years. Thereafter, skeletal age appears to keep pace with chronological age when compared to the Greulich & Pyle standards. For analytical purposes the skeletal ages were converted to logarithmic units and resulting values were compared through covariance using chronological age from 2 to 19 years as the covariate. As shown in Table 5.14, the highland males attained a greater adjusted skeletal age than their lowland counterparts. However, the highland and lowland girls attained equivalent skeletal ages. Furthermore, the regression coefficients indicate that the increase in skeletal age associated with an increase in chronological age is similar in both groups.

144

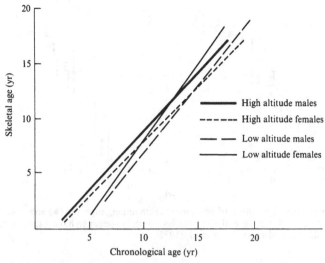

Fig. 5.15. Comparison of skeletal age of highland and lowland Ethiopian samples. The skeletal maturation among the highland appears to be more similar to those in the lowland. Adapted from Pawson (1971).

Table 5.14. *Comparison of mean $log_{10}$ skeletal age (adjusted for age by covariance analysis) of highland and lowland Ethiopian children*

|  | Adjusted mean $log_{10}$ skeletal age | Regression coefficient |
|---|---|---|
| Highland male | 0.93 | 0.06 |
| Lowland male | 0.85 | 0.06 |
| Significance level | $P = 0.001$ | NS |
| Highland female | 0.91 | 0.06 |
| Lowland female | 0.88 | 0.07 |
| Significance level | NS | NS |

Adapted from Clegg *et al.* (1972).

### Dental development

As shown in Fig. 5.16, under the age of 10 years the highland children appear to have a greater mean number of permanent teeth erupted. However, when tested by regression analysis no significant high-altitude group differences were found (Pawson, 1971).

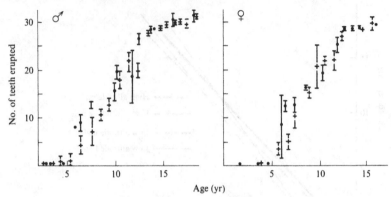

Fig. 5.16. Comparison of eruption of permanent teeth among highland (●) and lowland (+) Ethiopian samples. Under the age of 10 yr the highlanders appear to have a greater number of teeth erupted. Adapted from Clegg *et al.* (1972).

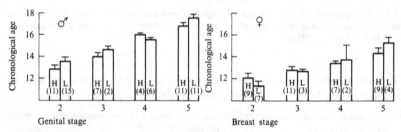

Fig. 5.17. Comparison of stages of pubertal maturation among Ethiopian samples. Unlike growth in height, secondary sexual maturation does not appear to show altitude-related differences. H, highland sample; L, lowland sample. Numbers in parentheses give the numbers in samples. Adapted from Clegg *et al.* (1972).

### Secondary sexual characteristics

Fig. 5.17 illustrates the pubertal ratings among Ethiopian children. The genital development ratings appear to be similar in both highland and lowland groups. In girls, breast development stages do not appear to show any defined difference.

### Conclusions

As indicated by the regression covariance and $\chi^2$-tests, growth and maturation of the highland children are advanced when compared to those of the lowland children. These differences, as Harrison *et al.* (1969) and Clegg *et al.* (1972) conclude, are probably related to the fact that the lowland environment is characterized by a higher incidence of infectious diseases such as malaria and intestinal parasitism. Consequently, it would appear

146

that these stresses have a greater negative effect on growth in the lowlands than any possible effects hypoxic stress may have on growth at high altitude (Clegg *et al.*, 1972). Furthermore, these differences may have a recent origin. This suggestion is based on two findings. (*a*) Among adult males there are no differences between altitudes in stature and weight although such differences occur among children. (*b*) Among adult females, contrary to the situation found in children, the lowlanders exceed in stature and weight their highland counterparts. In a recent investigation in Peru we also found that, due to recent deterioration in the ecological and socio-economic conditions, lowland tropical children, if the conditions remain the same, will probably not be as tall as their parents (Frisancho, Borkan & Klayman, 1975).

In terms of chest size, the children do not show any altitude-related differences. However, the adult highlanders, despite their equal statures, do show significantly greater measurements of chest size than the lowland natives. Furthermore, it must be pointed out that the difference between highland and lowland children is greater in terms of chest size (4.67% in chest width and 3.55% in chest circumference) than in terms of height (2.89%). Thus, it would appear that highland populations during adulthood do attain an increased chest size.

## Growth in the Himalayas

### Source of data

The data for this study come from investigations carried on among highland Sherpas and lowland Tibetans. The Sherpas inhabit the Himalayas of the Khumbu region. They live at altitudes that range from 3,475 to 4,050 m. According to Pawson (1974), in 1970 there were about 2,383 Sherpas living in the area, 1,123 males and 1,260 females, and 50% of the total population was under the age of 25 years. The subsistence pattern of the Sherpas is based upon a mixed economy of agriculture, herding and cash derived from the tourist trade. The staple foods are potatoes, rice and blackwheat; meat and milk of yaks are also consumed.

Following the imposition of direct Chinese political control over Tibet in 1959, refugee centers for Tibetans were established in the Kathmandu region, at an altitude of about 1,400 m. The economy of this population is directly tied to the production of woven carpets, which are then sold to visitors to the camp. In 1971, as part of the Thomas A. Dooley Foundation Program, daily distributions of milk and multiple vitamins were being made available to the children attending the camp school (Pawson, 1974). Furthermore, through a co-operative program of food-buying operated by refugee centers, animal protein in the form of water-buffalo meat was available to the children. According to Pawson (1974) the Tibetan children in Kathmandu were growing under more favorable

147

Table 5.15. *Anthropometric measurement of low-altitude (1,400 m) Tibetans and high-altitude (3,475–4,050 m) Sherpas*

| Age groups | Height (cm) High altitude N | Mean | S.D. | Low altitude N | Mean | S.D. | Weight (kg) High altitude N | Mean | S.D. | Low altitude N | Mean | S.D. | Chest circumference (cm) High altitude N | Mean | S.D. | Low altitude N | Mean | S.D. |
|---|---|---|---|---|---|---|---|---|---|---|---|---|---|---|---|---|---|---|
| **Males** | | | | | | | | | | | | | | | | | | |
| 0.0–1.9 | 9 | 69.3 | 5.92 | 8 | 68.7 | 4.85 | 9 | 7.6 | 1.88 | 13 | 7.0 | 1.09 | 9 | 45.0 | 3.12 | — | — | — |
| 2.0–3.9 | 20 | 81.6 | 5.35 | 16 | 76.6 | 5.99 | 19 | 11.0 | 1.53 | 17 | 9.5 | 3.11 | 19 | 49.7 | 2.06 | 1 | 45.3 | 0.0 |
| 4.0–5.9 | 8 | 93.2 | 8.65 | 31 | 90.4 | 5.91 | 7 | 13.3 | 1.76 | 31 | 12.7 | 2.00 | 7 | 53.2 | 3.65 | 22 | 52.5 | 3.37 |
| 6.0–7.9 | 21 | 101.8 | 7.53 | 23 | 102.8 | 6.26 | 16 | 15.7 | 2.51 | 23 | 15.9 | 1.35 | 21 | 56.3 | 2.79 | 23 | 55.2 | 2.08 |
| 8.0–9.9 | 23 | 109.5 | 9.60 | 12 | 109.8 | 4.76 | 23 | 17.9 | 3.97 | 12 | 18.2 | 2.29 | 23 | 58.2 | 3.47 | 12 | 58.4 | 3.42 |
| 10.0–11.9 | 29 | 121.0 | 6.33 | 8 | 120.5 | 8.69 | 29 | 21.5 | 2.11 | 8 | 21.6 | 3.78 | 29 | 61.8 | 2.77 | 8 | 60.8 | 2.77 |
| 12.0–13.9 | 21 | 126.9 | 6.75 | 6 | 126.4 | 12.09 | 21 | 24.7 | 3.17 | 6 | 27.4 | 4.29 | 21 | 64.4 | 3.29 | 6 | 64.2 | 2.86 |
| 14.0–15.9 | 22 | 134.3 | 6.44 | 2 | 144.0 | 15.63 | 22 | 29.4 | 3.40 | 4 | 34.2 | 6.26 | 18 | 67.2 | 2.85 | 2 | 68.9 | 0.85 |
| 16.0–17.9 | 15 | 145.2 | 7.17 | 4 | 145.8 | 11.55 | 15 | 36.0 | 5.89 | 3 | 36.1 | 8.15 | 8 | 71.2 | 4.11 | 4 | 71.6 | 6.71 |
| 18.0–19.9 | 4 | 147.5 | 9.57 | 3 | 154.8 | 2.39 | 4 | 39.2 | 5.11 | 3 | 45.6 | 1.13 | 1 | 73.2 | 0.0 | 3 | 79.1 | 1.88 |
| 20.0–22.0 | 5 | 152.6 | 4.54 | 2 | 160.8 | 3.05 | 5 | 46.3 | 3.60 | 2 | 52.9 | 5.46 | 2 | 78.2 | 2.97 | 2 | 82.4 | 1.20 |
| **Females** | | | | | | | | | | | | | | | | | | |
| 0.0–1.9 | 10 | 65.7 | 2.69 | 5 | 66.0 | 4.11 | 10 | 7.2 | 0.85 | 12 | 5.8 | 1.10 | 9 | 43.9 | 1.77 | — | — | — |
| 2.0–3.9 | 15 | 79.1 | 6.77 | 7 | 75.3 | 0.82 | 13 | 10.6 | 1.72 | 12 | 8.0 | 1.30 | 13 | 49.9 | 2.66 | — | — | — |
| 4.0–5.9 | 19 | 9.36 | 6.72 | 25 | 87.0 | 5.13 | 18 | 13.6 | 1.75 | 25 | 13.0 | 6.16 | 18 | 53.0 | 2.37 | 19 | 51.3 | 2.45 |
| 6.0–7.9 | 16 | 101.5 | 6.34 | 12 | 100.8 | 3.77 | 16 | 15.1 | 1.99 | 12 | 15.3 | 1.36 | 15 | 54.6 | 1.52 | 12 | 55.8 | 2.37 |
| 8.0–9.9 | 24 | 109.6 | 6.64 | 9 | 114.8 | 8.57 | 24 | 17.7 | 2.26 | 9 | 19.9 | 3.70 | 21 | 57.5 | 2.53 | 7 | 56.3 | 2.09 |
| 10.0–11.9 | 28 | 117.7 | 6.70 | 12 | 125.1 | 7.41 | 28 | 20.0 | 3.04 | 12 | 23.7 | 3.37 | 24 | 59.5 | 2.84 | 11 | 59.6 | 3.03 |
| 12.0–13.9 | 20 | 126.4 | 5.97 | 14 | 131.9 | 11.57 | 20 | 24.1 | 3.42 | 14 | 29.9 | 5.74 | 12 | 62.9 | 3.75 | 11 | 65.0 | 4.23 |
| 14.0–15.9 | 15 | 138.5 | 8.91 | 4 | 140.8 | 5.29 | 15 | 32.6 | 7.00 | 4 | 30.7 | 4.00 | 4 | 66.0 | 3.76 | — | — | — |
| 16.0–17.9 | 5 | 147.1 | 10.52 | 5 | 150.3 | 5.83 | 5 | 43.2 | 11.76 | 1 | 40.0 | 7.57 | 1 | 67.8 | 0.0 | — | — | — |
| 18.0–19.9 | 1 | 153.0 | 0.0 | 6 | 149.6 | 4.83 | 6 | 48.5 | 0.0 | 6 | 41.9 | 4.18 | — | — | — | — | — | — |
| 20.0–22.0 | 7 | 150.5 | 6.54 | 1 | 149.1 | 0.0 | 7 | 49.3 | 5.43 | 1 | 47.7 | 0.0 | — | — | — | — | — | — |

Data provided by I. G. Pawson (personal communication).

conditions than the highland Sherpas. These observations indicated that Tibetan children were receiving meat regularly, whereas the Sherpas ate meat only sporadically.

During the period of April to October, 1971, Pawson (1974) studied a sample of 180 highland Sherpa boys and 165 highland Sherpa girls. In addition, 122 Tibetan boys and 111 Tibetan girls were studied. The age of the sample ranged from birth to 22 years. Since no birth records are kept either among the Sherpas or Tibetans, the ages of the children were derived from interviews with the mothers. It is assumed that any errors in reported age are comparable among both highland and lowland children.

### Body size and composition

#### Height, weight and chest size

Table 5.15 summarizes the anthropometric measurements of the highland and low-altitude samples. It is quite evident that the number of children in some age groups is too small for adequate comparisons. Therefore, all inferences are drawn taking this fact into account. These data, and those illustrated in Fig. 5.18, indicate that growth in stature in both the highland

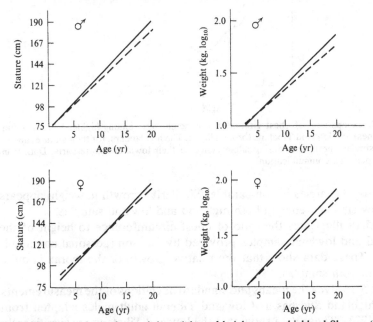

Fig. 5.18. Comparison of growth in weight and height among highland Sherpas (unbroken lines) and lowland Tibetans (broken lines). No major populational growth differences are evident from these data. Adapted from Pawson (1974) and regression equations from Pawson (personal communication).

149

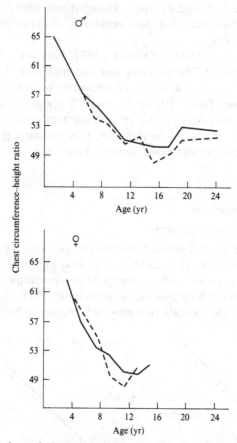

Fig. 5.19. Comparison of chest circumference–height ratio among highland Sherpas (unbroken lines) and lowland Tibetans (broken lines). The highlanders, for their stature, appear to have similar chest sizes during adolescence to their lowland counterparts. Data from Pawson (personal communication).

and lowland samples is comparable. Similarly, growth in weight appears to follow similar trends in both highland and lowland samples.

Fig. 5.19 illustrates the ratio of chest circumference to height of the highland and lowland samples provided by Pawson (personal communication). These data show that the relative growth of the thorax is quite similar in both samples.

Table 5.16 gives the means and standard deviations of the measurements of the highland Sherpas and lowland Tibetan adult males adapted from Weitz (1973). These data show that the lowland Tibetans are significantly ($P < 0.01$) taller and heavier than the highland Sherpas. However, the chest size in relation to height is comparable in both the highland Sherpas and the lowland Tibetans.

150

Table 5.16. *Comparison of anthropometric measurements of highland Sherpas and lowland Tibetan adult males*

| | High altitude (Sherpas, 3,475–4,050 M) | | | Low altitude (Tibetans, 1,400 m) | | | | | |
|---|---|---|---|---|---|---|---|---|---|
| | $\bar{x}$ | S.D. | N | $\bar{x}$ | S.D. | N | t | d.f. | P |
| Height (cm) | 162.17 | 7.05 | 62 | 166.94 | 4.87 | 23 | 2.99 | 83 | $\leqslant 0.01$ |
| Weight (kg) | 54.64 | 5.10 | 62 | 58.06 | 5.09 | 23 | 2.75 | 83 | $\leqslant 0.01$ |
| Chest circumference (cm) | 84.55 | 4.36 | 61 | 86.75 | 3.33 | 23 | 2.19 | 82 | $\leqslant 0.05$ |
| Chest–height ratio | 52.14 | — | 61 | 51.96 | — | 23 | — | — | — |
| Triceps skinfold (mm) | 4.18 | 1.12 | 62 | 4.91 | 1.76 | 23 | 2.27 | 83 | 0.05 |
| Subscapular skinfold (mm) | 6.85 | 1.69 | 62 | 7.24 | 2.03 | 23 | 0.89 | 83 | $> 0.05$ |
| Supra-iliac skinfold (mm) | 5.01 | 1.57 | 61 | 6.61 | 3.11 | 23 | 3.12 | 82 | $\leqslant 0.01$ |

Adapted from Weitz (1973).

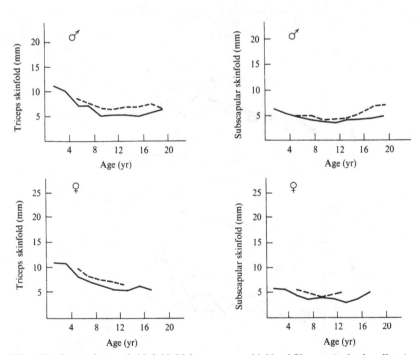

Fig. 5.20. Comparisons of skinfold thickness among highland Sherpas (unbroken lines) and lowland Tibetans (broken lines). The lowlanders are fatter than the highlanders. Adapted from Pawson (1974).

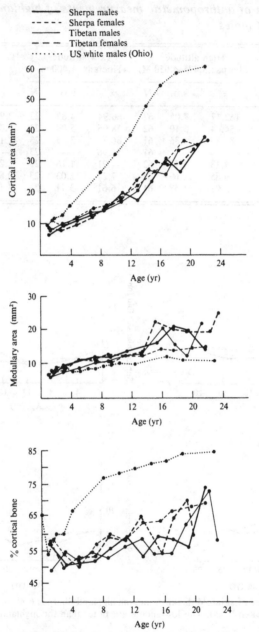

Fig. 5.21. Comparisons of cortical bone development at mid-shaft of the second metacarpal among highland Sherpas and lowland Tibetans. Both Himalayan samples have a markedly lower cortical bone due to an enhanced medullary area when compared to US standards. Adapted from Pawson (1974).

### Skinfolds

Fig. 5.20 illustrates the development of skinfold thickness at the triceps and subscapula. These data show that skinfold thicknesses among the lowland Tibetans at all ages are systematically higher than in the highland Sherpas. As shown in Table 5.16, the lowland adult Tibetans are also fatter than highland Sherpas.

### Cortical thickness

Fig. 5.21 depicts the percent cortical bone and medullary area at the mid-shaft of the second metacarpal. Both groups of Himalayan children have significantly less cortical bone than the US whites. These diminished amounts of cortical bone appear to be due to an increase in endosteal absorption as shown by the enlarged medullary area in the Himalayan children.

## Indices of maturation

### Skeletal development

Fig. 5.22 illustrates the assessments of skeletal age of the Sherpa and Tibetan children. Skeletal age was assessed with reference to the Greulich

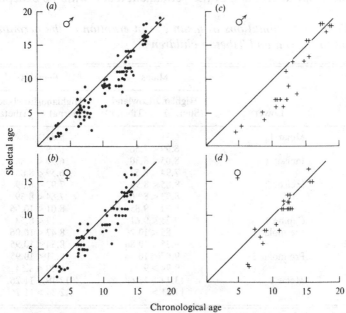

Fig. 5.22. Comparison of skeletal age of highland Sherpas (*a, b*) and lowland Tibetans (*c, d*). The highlanders, during childhood, are more delayed in skeletal age than the lowlanders. Adapted from Pawson (1974).

& Pyle (1959) US standards. From these data it is evident that in both highland and lowland children the skeletal age is significantly behind chronological age. The data also show that in both males and females the skeletal age of the highland Sherpas is systematically lower than that of the lowland Tibetans.

### Age at menarche

Pawson (1974) provided estimates of age at menarche derived by recall and radiographic assessment. The recall method was based upon a sample of sixty-one Sherpas and twenty-nine Tibetan women ranging in age from 12 to 22 years. The radiographic estimates were based upon the regression equations derived by Frisancho *et al.* (1969) and utilized a sample of forty-seven Sherpas and thirty-eight Tibetans. Surprisingly, both estimates give a mean age at menarche of 18.22 yr for the highland Sherpa women. The mean age at menarche (derived by radiographic estimation) for the lowland Tibetans equalled 16.12 yr, which is significantly ($P < 0.001$) lower than that of the highland Sherpas.

### Tooth eruption

Dental development in the Himalayas follows a reverse pattern to that seen in body size, skeletal maturation and age at menarche. As shown in Table 5.17 the mean age at eruption of the permanent teeth, with the exception

Table 5.17. *Probit estimations of mean age at eruption of the permanent dentition in Sherpa and Tibetan children*

| | | Males | Females |
|---|---|---|---|
| | | Highland/Lowland (Sherpa)    (Tibetan) | Highland/Lowland (Sherpa)    (Tibetan) |
| Upper | Lower | | |
| | Molar 1 | 6.31 | 5.83 |
| Molar 1 | | 6.99 < 7.56 | 6.09 |
| | Incisor 1 | 8.05 < 8.30 | 6.99 < 7.88 |
| Incisor 1 | | 7.94 < 8.57 | 6.59 < 8.03 |
| | Incisor 2 | 8.85 < 8.96 | 7.97 < 8.39 |
| Incisor 2 | | 8.93 < 8.96 | 7.64 < 8.39 |
| Pre-molar 2 | | 8.94 < 9.85 | 8.01 < 10.06 |
| | Canine 1 | 9.21 < 9.61 | 8.17 < 8.54 |
| | Pre-molar 1 | 8.85 < 10.26 | 8.47 < 10.06 |
| Pre-molar 2 | | 9.25 < 10.86 | 8.34 < 10.95 |
| | Pre-molar 2 | 9.07 < 10.86 | 8.51 < 10.95 |
| Canine | | 9.74 > 9.20 | 8.20 < 8.54 |
| | Molar 2 | 13.14 > 11.69 | 12.19 > 11.26 |
| Molar 2 | | 13.31 > | 12.34 > 11.26 |

No figures indicate insufficient data to estimate ages.
Adapted from Pawson (1974).

154

of the second premolar and second molars, is earlier among the highland Sherpas than in the lowland Tibetans.

## Conclusions

The anthropometric measurements indicate that growth in the highland Sherpas and low-altitude Tibetans is comparable. However, the estimates of skeletal maturation and age at menarche indicate a delayed maturation among the high-altitude Sherpas. In view of the fact that the low-altitude Tibetan children and adults have a greater amount of subcutaneous fat than the highland Sherpas, and assuming that variations in subcutaneous fat reflect differences in caloric reserves, it would appear that the Tibetans are living under more favorable conditions than the highland Sherpas. With regard to this point, Pawson (1974, p. 52) states 'after an examination of the general health and nutritional status of the children in the Tibetan refugee center it was apparent that they were growing under conditions that were more favorable than those of their Highland counterparts, the Sherpas. Although no detailed nutritional data were available for either group, it was apparent that the Tibetan children were receiving animal protein regularly, whereas the Sherpas only rarely eat meat.' In view of these differences, it is not clear at present to what extent the delayed growth of the highland Sherpas is related to high-altitude hypoxia or to the negative socio-economic conditions to which they are exposed.

On the other hand, the lack of differences between the highland Sherpas and lowland Tibetans may also be related to the fact that the low-altitude Tibetans are only recent migrants. As indicated by Pawson (1974), the adolescent Tibetans spent their intra-uterine months and early post-natal life in the highlands. Therefore, only those in the childhood period can be characterized as low-altitude natives. If this is the case, any differences in growth between highland and lowland samples at present cannot easily be established and must await further investigation. However, it is interesting to note that, among the adults, Weitz (1973) reports significantly taller stature and heavier weight for the lowland Tibetans. The reasons for this difference are not clear at present.

Surprisingly, however, the estimates of age of permanent dental eruption appear to occur at an earlier age among the Sherpas than in the low-altitude Tibetans. Another important finding of the Himalayan study is the markedly enlarged marrow cavity of both Sherpas and Tibetans, which results in a drastic decrease in percent cortical bone. Pawson (1974) suggests that the low cortical bone of the Himalayan samples is related to genetic factors, in view of the fact that Mongolic populations are characterized by thinner cortices (Garn, 1970).

155

# The biology of high-altitude peoples

## Growth in the Tien Shan mountains of the Soviet Union

*Source of data*

The data for this study come from investigations conducted by the Institute of Anthropology of Moscow State University among inhabitants of the Tien Shan mountains and low-altitude populations of the Kirov region. The present synthesis is based upon the publications of Miklashevskaia, Solovyeva & Godina (1972) and Psyzuk, Turusbekow & Brjancewa (1967).

The study was conducted among inhabitants of the Tien Shan mountains and the lowland region of Kirov between 1968 and 1970. In the Tien Shan mountains the villages of Kuzul-Djar, Sagunduk and Sidorov were included in the study. The villages are collectively referred to as Kuzul-Djar. These villages are located at altitudes that range from 2,300 to 2,800 m. However, during the summer, because of their herding economy, the inhabitants live at altitudes of 3,500 m. The inhabitants of these villages belong to the ethnic group of Kirghiz and arrived at their present location about 350 years ago. From these villages a sample of 900 children aged 8 to 18 years was studied.

In the lowland region of Kirov, located at about 700 m, a sample of 1,200 Kirghiz children aged 8 to 18 years was studied. Evaluations of PTC and dermatoglyphic data, along with anthroposcopic observations, revealed that both the highland and lowland groups belong to the same ethnic group or have similar genetic composition.

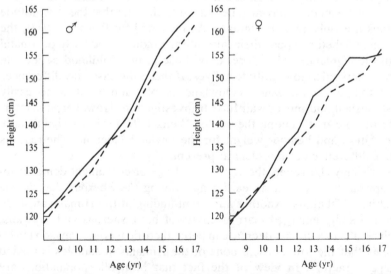

Fig. 5.23. Comparisons of growth in height of low-altitude (Kirov, unbroken lines) and high-altitude (Kuzul-Djar, broken lines) Russian populations. Age-for-age and especially in the males, highland children exhibit a delayed growth. Adapted from Miklashevskaia, Solovyeva & Godina (1972).

156

Table 5.18. *Comparison of anthropometric measurements of low-altitude* (*Kirov region* = *700 m*) *and high-altitude* (*Kuzul-Djar 2,300–2,800 m*) *Russian growth samples*

| Age group (yr) | N | Low altitude (Kirov region, 700 m) | | | | | | | |
|---|---|---|---|---|---|---|---|---|---|
| | | Height (cm) | | Transverse chest diameter (cm) | | Chest length (cm) | | Sum skinfold[a] (mm) | |
| | | Mean | S.D. | Mean | S.D. | Mean | S.D. | Mean | S.D. |
| *Males* | | | | | | | | | |
| 8 | 61 | 120.5 | 4.3 | 19.3 | 0.8 | 13.9 | 0.7 | 4.3 | 0.5 |
| 9 | 60 | 123.8 | 5.1 | 19.7 | 1.0 | 14.1 | 0.8 | 4.4 | 0.9 |
| 10 | 62 | 128.9 | 6.0 | 20.3 | 1.0 | 14.6 | 1.0 | 4.7 | 1.0 |
| 11 | 59 | 132.9 | 4.8 | 20.7 | 1.0 | 14.5 | 0.8 | 4.5 | 0.8 |
| 12 | 63 | 136.4 | 6.4 | 21.3 | 1.1 | 14.9 | 0.9 | 4.7 | 1.1 |
| 13 | 60 | 141.5 | 6.5 | 22.1 | 1.3 | 15.2 | 0.9 | 4.8 | 0.9 |
| 14 | 60 | 148.1 | 6.8 | 22.8 | 1.2 | 15.9 | 1.1 | 5.0 | 1.0 |
| 15 | 60 | 156.0 | 7.2 | 23.9 | 1.6 | 16.3 | 1.2 | 4.9 | 0.8 |
| 16 | 58 | 160.5 | 7.0 | 24.9 | 1.4 | 16.9 | 1.2 | 5.1 | 0.9 |
| 17 | 56 | 163.9 | 6.3 | 25.6 | 1.8 | 17.7 | 1.5 | — | — |
| *Females* | | | | | | | | | |
| 8 | 60 | 118.4 | 4.6 | 18.9 | 0.9 | 13.2 | 0.8 | 5.4 | 1.1 |
| 9 | 61 | 122.2 | 4.9 | 19.1 | 0.9 | 13.4 | 0.7 | 5.4 | 1.0 |
| 10 | 59 | 125.9 | 4.8 | 19.5 | 1.0 | 13.5 | 0.7 | 5.1 | 1.0 |
| 11 | 61 | 133.8 | 5.4 | 20.4 | 1.1 | 14.1 | 0.9 | 5.9 | 1.4 |
| 12 | 57 | 137.8 | 6.7 | 20.9 | 1.1 | 14.2 | 0.9 | 6.1 | 1.1 |
| 13 | 62 | 146.1 | 6.8 | 22.0 | 1.2 | 14.8 | 1.0 | 6.6 | 1.8 |
| 14 | 60 | 148.9 | 5.9 | 22.8 | 1.4 | 15.0 | 1.2 | 7.4 | 2.4 |
| 15 | 62 | 154.0 | 5.0 | 23.4 | 1.4 | 15.7 | 1.0 | 8.6 | 2.6 |
| 16 | 61 | 153.9 | 5.1 | 23.9 | 1.1 | 15.9 | 1.1 | 9.9 | 2.6 |
| 17 | 57 | 155.0 | 4.7 | 24.2 | 1.3 | 16.0 | 1.2 | — | — |
| High altidue (Kuzul-Djar, 2,300–2,800 m) | | | | | | | | | |
| *Males* | | | | | | | | | |
| 8 | 24 | 118.5 | 4.2 | 18.8 | 1.1 | 13.7 | 0.9 | 4.3 | 0.7 |
| 9 | 41 | 123.2 | 5.1 | 19.3 | 0.9 | 14.1 | 0.9 | 4.0 | 0.6 |
| 10 | 36 | 126.5 | 4.6 | 19.8 | 0.9 | 14.3 | 1.0 | 4.1 | 0.5 |
| 11 | 40 | 130.5 | 5.4 | 20.3 | 1.2 | 14.5 | 1.1 | 4.2 | 0.6 |
| 12 | 49 | 136.4 | 4.9 | 20.9 | 0.9 | 14.9 | 1.0 | 4.2 | 0.7 |
| 13 | 51 | 139.4 | 6.8 | 21.2 | 1.0 | 14.9 | 1.0 | 4.0 | 0.6 |
| 14 | 53 | 146.5 | 7.5 | 22.6 | 1.3 | 15.7 | 1.3 | 4.0 | 0.5 |
| 15 | 43 | 153.3 | 8.4 | 23.3 | 1.7 | 16.1 | 1.3 | 4.3 | 0.7 |
| 16 | 28 | 156.3 | 6.1 | 23.8 | 1.4 | 16.6 | 1.1 | 4.5 | 0.7 |
| 17 | 30 | 162.1 | 4.8 | 25.0 | 1.9 | 17.2 | 1.2 | — | — |
| *Females* | | | | | | | | | |
| 8 | 38 | 118.0 | 4.7 | 18.3 | 1.3 | 13.0 | 0.8 | 5.0 | 0.7 |
| 9 | 39 | 124.0 | 5.5 | 19.1 | 1.1 | 13.3 | 0.9 | 5.1 | 0.9 |
| 10 | 33 | 126.9 | 5.3 | 19.8 | 1.2 | 13.2 | 0.9 | 5.1 | 0.9 |
| 11 | 36 | 130.1 | 5.7 | 19.9 | 1.0 | 13.7 | 0.9 | 4.7 | 0.7 |
| 12 | 39 | 135.7 | 7.6 | 20.4 | 1.2 | 14.0 | 0.9 | 5.4 | 1.1 |
| 13 | 33 | 139.8 | 6.1 | 21.3 | 1.2 | 14.3 | 1.2 | 5.5 | 0.8 |
| 14 | 53 | 147.3 | 5.9 | 22.4 | 1.4 | 15.1 | 1.1 | 6.4 | 1.7 |
| 15 | 32 | 148.8 | 6.4 | 22.4 | 1.5 | 15.1 | 1.1 | 7.0 | 1.4 |
| 16 | 27 | 151.0 | 6.2 | 23.2 | 1.4 | 15.8 | 1.7 | 8.8 | 2.3 |
| 17 | 25 | 155.9 | 4.6 | 23.9 | 1.2 | 15.8 | 1.1 | — | — |

[a] Sum skinfold thickness = sum of subscapula, biceps, triceps and midaxillary divided by 4.
Adapted from Miklashevskaia, Solovyeva & Godina, 1972.

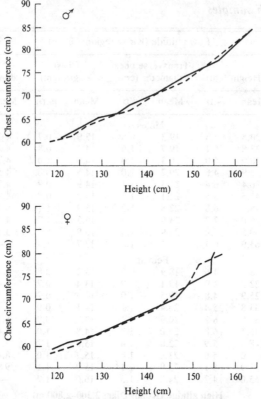

Fig. 5.24. Comparison of height in relation to chest circumference among low-altitude (unbroken lines) and high-altitude (broken lines) Russian populations. For their height the highland children exhibit similar chest circumferences to their lowland counterparts. Adapted from Miklashevskaia, Solovyeva & Godina (1972).

### Body size

Table 5.18 summarizes the anthropometric measurements of body size of the inhabitants of the Kirov and Tien Shan mountains. As illustrated in Fig. 5.23, the high-altitude boys at each age, excepting those at age 12, are systematically smaller than their low-altitude counterparts. Similarly, the high-altitude girls from 11 years onward are systematically shorter than the low-altitude samples. As shown in Fig. 5.24, for their stature, the high-altitude children attain the same chest circumference as their low altitude counterparts.

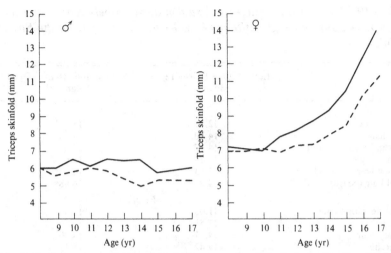

Fig. 5.25. Comparison of triceps skinfolds among low-altitude (unbroken lines) and high-altitude (broken lines) Russian populations. Age-for-age the highland children are leaner than their lowland counterparts. Adapted from Miklashevskaia, Solovyeva & Godina (1972).

### Skinfold thickness

The mean sum of skinfold thickness at the subscapula, biceps, triceps and mid-axillary is given in Table 5.18. From these data, and as illustrated in Fig. 5.25, it is evident that the high-altitude children have less subcutaneous fat than their low-altitude counterparts.

### Secondary sexual maturation

Table 5.19 gives the mean age for the attainment of stages of secondary sexual maturation. These data show that in the high-altitude males the voice change and axillary and facial hair appear about 6.5 months, 8 months, and 1 year and 3 months later, respectively, than in their low-altitude counterparts. Breast development in the high-altitude girls is delayed by about 3.5 months when compared to the low-altitude girls' breast development. The mean age at menarche in low-altitude samples occurs at about 14.4 years, while among high-altitude girls menarche occurs at about 15.2 years. In other words, the age at menarche at high altitude is delayed by about 1 year 3 months. Fig. 5.26 illustrates the percent, by age, of children having axillary hair at menarche. From these data it is evident that adolescent maturation in the high-altitude samples is delayed in both males and females.

159

Table 5.19. *Comparison of stages of sexual maturation between low-altitude (Kirov region) and high-altitude (Kuzul-Djar) samples from the Soviet Union*

|  | Low altitude (Kirov region) Mean (yr) | High altitude (Kuzul-Djar) Mean (yr) |
|---|---|---|
|  | Males | |
| Voice change | 12.71 | 13.25* |
| Pubic hair | 14.50 | 14.58 |
| Pubertal breast swelling | 14.17 | 13.83 |
| Adam's apple | 16.29 | 15.67 |
| Axillary hair | 15.83 | 16.50* |
| Facial hair (beard) | 14.83 | 16.08* |
|  | Females | |
| Breast development | 11.08 | 11.38 |
| Pubic hair | 13.58 | 13.50 |
| Axillary hair | 14.17 | 14.00 |
| Menarche | 14.42 | 15.17* |

\* $P < 0.05$.
Adapted from Miklashevskaia, Solovyeva & Godina (1972).

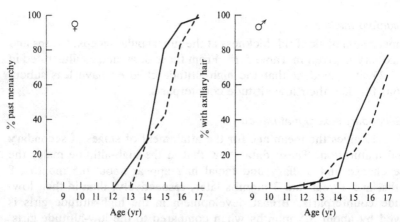

Fig. 5.26. Comparison of indices of sexual maturation among low-altitude (unbroken lines) and high-altitude (broken lines) Russian populations. At each age a greater percentage of lowland girls have attained menarche and a greater percentage of lowland boys exhibit axillary hair. Adapted from Miklashevskaia, Solovyeva & Godina (1972).

### Teeth eruption

As shown in Table 5.20, the mean number of teeth present at each age in the high-altitude children is comparable to that in their low-altitude counterparts. However, it is interesting to note that at 13 and 14 years

Table 5.20. *Comparison of mean number of permanent teeth by age among low-altitude (Kirov region) and high-altitude (Kuzul-Djar) children of the Soviet Union*

| Age group (yr) | Low altitude (Kirov region) | | | High altitude (Kuzul-Djar) | | |
|---|---|---|---|---|---|---|
| | N | Mean | S.D. | N | Mean | S.D. |
| | | | Males | | | |
| 8 | 60 | 10.2 | 2.3 | 24 | 11.0 | 3.4 |
| 9 | 60 | 13.1 | 2.6 | 35 | 13.1 | 2.4 |
| 10 | 62 | 16.3 | 4.1 | 40 | 16.7 | 4.5 |
| 11 | 59 | 20.0 | 4.6 | 40 | 20.5 | 4.7 |
| 12 | 63 | 23.2 | 4.8 | 48 | 23.4 | 4.6 |
| 13 | 60 | 26.3* | 3.0 | 50 | 24.1 | 4.2 |
| 14 | 60 | 27.6* | 1.0 | 50 | 26.6 | 2.6 |
| 15 | 60 | 27.9 | 0.4 | 43 | 27.6 | 1.1 |
| | | | Females | | | |
| 8 | 60 | 11.0 | 2.3 | 40 | 11.5 | 3.3 |
| 9 | 61 | 14.9 | 3.7 | 36 | 14.7 | 3.4 |
| 10 | 59 | 17.6 | 4.2 | 30 | 18.6 | 5.4 |
| 11 | 61 | 22.2 | 3.8 | 35 | 21.1 | 4.0 |
| 12 | 57 | 25.5 | 3.2 | 36 | 24.6 | 3.3 |
| 13 | 62 | 27.0 | 1.8 | 30 | 26.9 | 2.5 |
| 14 | 60 | 27.8 | 0.8 | 53 | 27.5 | 1.4 |
| 15 | 62 | 28.0 | 0.6 | 32 | 27.5 | 1.3 |

* $P < 0.01$.

of age the high-altitude boys have a significantly ($P < 0.01$) smaller number of teeth than their low-altitude counterparts.

### Respiratory system

Table 5.21 summarizes three measures of respiratory function conducted among highland inhabitants of the Nyryn region, located above 2,050 m, and lowland inhabitants from the Frunze region, both of the Soviet Union. These data show that in both males and females, at ages 8, 11 and 14 years, the highland subjects have a significantly ($P < 0.05$) greater tidal volume and respiratory volume per minute than their low-altitude counterparts. Significant differences in vital capacity are evident at 11 and 14 years in males and only at 14 years in females.

### Conclusions

As inferred from the anthropometric measurements and indices of puberty, growth and maturation are delayed among highland children from

## The biology of high-altitude peoples

Table 5.21. *Comparison of respiratory volumes among low-altitude (Frunze) and high-altitude (Nyryn) samples from the Soviet Union*

| | Low altitude (Frunze, 770 m) | | High altitude (Nyryn, 2,050 m) | | |
|---|---|---|---|---|---|
| | Mean | S.E. | Mean | S.E. | t |
| | | Males | | | |
| | | Age: 8 yr | | | |
| Tidal volume (ml/breath rate) | 185.0 | 5.3 | 239.6 | 10.7 | * |
| Respiratory volume (ml/min) | 3810.0 | 100.0 | 5240.0 | 240.0 | * |
| Vital capacity (ml) | 1761.0 | 20.0 | 1328.0 | 35.0 | NS |
| | | Age: 11 yr | | | |
| Tidal volume (ml/breath rate) | 180.4 | 14.4 | 281.8 | 13.3 | * |
| Respiratory volume (ml/min) | 3450.0 | 210.0 | 5000.0 | 220.0 | * |
| Vital capacity (ml) | 1861.0 | 81.0 | 1928.0 | 33.0 | * |
| | | Age: 14 yr | | | |
| Tidal volume (ml/breath rate) | 222.4 | 16.7 | 318.8 | 13.0 | * |
| Respiratory volume (ml/min) | 4320.0 | 180.0 | 6870.0 | 200.0 | * |
| Vital capacity (ml) | 2406.0 | 95.0 | 2516.0 | 47.0 | NS |
| | | Females | | | |
| | | Age: 8 yr | | | |
| Tidal volume (ml/breath rate) | 216.0 | 9.4 | 249.3 | 19.5 | * |
| Respiratory volume (ml/min) | 3480.0 | 150.0 | 5250.0 | 240.0 | * |
| Vital capacity (ml) | 1280.0 | 28.0 | 1143.0 | 33.0 | NS |
| | | Age: 11 yr | | | |
| Tidal volume (ml/breath rate) | 176.0 | 14.3 | 251.7 | 11.7 | * |
| Respiratory volume (ml/min) | 3510.0 | 140.0 | 5170.0 | 220.0 | * |
| Vital capacity (ml) | 1606.0 | 55.0 | 1636.0 | 31.0 | NS |
| | | Age: 14 yr | | | |
| Tidal volume (ml/breath rate) | 248.0 | 6.5 | 316.8 | 22.7 | * |
| Respiratory volume (ml/min) | 4880.0 | 190.0 | 6690.0 | 410.0 | * |
| Vital capacity (ml) | 2406.0 | 95.0 | 2516.0 | 47.0 | * |

* $P < 0.05$.
Adapted from Pyzuk, Turusbekow & Brjancewa (1967).

the Tien Shan mountains when compared to low-altitude counterparts from the Kirov region. Based on the findings from Andean populations and evidence from experimental animals, Miklashevskaia, Solovyeva & Godina (1972) postulate that the delayed linear growth of highland inhabitants reflects the influences of the highland environment. However, in view of the fact that highlanders had significantly lower subcutaneous fat than the lowlanders, it would appear that part of the delayed growth may be related to nutritional factors as well.

Despite their lower growth in body weight the high-altitude samples exhibit a chest size (measured by antero-posterior chest diameter) that is comparable to that of their taller low-altitude counterparts. However, evaluations of respiratory function indicate that even at 2,000 m growth in the highlands is associated with increased tidal and respiratory volume.

## Overview

The Andean samples, either Quechua or Mestizo and irrespective of location (central or southern highlands), exhibit a slow growth in body size, delayed skeletal maturation, and late onset of secondary sexual maturation, which do not seem entirely due to nutritional factors. This conclusion is inferred from the facts that dietary intakes of selected highland samples do not show meaningful relationships to growth, and that intra-population highland differences in weight, subcutaneous fat and arm muscle do not parallel differences in growth in height. However, in view of the difficulties of obtaining adequate information on dietary factors and the socio-economic conditions of highland populations, the possible role of nutritional factors must also be considered in explanations of the delayed patterns of growth in the highlands.

Another way to clarify the roles of hypoxic and nutritional stress is to compare highland and lowland children matched for skinfold thickness (as a measure of caloric reserves) and body muscle (as a measure of protein reserves) in order to determine whether differences in linear growth continue to appear. This procedure has proved to be quite useful in ascertaining the influence of nutritional limitation on Central American populations (Frisancho & Garn, 1971; Frisancho *et al.*, 1971).

Evaluations of Ethiopian highland and lowland children show that the highland children appear to grow and mature at a faster rate than their lowland counterparts. These differences are probably related to the fact that the environmental conditions in the lowlands, as shown by the high incidence of infectious diseases (malaria) and intestinal parasitism, are worse than those in the highlands. The fact that the highland and lowland adults do not differ in stature would suggest that the retarded growth of the lowlanders is probably of recent origin and not a permanent characteristic. Indeed, as previously discussed, it is suggested that the low-altitude Ethiopian children, like the lowland Peruvian children, are going through a reverse secular trend; if the present conditions persist, the new generation will not be as tall as their parents.

Measurements of chest size among adults suggest that despite the lack of differences in stature, highlanders tend to have an increased chest size when compared to lowlanders. Evaluations of migrants also suggest that the increased thorax of highlanders can be acquired by lowland natives who migrate to the highlands. Furthermore, as indicated by Harrison *et al.* (1969), whether stature or weight is taken as the measure of body size, the highlanders have a greater respiratory capacity than the lowlanders, regardless of the size variation. In view of the fact that the highland and lowland samples do not exhibit any genetic divergence in terms of blood groups and other marked frequencies, Harrison *et al.* (1969) suggest that most of the highland and lowland differences can be viewed as arising from phenotypic flexibility. On the other hand, Hoff (1974) noted that among children born at low altitudes but whose parents were highland Quechua natives, the chest size and forced expiratory volume were equal to those attained by highland Quechua children. Accordingly, Hoff (1974) postulated that these traits were under genetic control.

Growth of body size among Himalayan Sherpas is comparable to that of the lowland Tibetans. On the other hand, in terms of measurements of subcutaneous fat, and measurements of maturation such as skeletal age, tooth eruption and age at menarche, the highland Sherpas were more retarded than the lowland Tibetans. Furthermore, the adult highland Sherpas also appear to be significantly shorter in stature, heavier in weight, and leaner (thin skinfolds) than the adult lowland Tibetans. Thus it appears that body size in the highland Sherpas during adulthood is smaller than in the lowland Tibetans. The sources of these differences are difficult to evaluate. First, the adult lowland Tibetans spent their intra-uterine months and first post-natal years in the highlands (Pawson, 1974). Therefore, only the younger lowland Tibetan children can be characterized as true native lowlanders. In view of the fact that both Sherpas and lowland Tibetans come from the same gene pool, Pawson (1974) postulates that retarded growth of the Himalayan population has a genetic basis, and that the greater stature of the adult Tibetans results from the removal of hypoxic stress at a critical point in their development, in this manner resulting in a longer or more rapid spurt of growth. Secondly, the socio-economic conditions for the Sherpas were found to be less favorable than those in the lowlands. Assuming that variations in skinfold thickness reflect differences in caloric reserves, it would appear that the less favorable socio-economic conditions of the Sherpas are also reflected by their lower caloric reserves (thinner skinfolds) than those of the Tibetans. For these reasons, it is difficult to determine the extent to which the retarded growth of the highland Sherpas reflects either the growth-retarding effect of high-altitude hypoxia or genetic factors. Hence, further research is required in this area. In the same manner, while the delayed growth of the inhabitants of the Tien Shan mountains does not appear to be related to genetic differences, the role of nutritional factors is difficult to separate from that of hypoxia.

In conclusion, the growth-retarding effects of high-altitude hypoxia can be inferred only among the Andean populations, while among the Ethiopian, Himalayan, and Tien Shan populations any possible hypoxic effects are obscured by socio-economic and ecological factors. As previously indicated, high-altitude hypoxia can only partially explain the delayed growth of Andean populations, and consequently the patterns of growth of highland Peruvian populations reflect the influence of the complex highland environment, which includes hypoxia, temperature, and nutritional stresses. The hypothesis that hypoxia affects human growth is supported by experimental studies on animals. Several investigations have demonstrated that rats and guinea-pigs born and raised in hypoxic environments and with favorable conditions of nutrition and temperature at all ages have a slower growth rate than their sea-level counterparts (Gordon *et al.*, 1943; Cohn & D'Amour, 1951; Krum, 1957; Hale *et al.*, 1959; Timiras, 1965; Delaquerriere-Richardson, Forbes & Valdivia, 1965; Weihe, 1966; Timiras & Wooley, 1966; Naeye, 1966; Cheek, Graystone & Roe, 1969; Schnakenberg & Burlington, 1970; Burri & Weibel, 1971*a*, *b*; Bartlett & Remmers, 1971; Petropoulos, Dabal & Timiras, 1972; McGrath *et al.*, 1973).

These depressed growth rates are said to be related to the anorexia experienced by experimental animals (Schnakenberg, Krabill & Weiser, 1971) and decreased efficiency of intestinal food absorption (Van Liere *et al.*, 1948; Chinn & Hannon, 1969). However, these factors can not explain all the growth retardation seen among animals exposed to high-altitude hypoxia. As indicated by Naeye (1966) growth retardation resulting from hypoxic stress is related to a low number of cells, while retardation under nutritional deficiency is due to a low mass of cytoplasm. In fact, Cheek *et al.* (1969) demonstrated that hypoxia caused a failure of brain DNA and protein content to increase. In other words, it appears that high-altitude hypoxia interferes with cell multiplication. From the studies of Timiras (1965), the growth-retarding effects of high-altitude hypoxia continue even after several generations of exposure to hypoxia. Thus, it is quite possible that part of the growth-retardation of highland children is related to a lower rate of cell multiplication.

Studies of pulmonary function in Andean and Ethiopian populations indicate that both immigrants and highland natives have an increased respiratory capacity. In view of the importance of the respiratory system under conditions of low oxygen availability, we postulate that the increased respiratory capacity of the highland resident is an adaptive response to high-altitude hypoxia. Furthermore, on the basis of studies of sea-level subjects who migrated to high altitudes during the developmental period, and supported by experimental studies on rats, it is concluded that the enlarged lung volume of high-altitude residents is developmental in nature (Frisancho *et al.*, 1973*c*; Frisancho, 1975).

The influence of developmental factors in man's functional adaptation to high altitude is also inferred from studies on aerobic capacity (Frisancho

## The biology of high-altitude peoples

et al., 1973a; Frisancho, 1975b). In previous studies we have demonstrated that lowland natives acclimatized to high altitude during childhood and adolescence attained a maximal work aerobic capacity and ventilation that was equal to that of the highland natives. In contrast, lowland natives (Peruvian and US subjects) when acclimatized to high altitudes as adults attained significantly lower aerobic capacities and higher pulmonary ventilation than the highland natives. The extent to which developmental factors influence the attainment of aerobic capacity at high altitudes has been described by Frisancho et al. (1973a). These data show that among lowland natives acclimatized to high altitudes during growth and development, the attainment of aerobic capacity is directly related to age at migration and length of residence. In contrast, when subjects were acclimatized to high altitudes as adults, age at migration and length of residence did not influence the attainment of aerobic capacity. In other words, from these investigations it appears that the attainment of normal aerobic capacity at high altitudes is influenced by adaptation occurring during the developmental period (Frisancho et al., 1973b; Frisancho, 1975b). In order to test the hypothesis that the small body size of the highland native is a by-product of adaptive responses to the hypoxic highland environment, we tested the maximal aerobic capacity of highland Quechua natives (Frisancho, Velasquez & Sanchez, 1975). This study demonstrated that subjects of short stature attained a higher maximal aerobic capacity than their counterparts of large body size when tested under identical conditions. Taking into account the importance of developmental factors, the patterns of growth and development of high-altitude populations may be viewed in part as a developmental adaptive process than enables man to survive under conditions of low oxygen availability.

Results of the studies in the Himalayas and Tien Shan mountains demonstrate that the enlargement of the thorax is not a general characteristic of all highland populations. Hence, it would appear that the factors associated with an increase in chest size in the Andean regions are not the same as those in Asiatic populations. The cause of such different effects and different responses must be answered by future research.

This work was supported in part by grant HB-13805 from the National Institutes of Health and by grant GS-37542X from the National Science Foundation. Various phases of the presented investigations were conducted in co-operation with the Institute of Andean Biology of the National University of San Marcos of Peru, and the International Biological Program of the United States.

166

## References

Baker, P. T., Frisancho, A. R. & Thomas, R. B. (1966). A preliminary analysis of human growth in the Peruvian Andes. In *Human adaptability to environments and physical fitness*, ed. M. S. Malhotra, pp. 259–69. Madrawe, India: Defense Institute of Physiology and Allied Sciences.

Baker, P. T., Hunt, E. B. & Sen, S. T. (1958). The growth and interrelations of skinfolds and brachial tissues in man. *American Journal of physical Anthropology*, **16**, 39–58.

Bartlett, D. (1972). Postnatal development of the mammalian lung. In *Regulation of organ and tissue growth*, ed. R.Goss, pp. 197–209. New York & London: Academic Press.

Bartlett, D. & Remmers, J. E. (1971). Effects of high-altitude exposure on the lungs of young rats. *Respiratory Physiology*, **13**, 116–25.

Beall, C. M., Baker, P. T., Baker, T. S. & Haas, J. D. (1977). The effects of high altitude on adolescent growth in Southern Peruvian Amerindians. *Human Biology*, **49**, 109–24.

Bharadwaj, J., Singh, A. P. & Malhotra, M. S. (1973). Body composition of the high-altitude natives of Ladakh – a comparison with sea-level residents. *Human Biology*, **45**, 423–34.

Bjure, J. (1963). Sperometric studies in normal subjects. IV. Ventilatory capacities in healthy children 7–17 years of age. *Acta Paediatrica*, **52**, 232–40.

Bouloux, C. J. (1968). *Contribution a l'étude biologique des phénomènes pubertaires en très haute altitude (La Paz)*. Toulouse, France: Centre d'Hématypologie du Centre National de la Recherche Scientifique. Centre Régional de Transfusion sanguine et d'Hématologie.

Boyce, A. J., Haight, J. S. J., Rimmer, D. B. & Harrison, G. A. (1974). Respiratory function in Peruvian Quechua Indians. *Annals of human Biology*, **1**, 137–48.

Burri, P. H. & Weibel, E. R. (1971a). Morphometric estimation of pulmonary diffusion capacity. II. Effect of $P_{O_2}$ on the growing lung. Adaptation of the growing rat lung to hypoxia and hyperhypoxia. *Respiratory Physiology*, **11**, 247–64.

Burri, P. H. & Weibel, E. R. (1971b). Morphometric evaluation of changes in lung structure due to high altitude. In *High-altitude physiology. Cardiac and respiratory aspects*, ed. R. Porter & J. Knight, pp. 15–30. Edinburgh & London: Churchill, Livingstone.

Cheek, D., Graystone, J. A. & Rowe, R. A. (1969). Hypoxia and malnutrition in newborn rats: effects of RNA, DNA and protein tissues. *American Journal of Physiology*, **217**, 642–5.

Chinn, K. S. K. & Hannon, J. P. (1969). Efficiency of food utilization at high altitude. *Federation Proceedings*, **28**, 944–7.

Clegg, E. J., Pawson, I. G., Ashton, E. H. & Flinn, R. M. (1972). The growth of children at different altitudes in Ethiopia. *Philosophical Transactions of the Royal Society of London*, **264B**, 403–37.

Cohn, E. W. & D'Amour, R. E. (1951). Effects of high altitude and cobalt on growth and polycythemia in rats. *American Journal of Physiology*, **166**, 394–9.

Comroe, J. H., Forster, R. E., Dubois, A. B., Briscoe, W. A. & Carlsen, E. (1955). *The lung: clinical physiology and pulmonary function tests*. Chicago: Yearbook Publishers, Inc.

Consolazio, C. F., Johnson, R. E. & Peccora, L. T. (1963). *The physiological measurement of metabolic function in man*. New York: McGraw-Hill.

167

## The biology of high-altitude peoples

Conway, D. L. & Baker, P. T. (1972). Skin reflectance of Quechua Indians: the effects of genetic admixture, sex and age. *American Journal of physical Anthropology*, **36**, 267–383.

Cotes, J. E. (1965). *Lung function*. Oxford: Blackwell.

Delaquerriere-Richardson, L., Forbes, E. S. & Valdivia, E. (1965). Effect of simulated high altitude on the growth rate of albino guinea pigs. *Journal of applied Physiology*, **20**, 1022–5.

Donayre, J. (1966). Population growth and fertility at high altitude. In *Life at high altitudes*. Washington, DC: Pan American Health Organization. Scientific Publication No. 140, pp. 74–9.

Dunnill, M. S. (1962). Postnatal growth of the lung. *Thorax*, **17**, 329–33.

Frisancho, A. R. (1966). *Human growth in a high-altitude population. Andean bio-cultural studies*. The Pennsylvania State University.

Frisancho, A. R. (1969a). Human growth and pulmonary function of a high-altitude Peruvian Quechua population. *Human Biology*, **41**, 365–79.

Frisancho, A. R. (1969b). Growth, physique and pulmonary function at high altitude: a field study of a Peruvian Quechua population. PhD dissertation, Pennsylvania State University.

Frisancho, A. R. (1976a). Growth and functional development at high altitude. In *Man in the Andes: A Multidisciplinary Study of High Altitude Quechua*, ed. P. T. Baker & M. A. Little, pp. 180–207. Stroudsberg, Pa.: Dowden, Hutchinson & Ross.

Frisancho, A. R. (1975b). Functional adaptation to high-altitude hypoxia. *Science, Washington*, **187**, 313–19.

Frisancho, A. R. & Baker, P. T. (1970). Altitude and growth: a study of the patterns of physical growth of a high-altitude Peruvian Quechua population. *American Journal of physical Anthropology*, **32**, 279–92.

Frisancho, A. R., Borkan, G. A. & Klayman, J. F. (1975). Pattern of growth of lowland and highland Peruvian Quechua of similar genetic composition. *Human Biology*, **47**, 233–43.

Frisancho, A. R. & Garn, S. M. (1971). The implications of skinfolds and muscle size to developmental and nutritional status of Central American children. III. Guatemala. *Tropical Geography and Medicine*, **23**, 167–72.

Frisancho, A. R., Garn, S. M. & Ascoli, W. (1970). Childhood retardation resulting in reduction of adult body size due to lesser adolescent skeletal delay. *American Journal of physical Anthropology*, **33**, 325–36.

Frisancho, A. R., Garn, S. M. & McCreery, L. D. (1971). Relationship of skinfolds and muscle size to growth of children. I. Costa Rica. *American Journal of physical Anthropology*, **35**, 85–90.

Frisancho, A. R., Garn, S. M. & Rohmann, C. G. (1969). Age at menarche: a new method of prediction and retrospective assessment based on hand X-rays. *Human Biology*, **41**, 42–50.

Frisancho, A. R., Martinez, C., Velasquez, T., Sanchez, J. & Montoye, H. (1973a). Influence of developmental adaptation on aerobic capacity at high altitude. *Journal of applied Physiology*, **34**, 176–80.

Frisancho, A. R., Sanchez, J., Pallardel, D. & Yanez, L. (1973b). Adaptive significance of small body size under poor socio-economic conditions in Southern Peru. *American Journal of physical Anthropology*, **39**, 255–62.

Frisancho, A. R., Velasquez, T. & Sanchez, J. (1973c). Influences of developmental adaptation on lung function at high altitude. *Human Biology*, **45**, 525–94.

168

Frisancho, A. R., Velasquez, T. & Sanchez, J. (1975). Possible adaptive significance of small body size in the attainment of aerobic capacity among high-
altitude Quechua natives. In *Biosocial interrelations in population adaptation.
Proceedings of the annual meeting of the society of Human Biology*, ed. E.
Watts, F. Johnston & G. Lasker, pp. 55–67. Chicago: Mouton Publishers.

Garn, S. M. (1970). *The earlier gain and later loss of cortical bone*. Springfield,
Illinois: Charles L. Thomas.

Garn, S. M. & Haskell, J. A. (1960). Fat thickness and developmental status in
childhood and adolescence. *Journal of Diseases of Childhood*, **99**, 746–51.

Garn, S. M. & Rohmann, C. G. (1966). Interaction of nutrition and genetics in the
timing of growth and development. *Pediatric Clinics of North America*, **13**,
353–79.

Garruto, R. (1969). Pulmonary function and body morphology: selected relationships studied at high altitudes. MA thesis in Anthropology, The Pennsylvania
State University.

Gordon, A. S., Tornetta, F. J., D'Angelo, S. A. & Charipper, H. A. (1943). Effects
of low atmospheric pressures on the activity of the thyroid, reproductive
system and anterior lobe of the pituitary in the rat. *Endocrinology*, **33**, 366–83.

Greulich, W. W. & Pyle, S. I. (1959). *Radiographic atlas of skeletal development
of the hand and wrist*, 2nd ed. Stanford, California: Stanford University
Press.

Guerra-Garcia, R. (1971). Testosterone metabolism in men exposed to high
altitude. *Acta Endocrinologica, Panama*, **2**, 55.

Gursky, M. (1969). A dietary survey of three Peruvian highland communities. MA
thesis in Anthropology, The Pennsylvania State University.

Hale, H. B., Mefferd, R. B., Waters, G., Forster, G. E & Criscuolo, D. (1959).
Influence of long-term exposure to adverse environments on organ weights
and histology. *American Journal of Physiology*, **196**, 520–4.

Harrison, G. A., Küchemann, G. F., Moore, M. A. S., Boyce, A. J., Baju, T.,
Mourant, A. E., Godber, M. J., Glasgow, B. G., Kopeć, A. C., Tills, D. &
Clegg, E. J. (1969). The effects of altitudinal variation in Ethiopian populations. *Philosophical Transactions of the Royal Society of London*, **256B**,
147–82.

Hoff, C. (1974). Altitudinal variations in the physical growth and development of
Peruvian Quechua. *Homo*, **24**, 87–99.

Hunt, R. & Schraer, H. (1965). Skeletal response of rats exposed to reduced
barometric pressure. *American Journal of Physiology*, **208**, 1217–21.

Hurtado, A. (1932). Respiratory adaptation in the Indian natives of the Peruvian
Andes. Studies at high altitude. *American Journal of physical Anthropology*,
**17**, 137–61.

Hurtado, A. (1964). Animals in high altitude: resident man. In *Handbook of
Physiology, section 4, Adaptation to the Environment*, ed. D. B. Dill, E. F.
Adolph & C. G. Wilber, pp. 843–60. Washington, DC: American Physiological
Society.

Hurtado, A., Velasquez, T., Reynafarje, C., Lozano, R., Chaves, R., Aste-Salazar,
H., Reynafarje, B., Sanchez, C. & Munoz, J. (1956). *Mechanisms of natural
acclimatization. Studies on the native residents of Morococha, Peru, at an
altitude of 14,900 feet*. Randolph AFS, Texas: USAF School of Aviation
Medicine Report 56–1.

Krum, A. A. (1957). Reproduction and growth of laboratory rats and mice at
altitude. PhD thesis, University of California, Berkeley.

169

Llerena, L. A. (1973). *Determinacion de hormona luteinizatante por radioimmuno-ensayo. Universidad Peruana cayetano Heredia*. Lima, Peru: Instituto de investigaciones de altura.

Lloyd, J. K., Wolff, O. H. & Whelan, S. (1961). Childhood obesity. A long term study of height and weight. *British medical Journal*, 15, 145–8.

McGrath, J. J., Prochazka, J., Pelouch, V. & Ostadal, B. (1973). Physiological responses of rats to intermittent high-altitude stress: effects of age. *Journal of applied Physiology*, 34, 289–93.

Mazess, R. B. & Baker, P. T. (1964). Diet of Quechua Indians living at high altitude: Nuñoa, Peru. *American Journal of clinical Nutrition*, 15, 341–51.

Merino, C. & Reynafarje, C. (1949). Bone marrow studies in polycythencia of high altitudes. *Journal of laboratory and clinical Medicine*, 34, 637–47.

Miklashevskaia, N. N., Solovyeva, V. S. & Godina, E. Z. (1972). Growth and development in high-altitude regions of Southern Kirghizia, USSR. *Vopros Anthropologii*, 40, 71–91. (Transl. B. Honeyman Heath.)

Naeye, R. L. (1966). Organ and cellular development in mice growing at simulated high altitude. *Laboratory Investigation*, 15, 700–6.

Pawson, I. G. (1971). The effects of altitudinal variation on growth in an Ethiopian population. MA thesis, The Pennsylvania State University.

Pawson, I. G. (1974). The growth and development of high altitude children with special emphasis on populations of Tibetan origin in Nepal. PhD thesis, The Pennsylvania State University.

Peñaloza, J. B. (1971). Crecimiento y desarrollo sexual del adolescente andino. PhD thesis, Universidad Nacional Mayor de San Marcos, Lima, Peru.

Peñaloza, D., Gamboa, R., Dyer, J., Echevarria, M. & Marticorena, E. (1960). The influence of high altitudes on the electrical activity of the heart. I. Electrocardiographic and vectorcardiographic observations in the newborn, infants and children. *American Heart Journal*, 59, 111–28.

Petropoulos, E. A., Dabal, K. B. & Timiras, P. S. (1972). Biological effects of high altitude on myelinogenesis in brain of the developing rat. *American Journal of Physiology*, 223, 951–7.

Picón-Reátegui, E. (1976). Nutrition. In *Man in the Andes: A Multidisciplinary Study of High Altitude Quechua*, ed. P. T. Baker & M. A. Little, pp. 208–36. Stroudsburg, Pa.: Dowden, Hutchinson & Ross.

Preto, J. C. & Calderon, M. (1947). *Estudios bioantropometricos en los escolares Peruanos*. Boletin del Instituto Psicopedagico Nacional 2, Lima, Peru.

Pyzuk, M., Turusbekow, B. T. & Brjancewa, L. A. (1967). Certain properties of the respiratory system in school children in various altitudes and climatic conditions. *Human Biology*, 39, 35–52.

Reynolds, E. L. (1950). The distribution of subcutaneous fat in childhood and adolescence. *Monographs of the Society for Research in Childhood Development*, 15, 189.

Rothhammer, F. & Spielman, R. (1972). Anthropometric variation in the Aymara: genetic, geographic and topographic contributions. *American Journal of human Genetics*, 24, 371–80.

Schnakenberg, D. D. & Burlington, R. F. (1970). Effect of high carbohydrate, protein and fat diets and high altitude on growth and caloric intake of rats. *Proceedings of the Society for experimental Biology and Medicine*, 134, 905.

Schnakenberg, D. D., Krabill, L. F. & Weiser, P. C. (1971). The anorexic effect of high altitude on weight gain, nitrogen retention and body composition of rats. *Journal of Nutrition*, 101, 787–96.

Tanner, J. M. (1962). *Growth at adolescence*, 2nd ed. Oxford: Blackwell Scientific Publications.

Tanner, J. M., Whitehouse, R. H. & Takaishi, M. (1966). Standards from birth to maturity for height, weight, height velocity and weight velocity: British children, 1965. *Archives of Diseases of Childhood*, **41**, 454–71, 613–35.

Thomas, R. B. (1972). Human adaptation to a high Andean energy flow system. PhD thesis, The Pennsylvania State University.

Timiras, P. S. (1965). High altitude studies. In *Methods of animal experimentation*, vol. 2, ed. W. I. Gay, pp. 333–69. New York & London: Academic Press.

Timiras, P. S. & Wooley, D. E. (1966). Functional and morphological development of brain and other organs at high altitude. *Federation Proceedings*, **25**, 1312–20.

Van Liere, E. J., Crabtree, W. V., Northrup, D. W. & Stickney, J. C. (1948). Effect of anoxic anoxia on propulsive motility of small intestine. *Proceedings of the Society for experimental Biology and Medicine*, **67**, 331.

Weihe, S. H. (1966). Influence of age, physical activity and ambient temperature on acclimatization of rats to high altitude. *Federation Proceedings*, **25**, 1342.

Weitz, C. A. (1973). The effects of aging and habitual activity pattern on exercise performance among a high altitude Himalayan population. PhD thesis, The Pennsylvania State University.

Wintrobe, M. M. (1967). *Clinical Hematology*. Philadelphia: Lea & Febiger.

# 6.  Work capacity of high-altitude natives

E. R. BUSKIRK

The effects of chronic hypoxia on physical performance and the physio-logical correlates of performance have been examined in considerable detail during the past two decades. Several resident population groups that reside at high altitude have been investigated, including those in the USA, Bolivia, Chile, Peru, Ethiopia, USSR and Nepal. One common feature of people who live at altitudes of about 3,000 m or above (exception US population at Leadville, Colorado) is their relatively small stature. In order to compare appropriately physical performances and associated physiological differences that might explain differences in performance, it is imperative to consider age, stage of growth or maturation, body size and composition, genetic background, environmental conditions, nutri-tion, physical condition and prevalence of compromised health or presence of disease. Appropriate control of all confounding variables has rarely been accomplished in the investigations reported to date. Nevertheless, certain comparative interpretations can be made concerning the work capacity of high-altitude natives largely as reflected by aerobic capacity. Inclusion of consideration of some of the impacting variables mentioned above could modify conclusions drawn from investigations that have been reported, but such clarification remains for the future.

It is apparent from visual inspection of high-altitude natives at work and at play that performance of both high intensities and durations of work, e.g. Sherpas load-carrying for expeditions into the Himalayas and par-ticipation by Peruvian and Bolivian highlanders in strenuous games, such as soccer, is not only possible but engaged in regularly. These feats leave little doubt that high-altitude natives can lead physically active existences and accomplish tasks that are most difficult for all but well-conditioned newcomers who are partially altitude acclimatized. Thus, there is no doubt that man can adapt to intermediate terrestrial altitude, i.e. to altitudes between about 2,000 and 5,000 m. Such a conclusion is strengthened by the ample evidence from cultures thriving for many centuries at these heights.

A number of excellent reviews have been prepared by several authors and the interested reader should consult these for appropriate background as well as for additional information. These references are: Baker (1969); Balke (1964); Buskirk (1971, 1976); Dill, Adolph & Wilber (1964); Hurtado (1964); Lahiri (1974); Kollias & Buskirk (1974); Mazess (1970). Of further

173

## The biology of high-altitude peoples

general interest is the recent book *Man in the Andes: a multidisciplinary study of high-altitude Quechua*, edited by Baker & Little, 1976. This treatise is probably the most comprehensive one ever prepared about a group of native highlanders.

### Aerobic capacity

Numerous studies have shown that the aerobic capacity or the ability to deliver oxygen to working muscles is comparable in the highland native and the lowland native of comparable genetic background (Andersen, 1973; Grover *et al.*, 1967; Mazess, 1969). This statement is true for groups tested at the altitude at which they habitually reside, but if the lowland native goes to altitude then his aerobic capacity is reduced and remains so for an extended period. An exception to this was found by Mazess (1969) for university students from the lowland region who attended a university in the highlands at 3,000 m and were tested at 3,840 m. The lowlanders had essentially the same aerobic capacity as their highland counterparts. Comparable data to those of Mazess have been obtained on children by Andersen (1973). He found essentially no difference in the aerobic capacities of children who were tested at 1,500 m and 3,000 m.

There is no evidence that the energy costs of doing specific amounts of submaximal work during walking and load-carrying are any different for the highland as compared to the lowland native (Hurtado, 1964; Kollias & Buskirk, 1974). The proportionate relationship between oxygen consumption and work load is unchanged at altitude, at least during walking. In studies on the cycle ergometer, highland natives were found to utilize more oxygen than their sea-level counterparts, but this has been regarded as an artifact of cycle ergometry, the performance of which requires both acquired skill and considerable leg strength (Kollias *et al.*, 1968).

The highland native does not hyperventilate at altitude during rest or the performance of exercise as does the newcomer to the highlands (Frisancho *et al.*, 1973; Lahiri *et al.*, 1976). The native highlander's circulatory system and blood oxygen-carrying capacity is presumably adequate to alleviate the need for hyperventilation required by the un-acclimatized newcomer to altitude.

A summary of several studies in the literature on aerobic capacity is presented in Fig. 1. In Fig. 1 the solid lines represent data accumulated from a number of studies by Shephard (1966) in an attempt to provide representative values for aerobic capacity as an assessment of cardio-respiratory performance in several populations. Shephard's groupings were: Canada, US, Scandinavia and other countries. The two extremes for the data were: Scandinavia, high values and the United States, low

174

## Work capacity of high-altitude natives

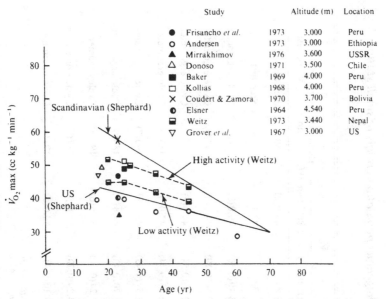

| Study | | | Altitude (m) | Location |
|---|---|---|---|---|
| ● | Frisancho *et al.* | 1973 | 3,000 | Peru |
| ○ | Andersen | 1973 | 3,000 | Ethiopia |
| ▲ | Mirrakhimov | 1976 | 3,600 | USSR |
| △ | Donoso | 1971 | 3,500 | Chile |
| ■ | Baker | 1969 | 4,000 | Peru |
| □ | Kollias | 1968 | 4,000 | Peru |
| ✕ | Coudert & Zamora | 1970 | 3,700 | Bolivia |
| ◐ | Elsner | 1964 | 4,540 | Peru |
| ◘ | Weitz | 1973 | 3,440 | Nepal |
| ▽ | Grover *et al.* | 1967 | 3,000 | US |

Fig. 6.1. Relationship of maximal oxygen uptake to age. Data obtained from several studies on native highlanders are plotted in relation to average values for groups from Scandinavia and US. The empirical regression lines were plotted from tabular data accumulated by Shephard (1966).

values. Note that most of the data obtained from highland natives fall between the Scandinavia and USA regression lines. The exceptions, all of which are on the low side, are from Andersen (1973) who studied Ethiopian natives from Debarech, Mirrakhimov (Chapter 10) who studied residents of Nurhab in the Pamirs and Elsner, Bolstad & Forno (1964), who studied Peruvian highlanders at Morococha. The reason for these low values may reside either in the testing procedures or in the relatively low fitness levels of the subjects studied.

A mean value for aerobic capacity based on data from highland natives within the age range 18–29 years was calculated to be 45.6 cc kg$^{-1}$ min$^{-1}$ – a value essentially equal to that found at sea level or low altitudes for other indigenous population groups of small stature and body size. Data from the latter groups appear in Table 1.

It is apparent from Fig. 1 that both age and habitual physical activity, the latter providing superior physical condition, influence aerobic capacity, age negatively and physical activity positively. The age trend is clearly documented in the studies by Andersen (1973) with residents of two communities in Ethiopia (Fig. 2). The effect of difference in aerobic capacity as related to physical activity is re-emphasized in Fig. 3. The data are from Weitz's study of Nepalese Sherpas from Namche Bazar

175

Table 1. *Maximum oxygen uptake and related measurements of indigenous population groups of small stature, i.e. less than 170 cm*

| Group | Reference | N | Age | Height (cm) | Weight (kg) | $\dot{V}_{O_2}$ (l. min$^{-1}$) | $\dot{V}_{O_2}$ (cc kg$^{-1}$ min$^{-1}$) | Heart rate (beats min$^{-1}$) | Work test |
|---|---|---|---|---|---|---|---|---|---|
| South African Bantu | Wyndham & Heyns (1969) | 23 | 18–25 | 166(6) | 59(6) | 2.85 | 48.0(2) | 180 | Step test |
| Australian Kalahari Bushmen | Wyndham et al. (1963) | 3 | Young adult | 161(6) | 51(5) | 2.39 | 47.1 | – | Step test |
| Arctic Indians | Andersen et al. (1960) | 8 | 29(8) | 166(6) | 63(3) | 3.10(0.50) | 49.0(8) | – | Cycle ergometer |
| Eskimos | Andersen & Hart (1963) | 8 | 23(4) | 162(13) | 60(8) | 2.56(0.19) | 44.0(5) | 173(9) | Cycle ergometer |
| Eskimos | Erickson (1957) | 10 | 20(3) | 167(4) | 66(6) | 2.79(0.54) | 44.1(7) | – | Cycle ergometer |
| Ainus | Ikai et al. (1971) | 21 | 27(8) | 162(6) | 57(8) | 2.43(0.44) | 42.0(6) | 188(9) | Cycle ergometer |
| Japanese farmers | Ikai et al. (1971) | 25 | 29(9) | 163(6) | 62(8) | 2.57(0.45) | 42.5(8) | 190(7) | Cycle ergometer |
| Japanese, Honshu and Hokkaido | Ikai et al. (1971) | 48 | 24(3) | 166(5) | 63(8) | 2.70(0.43) | 43.4(7) | 188(8) | Cycle ergometer |
| Kurds | Davies et al. (1972) | 12 | 26(4) | 169(7) | 64(6) | 3.24(0.38) | 50.6(6) | 193(6) | Cycle ergometer |
| Yemenites | Davies et al. (1972) | 15 | 25(4) | 163(5) | 64(10) | 3.29(0.70) | 51.4(5) | 187(7) | Cycle ergometer |
| Yoruba | Davies et al. (1972) | 11 | 25(2) | 169(5) | 62(5) | 3.10(0.44) | 50.0(5) | 191(6) | Cycle ergometer |

Mean    46.6
Range   42–57

The numbers in parentheses are the standard deviations.

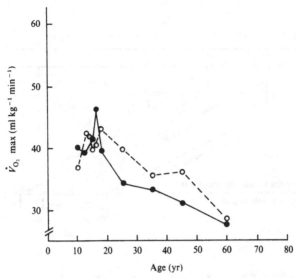

Fig. 6.2. Maximal oxygen uptake in relation to age in two groups of highland natives from Ethiopia. Data replotted from studies of Andersen (1973). Filled circles, 3,000 m Debarek; open circles, 1,500 m Adi-Arkai.

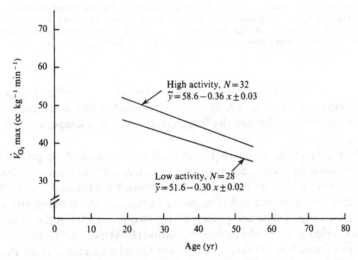

Fig. 6.3. Maximal oxygen uptake in relation to age in highland natives from Namche Bazar, Nepal, 3,440 m. Regression lines are given for two groups of subjects who differed in habitual physical activity. The data are from Weitz (1973).

177

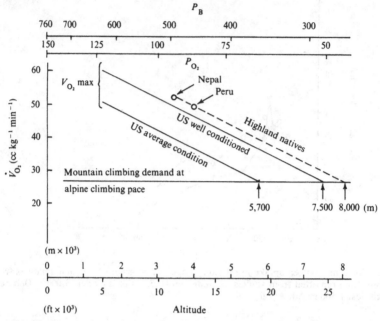

Fig. 6.4. The energy demands in terms of oxygen utilization for mountain climbing are indicated in relation to the progressive decrease in maximal oxygen uptake with altitude. The greater aerobic capacity of the highland native provides an additional functional capacity for climbing in comparison to unacclimatized men. Data are from Kollias *et al.* (1968).

(3,440 m). The respective regression lines for the dissimilar physical activity groups are shown. Those who were most active were porters, herders and farmers, whereas the less active were storekeepers, tradesmen and artisans.

That the highland native can perform hard work at altitude is probably best documented by their performance as porters on mountain climbing expeditions. One of the reasons for their superior performance may be found by inspecting their aerobic capacity in relation to climbing energy expenditure. In Fig. 4 data are presented for two groups from the United States who varied in level of physical condition (Buskirk, 1976), compared with Sherpas from Nepal (Weitz, 1973) and Quechua Indians from Peru (Kollias *et al.*, 1968). Assuming that there is a linear negative regression for aerobic capacity in relation to increasing hypoxia in all groups studied, it is apparent that mountain climbing at an average alpine pace would enable the highland Sherpa or Quechua to climb to about 8,000 m before reaching his aerobic capacity, whereas the non-native climber in excellent condition could climb to 7,500 m and the non-native climber in average

178

condition would reach only 5,700 m. Thus, the highland native can work well within his aerobic-capacity limit while climbing to substantially higher terrestrial heights than his non-native counterpart. In terms of load carrying, the Sherpa has demonstrated that he can also use his high aerobic capacity to advantage. Pugh (1966) has reported that the Sherpa can carry lods of 63 kg to about 3,000 m, loads of 36 kg to 3,500 m, 27 kg above 3,500 and 18 kg near 7,000 m.

In recent years the relatively untapped athletic talent in Africa, particularly for running, has been recognized by world-class performances in the Olympics and world games. A number of world-class distance runners have come from highland areas in Africa including M. Eftir from Ethiopia and N. Temu from Kenya. Unfortunately, neither was allowed to compete in the 1976 Olympics for political reasons. It is speculated that other world-class athletes exist in other highland areas around the world, but their latent performance capacities go unrecognized for lack of organized programs and training facilities. Thus, the present physical conditioning of highland natives will continue to be in such recreational activities as dancing and soccer and such occupational activities as walking, load carrying, lifting and digging. World-class athletic performance remains for the future among indigenous highland groups and hinges on an interest being shown in these people in order to provide them with appropriate instruction, training and conditioning. Evidence, to date, suggests that beneficial training intensities for running can be undertaken at altitudes up to 3,500 m. L. Virren, who won both the 5,000 and 10,000 m runs in the 1976 Olympics and did well in the marathon, trained extensively in the Pyrenees. M. Gamudi from Tunisia, an outstanding marathon runner, has utilized the Atlas mountains for his training.

**The effects of migration on aerobic capacity**

Velasquez (1964) demonstrated that work performance based on time to exhaustion when running on a treadmill can be improved by residence at 4,000 m for 12 months (Fig. 5). Running speed was 133 m min$^{-1}$ and the treadmill grade was 11%. A concomitant increase in aerobic capacity was found. Studies conducted since have demonstrated that acclimatization to altitude occurs very slowly and is difficult to demonstrate with physical performance assessment or related physiological variables for periods of 6 months or less (Kollias & Buskirk, 1974; Baker, 1976).

Active children tend to have relatively high aerobic capacities compared to older adults and their aerobic capacities are affected by physical conditioning to a lesser extent than adults. The higher values in children in relation to those in adults are demonstrated in Andersen's (1973) results (see Fig. 2). Frisancho *et al.* (1973) have stated that being born in the

179

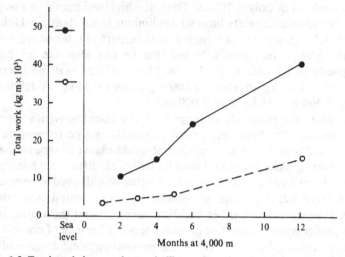

Fig. 6.5. Total work done on the treadmill at various times by subjects from Lima (sea-level) who spent 12 months at Morococha, Peru, 4,000 m. Each line is for one subject. Although major interindividual differences were observed, all subjects showed some acclimatization to hypoxia. Data are from Velasquez (1964).

highlands or migrating there at an early age leads to a high aerobic capacity at high altitude. The results reported are suggestive but not conclusive. Data from the study of Frisancho *et al.* are replotted in Fig. 6. The regression line relating aerobic capacity to age is provided which indicates about a 1 cc $kg^{-1}$ $min^{-1}$ loss in maximal oxygen uptake per year of greater age at migration. In the left portion of Fig. 6 are plotted data from Frisancho *et al.* for the highland control subjects, i.e. those born at 3,000 m and the lowland newcomers to 3,000 m. Note that the intercept for the regression line predicts a $\dot{V}_{O_2}$ max of approximately 54 cc $kg^{-1}$ $min^{-1}$ for the native highlanders – a value about plus two standard deviations above the mean measured value of 46.3 cc $kg^{-1}$ $min^{-1}$. This discrepancy suggests that group matching with respect to physical condition or some other modifying factor was unequal. On the right side of Fig. 6, data from Mazess's (1969) study of similar subjects suggests that students who migrate to the highlands to attend school have an aerobic capacity equal to the indigenous highland students. Mazess's data do not fit the regression line of Frisancho *et al.* Thus, there appears ample reason to explore further the temporal effect of migration in relation to the stage of growth, development or maturation in groups matched for level of physical condition. Baker (1976) has also written a critique of the study reported by Frisancho *et al.* in which he suggests a curvilinear regression based on a critical age at which migration must occur or a critical duration of

180

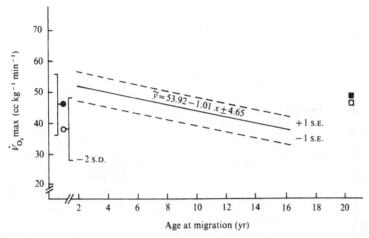

Fig. 6.6. Decrease in maximal oxygen uptake in relation to age at migration to 3,000 m. Data from Frisancho *et al.* (1973) (circles) also include the mean and ±2 s.d. for a group of highland natives (solid symbols) and groups of newcomers to 3,000 m (open symbols). Comparable data from a study by Mazess (1969) (squares) are included but do not fit the regression plot.

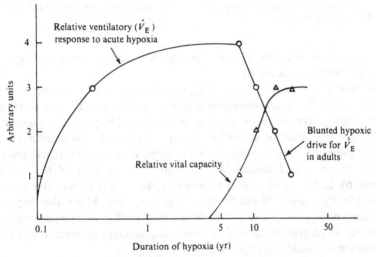

Fig. 6.7. Respiratory acclimatization to hypoxia as found by Lahiri *et al.* (1976). Figure plotted from tabular data. A log scale was utilized for the abscissa. The concept of developmental adaptation is emphasized.

181

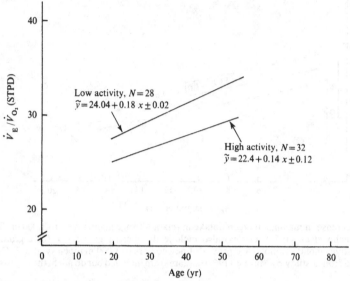

Fig. 6.8. Ventilatory exchange ratio in two groups of native highlanders from Namche Bazar, Nepal. The groups varied in habitual physical activity. Those who were less active had significantly higher ventilatory exchange ratios. STPD, standard temperature and pressure, dry. Data from Weitz (1973).

exposure to high altitude. Baker also suggests that the younger migrants may have been unusually fit.

Further support for the concept that environmental acclimatization to hypoxia occurs has been demonstrated by Lahiri *et al.* (1976) who found development of hypervital capacity with duration of hypoxic exposure (see Fig. 7). Another feature is further documentation of a blunted hypoxic drive for pulmonary ventilation after 6 or more years' exposure to hypoxia in older children and adults. A schematic presentation of the results obtained by Lahiri *et al.* (1976) is given in Fig. 7. It is interesting to note that relatively better physical condition may also blunt the hypoxic ventilatory drive, for Weitz (1973) found that $\dot{V}_E/\dot{V}_{O_2}$ ratio was considerably lower in physically active Nepalese highlanders compared to those who were more sedentary (see Fig. 8).

Pulmonary ventilation data also suggest that newcomer Quechua to the highlands in Peru have a partial blunting of their hypoxic drive for pulmonary ventilation (Kollias & Buskirk, 1974; Buskirk, 1976). The respective data are plotted in Fig. 9. The highland Quechua have essentially the same ventilatory response to increased exercise as do sea-level residents tested at sea level.

## Work capacity of high-altitude natives

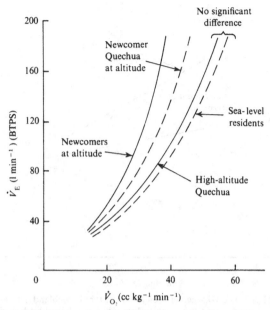

Fig. 6.9. Relationship between exercise ventilation and oxygen uptake in Quechua and US subjects abruptly moved to Nuñoa, Peru (4,000 m) and in native highland Quechua. The US subjects were also tested at sea level. BTPS, barometric temperature and pressure, saturated. Data from Buskirk (1976), Kollias *et al.* (1968) and Kollias & Buskirk (1974).

### Physiological factors related to aerobic capacity and performance

Probably the most important factors related to the relatively large aerobic capacity of physically active highland natives are their greater diffusing capacity for oxygen, their larger right ventricular mass, the greater oxygen-carrying capacity of their blood, the greater capillarity of their skeletal muscles and lessened diffusion distances for oxygen, the presence of a modified Bohr effect that favors oxygen unloading from hemoglobin in the periphery including skeletal muscle, as well as the greater number of mitochondria in skeletal muscle, plus an increased concentration of oxidative chain enzymes.

Aerobic capacity may be regarded as an integrated measure of oxygen intake, transport and utilization. Any step in the oxygen transfer process can be limiting under hypoxic conditions. For example, transfer of oxygen across the alveolar capillary membrane can be an important limiting factor. The large diffusing capacity of the lungs of native highlanders preserves the essential transfer of oxygen (Velasquez, 1956). The newcomer pays the price of having to support hyperventilation during exercise and the oxygen cost of the increased breathing partially reduces aerobic capacity.

183

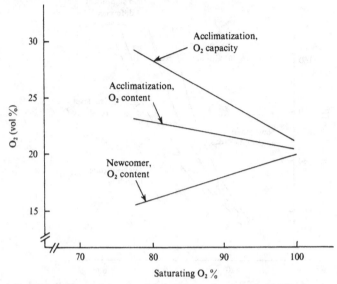

Fig. 6.10. Effect of hypoxia on oxygen capacity and content in highland natives. The oxygen content for the newcomer (acute exposure to hypoxia ) was based by Lahiri (1974) on an oxygen capacity of 20 ml $O_2$/100 ml blood and shows a linear decrease as the oxygen saturation of hemoglobin decreases. The data for chronic hypoxia were obtained by Lahiri from several references. Lahiri states 'The $O_2$ capacity response to chronic hypoxia is such that $O_2$ content is somewhat greater than the sea-level value at a given altitude'.

Presumably, maximal cardiac output per unit body mass is roughly equivalent in those acclimatized to hypoxia and the sea-level resident – at least for those of comparable physical condition. Extensive studies of cardiac output response to exercise have not been made in highland natives. The mass of the right ventricle is increased presumably because of the increased pulmonary arterial pressure against which the ventricle must pump. The greater capabilities of the right ventricle insure pulmonary perfusion and create conditions to support the greater diffusion capacity found in the highland native (Kollias & Buskirk, 1974).

Evidence for differences in oxygen-carrying capacity and oxygen content of arterial blood in highland natives has been presented by Lahiri (1974) and Fig. 10 has been adapted from their presentation. A higher hemoglobin concentration and a larger red-cell mass have been routinely found in the native highlander compared to the sea-level resident or the newcomer to altitude (Garruto, 1976). The polycythemia is only moderate but of distinct advantage for relatively increased arterial oxygen-carrying capacity for the performance of hard work. Because of the larger number of red blood cells and the greater hemoconcentration, the arterial oxygen content at equal percentages of oxygen saturation is greater in the native

184

highlanders than in newcomers to altitude (Garruto, 1976). Thus, the chronic erythropoietic response to hypoxia is adequate for minimization of tissue hypoxia including skeletal muscle.

A larger Bohr effect in highland natives compared to Europeans studied at Cerro de Pasco, Peru (4,350 m) has been reported by Morpurgo *et al.* (1970). The respective typical positions of the oxygen dissociation curves of hemoglobin deviate to the right in the highland native both at pH 7.4 and at pH 6.7. The migrant Europeans were presumably well acclimatized to hypoxia and had lived at high altitude for years. The shift to the right of the oxygen dissociation curve favors the unloading of oxygen from hemoglobin in the periphery and presumably in skeletal muscles during the performance of hard work.

Increased vascularity (proliferation of small blood vessels including capillaries) and greater perfusion of skeletal muscle (associated with decreased peripheral resistance) under hypoxic conditions is an asset to the native highlander while this perfusion may be limiting in the newcomer to altitude. Velasquez (1976) has stated 'More oxygen in the capillaries, changes in affinity of hemoglobin for oxygen and slower capillary blood flow are some of the important mechanisms employed to maintain high (oxygen) pressure in the capillaries. The net effect of these adaptive mechanisms at high altitude is an efficient and normal oxygen delivery to the tissues'.

Lahiri (1974) cites evidence that mitochondrial concentration and membrane surface per unit mass tissue and cells are increased in the altitude-adapted, as are the enzymes of the respiratory oxidative chain. Thus, diffusion distance may be shortened for oxygen in the high-altitude native not only because of increased capillarity of skeletal muscle but because of increased mitochondrial surface availability as well.

While the answers are not all in, the functional adaptations associated with acclimatization to hypoxia are being uncovered to reveal the physiological reasons for the native highlanders' superior performance capacities at high altitude.

## References

Andersen, K. L. (1973). The effect of altitude variation on the physical performance capacity of Ethiopian men. Part I. Maximal oxygen uptake and some related circulatory and respiratory functions in relation to aging. Part II. Development of physical performance capacity during adolescence. In *Physical Fitness*, ed. V. Seliger, pp. 13–33, 34–46. Prague: Universita Karlova.

Andersen, K. L., Bolstad, A., Loyning, Y. & Irving, L. (1960). Physical fitness of Arctic Indians. *Journal of applied Physiology*, **15**, 645–8.

Andersen, K. L. & Hart, J. S. (1963). Aerobic working capacity of Eskimos. *Journal of applied Physiology*, **18**, 764–8.

The biology of high-altitude peoples

Baker, P. T. (1969). Human adaptation to high altitude. *Science, Washington,* **163,** 1149–56.
Baker, P. T. (1976). Work performance of highland natives. In *Man in the Andes: A Multidisciplinary Study of High Altitude Quecha,* ed. P. T. Baker & M. A. Little, pp. 300–14. Stroudsburg, Pa.: Dowden, Hutchinson & Ross.
Balke, B. (1964). Work capacity and its limiting factors at high altitude. In *The Physiological Effects of High Altitude,* ed. W. H. Weihe, pp. 230–40. New York: Macmillan.
Buskirk, E. R. (1971). Work and fatigue in high altitude. In *Physiology of Work Capacity and Fatigue,* ed. E. Simonson, pp. 312–22. Springfield, Illinois: Charles C. Thomas.
Buskirk, E. R. (1976). Work performance of newcomers to the Peruvian highlands. In *Man in the Andes: A Multidisciplinary Study of High Altitude Quecha,* ed. P. T. Baker & M. A. Little, pp. 283–99. Stroudsburg, Pa.: Dowden, Hutchinson & Ross.
Coudert, J. & Zamora, M. P. (1970). *Estudio del consumo de oxigeno en La Paz (3,700 m) sobre un grupo de atleas nativos de la altura,* pp. 109–13. Instituto Boliviano de Biologia de Altura Anuario.
Davies, C. T. M., Barnes, C., Fox, R. H., Ojikutu, R. O. & Samueloff, A. S. (1972). Ethnic differences in physical working capacity. *Journal of applied Physiology,* **33,** 726–32.
Dill, D. G., Adolph, E. F. & Wilber, C. G. (eds.) (1964). *Handbook of Physiology: Adaptation to the Environment.* Washington, DC: American Physiological Society.
Donoso, H. P., Apud, E., Sanudo, M. C. & Santolaya, R. (1971). Capacidad aerobica como indice de adecuadad fisica en muestras de poblaciones (Urbanas y Nativas De La Altura) y en un grupo de atletas de seleccion. *Revista Medica de Chile,* **99,** 719–31.
Elsner, R. W., Bolstad, A. & Forno, C. (1964). Maximum oxygen consumption of Peruvian Indians native to high altitude. In *The Physiological Effects of High Altitude,* ed. W. H. Weihe, pp. 217–33. New York: Macmillan.
Erikson, H. (1957). The respiratory response to acute exercise of Eskimos and Whites. *Acta physiologica scandinavia,* **41,** 1–11.
Frisancho, A. R., Martinez, C., Velasquez, T., Sanchez, J. & Montoye, H. (1973). Influences of developmental adaptation on aerobic capacity at high altitude. *Journal of applied Physiology,* **34,** 176–80.
Garruto, R. M. (1976). Hematology. In *Man in the Andes: A Multidisciplinary Study of High Altitude Quecha,* ed. P. T. Baker & M. A. Little, pp. 261–82. Stroudsburg, Pa.: Dowden, Hutchinson & Ross.
Grover, R. F., Reeves, J. T., Grover, E. B. & Leathers, J. S. (1967). Muscular exercise in young men native to 3,100 m altitude. *Journal of applied Physiology,* **22,** 555–64.
Hurtado, A. (1964). Animals in high altitudes: resident man. In *Handbook of Physiology, Section 4. Adaptation to Environment,* pp. 843–60. Washington, DC: American Physiological Society.
Ikai, M., Ishi, K., Miyamura, M., Kusano, K., Bar-Or, O., Kollias, J. & Buskirk, E. R. (1971). Aerobic capacity of Ainu and other Japanese on Hokkaido. *Medicine and Science in Sports,* **3,** 6–11.
Kollias, J. & Buskirk, E. R. (1974). Exercise and altitude. In *Medicine and Science of Exercise and Sport,* 2nd edn, ed. W. R. Johnson & E. R. Buskirk, pp. 211–27. New York: Harper & Row.

# Work capacity of high-altitude natives

Kollias, J., Buskirk, E. R., Akers, R. F., Prokop, E. K., Baker, P. T. & Picón-Reátegui, E. (1968). Work capacity of long time residents and newcomers to altitude. *Journal of applied Physiology*, **24**, 792-9.

Lahiri, S. (1974). Physiological responses and adaptations to high altitude. In *Environmental Physiology*, ed. D. Robertshaw, pp. 273–311. Baltimore: University Park Press.

Lahiri, S., DeLaney, R. G., Brody, J. S., Simpser, M., Velasquez, T., Motoyama, E. K. & Polgar, C. (1976). Relative role of environmental and genetic factors in respiratory adaptation to high altitude. *Nature, London*, **261**, 133-5.

Mazess, R. B. (1969). Exercise performance of Indian and White high altitude residents. *Human Biology*, **41**, 494-518.

Mazess, R. B. (1970). Cardiorespiratory characteristics and adaptation to high altitudes. *American Journal of physical Anthropology*, **32**, 267-78.

Morpurgo, G., Battaglia, P., Bernini, L., Paolucci, A. M. & Modanio, G. (1970). Higher Bohr effect in Indian natives of Peruvian highlands as compared with Europeans. *Nature, London*, **227**, 387-8.

Pugh, L. G. C. E. (1966). A programme for physiological studies of high altitude peoples. In *The Biology of Human Adaptability*, ed. P. T. Baker & J. S. Weiner. Oxford: The Clarendon Press.

Shephard, R. J. (1966). World standards of cardiorespiratory performance. *Archives of environmental Health*, **13**, 664-72.

Velasquez, T. (1956). *Maximal diffusing capacity of the lungs at high altitudes.* Report 56-108 School of Aviation Medicine, USAF Randolph Field, Texas.

Velasquez, T. (1964). Response to physical activity during adaptation to altitude. In *The physiological effects of high altitude*, ed. W. H. Weihe, pp. 289-99. New York: Macmillan.

Velasquez, T. (1976). Pulmonary function and oxygen transport. In *Man in the Andes: A Multidisciplinary Study of High Altitude Quecha*, ed. P. T. Baker & M. A. Little, pp. 287-60. Stroudsburg, Pa.: Dowden, Hutchinson & Ross.

Weitz, C. A. (1973). The effects of aging and habitual activity pattern on exercise performance among a high altitude Nepalese population. PhD Thesis, The Pennsylvania State University.

Wyndham, C. H. & Heyns, A. J. (1969). Determinants of oxygen consumption and maximum oxygen intake of Bantu and Caucasian males. *Internationale Zeitschrift für angewandte Physiologie*, **27**, 51-75.

Wyndham, C. H., Strydom, N. B., Morrison, J. F., Peter, J., Williams, C. G., Gredell, G. A. G. & Joffe, A. (1963). Differences between ethnic groups in physical working capacity. *Journal of applied Physiology*, **18**, 361-6.

# 7. The haematological characteristics of high-altitude populations

J. C. QUILICI & H. VERGNES

The fundamental part played by blood in all processes of exchanges between the outside environment and the inside environment of vertebrates accounts for the fact that all biologists who have studied the problem of adaptation to altitude have found it necessary to carry out haematological observations. As a consequence, the literature dealing with haematology and altitude is particularly rich in information.

F. Viault, in 1890, was probably the first to draw attention to the polycythaemia found at altitude. Since then many studies have treated haematology at altitude, but the major interest of the scientists in this problem may be said to have begun in the 1920s. Four names must be cited, Barcroft (1925), Harrop (1922), Haldane (1926) and finally, undoubtedly the most important, C. Monge (1929), the true founder of haematology at altitude. Since then, numerous studies have been undertaken in various aspects of the topic. Though important progress has been made towards a better knowledge of the haematological phenomena of adaptation to altitude, it must be pointed out from the start that knowledge on the subject is still very fragmentary.

This paradox may be explained by several problems, often common to altitude-related topics other than haematology. Briefly stated the problems are:

(1) Altitude in itself does not create a uniform ecological situation. On the contrary, owing to geographical and climatic factors, the environment displays substantial variations in cold (Dollfus, 1973; Dollfus & Lavallee, 1974), humidity and natural radiation levels. On the other hand, the single variable which remains proportional to altitude, whatever the latitude may be, is the partial oxygen pressure. It is also, as Ruffie (1972) stressed, the only factor in the environment against which man has been unable to find any sort of cultural protection. To survive in regions of high altitude, he has therefore had to rely solely on his biological capacities to adapt.

Altitude also creates two fundamentally distinct situations according to the zones of human settlement: on the one hand, the high plateaus and on the other, the high valleys. In the former, man's activity takes place mainly at a constant altitude; in the latter, his activity exposes him to varying altitudes. Commonly these variations involve the altitude range 1,500–2,000 m (e.g. central Andean valleys of Peru, Nepalese valleys).

189

*The biology of high-altitude peoples*

(2) A second major difficulty arises from the length of stay of men at altitude. This problem appears to be one of the most important ones. To overlook it leads to errors in explaining the observed phenomena; errors that delayed for a long time, and still often confound, our understanding of altitudinal biology.

Here we find it necessary to distinguish three fundamental situations:

(a) Men born at high altitude and permanently residing at that same altitude: these individuals we shall call, as Vellard (1965) does, 'Altiplanide'.*

(b) Men born at altitude but undergoing more or less prolonged variations of altitude: we shall designate such individuals 'Andide'.*

(c) Men born at low altitudes who have been residing at high altitude for more or less prolonged time.

Between these basic groups, every intermediate situation may be met: this fact complicates the interpretation of available data still more. The great confusion which still exists in the case of the term 'native of altitude' seems to us to bear the main responsibility for the difficulty encountered in interpreting the often contradictory results of various authors.

The literature that we reviewed prior to the writing of this chapter made it obvious to us that the difficulties just described, though fundamental, were not always considered by researchers.

**Altitude polycythaemia**

Most of the published reports on haematology and altitude are concerned with altitude polycythaemia. This clinical concept, described in detail by C. Monge (1929) and M. Monge & C. Monge (1966), has led to a certain confusion which we find it necessary to clarify.

Hypoxia occurs at the tissue level as soon as more oxygen is needed than the blood can supply (Rørth, 1972). The gaseous exchanges in man, as in all species of vertebrates leading a normal aerobic life, take place in four stages: (1) pulmonary ventilation, when the molecules of gas are transferred from the external environment to the pulmonary alveoli; (2) alveolar exchanges, when the oxygen of the air in the alveoli diffuses to the pulmonary capillaries; (3) circulatory stage, when the oxygen dissolved in the plasma and fixed on the haemoglobin is carried by the blood circulation; and (4) a stage of exchange at the tissue level when the oxygen diffuses from the capillary blood vessels to the tissues and cells.

To improve the capacities of oxygen delivery to the tissues, the organism

* These terms apply well to the Andean region only. It must also be noted that we do not intend to imply the same anthropological meaning as Vellard.

190

must alter two functions. First, it must increase the efficiency of the transportation by the blood, and second, it must raise the rate of delivery at the tissue level by bringing into play regulating systems affecting the affinity of oxygen to haemoglobin. A rise in oxygen delivery is accomplished by two mechanisms which are activated by a common stimulus (the lowering of the $P_{O_2}$ to a critical level). These mechanisms intervene in close co-ordination to maintain homeostasis. The mechanisms are reactional polycythaemia through acceleration of the erythropoiesis, and metabolic alterations of glycolysis leading to a modulation of oxygen affinity for the normal respiratory pigment of the haemoglobin.

Under experimental or natural (high-altitude) conditions of hypoxia, the first compensating mechanism is manifested by an increasing level of functional haemoglobin and circulating erythrocytes. This rise constitutes the classical phenomena called polycythaemia.

*Experimental results*

The processes leading to altitude polycythaemia have been revealed by many investigators. These studies indicate that the response of the haematopoietic system to the hypoxic stimulus is extremely complicated. Partially for this reason, fundamental research on the topic has been mostly confined to non-human studies.

The most comprehensive experimental studies carried out on animals are those of Lord & Murphy (1973) on mice. The authors have analysed the kinetics of erythropoietic production in a few strains of mice undergoing an exposure to a simulated altitude of 6,680 m for periods ranging from 1 to 15 days in succession. The observed changes were progressive over the time of observation. A striking finding related to the early exposure period corresponding to the first hours of exposure. During this time rapid changes took place, mainly at the haematocrit level (Fig. 7.1) and in the number of haematopoietic stem cells in the femoral bone marrow (Fig. 7.2). As the process continued there was a progressive rise in haematocrit. In the animals studied, Lord & Murphy (1973) noted that the haematocrit reached very high equilibrium values of between 70% and 75%. The number of haematopoietically active cells gradually increased until they reached a plateau 6 or 7 days after the beginning of the experiment. Thereafter the number remained nearly constant.

The authors also explored the turnover of the erythropoietic 'islets' in the femoral bone marrow and in the spleen. Figs. 7.3 and 7.4 show the differences in cellular kinetics in response to hypoxia for both tissues. It is interesting to note the cycling nature of the response, which is particularly clear in the bone marrow. The exact mechanisms which start the reaction of bone marrow hypercellularity are not yet fully explained.

191

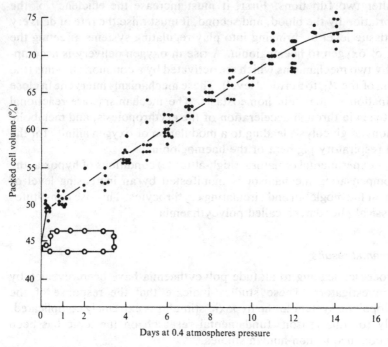

Fig. 7.1. Variation of red cell volume in mice during simulated altitude. O, control animals.

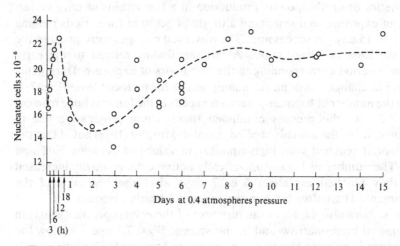

Fig. 7.2. Variation of haematopoietic stem cells in femoral bone marrow in mice.

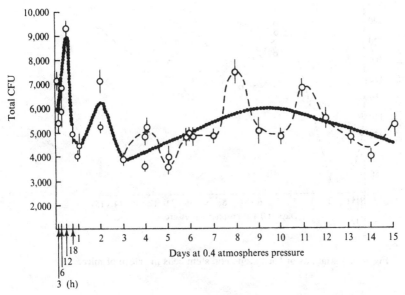

Fig. 7.3. Variation of absolute number of CFU in the femoral shaft bone marrow in mice. Dotted line indicates the general trend and the broken line the oscillatory pattern.

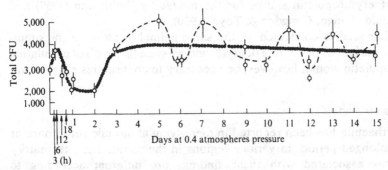

Fig. 7.4. Variation of absolute number of CFU in the whole spleen, in mice. The dotted line indicates the general trend and the broken line the oscillatory pattern.

Schooley, Garcia, Cantor & Havens (1968) studied mice which were injected daily with anti-erythropoietin while being exposed for 5 days to an altitude of 5,400 m. They found an almost complete inhibition of the polycythaemia response. This study suggests that the polycythaemia of altitude must be linked with a hypersecretion of erythropoietin (Stohlmann, 1959; Albrecht & Littell, 1972).

The research carried out by Lord & Murphy (1973) provided additional support for this hypothesis because the cyclic character of the turnover

193

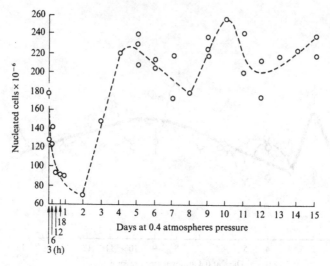

Fig. 7.5. Variations of haematopoietic stem cells in spleen of mice.

of the haematopoietic 'islets' (Fig. 7.5) in the animals undergoing experimental hypoxia may well be closely related to fluctuations in the plasma level of erythropoietin shown for the mouse by Stohlmann (1959) and McDonald, Lange, Congdon & Toya (1970).

Further studies conducted on the same animal to measure the simultaneous variations of the rate of erythropoiesis and the levels of circulating erythropoietin would, however, be necessary to confirm this suggestion.

### Studies on man

Polycythaemia has been reported in men living at altitude for a more or less prolonged period, in various regions in the world. The explanatory problems associated with these findings are different according to whether the polycythaemia reaction is observed in a normal subject or in individuals with other health problems (cardiovascular complaints in altitude, acute or chronic bronchopulmonary diseases). Owing to this fundamental distinction, it is advisable from the start to distinguish clearly the case of subjects apparently in good health from those with medical problems.

We shall consider the principal results relating to altitude polycythaemia by various authors, according to geographical regions: (1) Africa (the high plateau of Ethiopia), (2) Asia (data gathered from Indian or Nepalese groups), (3) Latin America (research carried out in Mexico and in various countries of the Andean region, principally reports by Peruvian and

194

Bolivian researchers), and (4) North America (studies carried out in the Rockies).

### Studies carried out in Africa (Ethiopian highlands)

In 1969, Harrison *et al.* (1969) published the results of an anthropological and medical survey of the inhabitants of three localities situated in the northern part of the Ethiopian plateau. This survey was conducted as a part of the British IBP. Its object was the analysis of the biological factors relating to human adaptation to high altitude among natives living in the highlands of East Africa. Ethiopia offered a particularly interesting area for investigation since the greater part of the country is situated at an average altitude of 2,500 m, with a few inhabited areas over 3,000 m.

The surveys were conducted in three villages of northern Ethiopia (Simien zone): Debarech, Adi-Arkai and Geech. Debarech stands at the edge of the plateau at an altitude of 3,000 m. Adi-Arkai, farther to the north, is built on the foothills of the plateau at a height of 1,500 m, and Geech is a small community in the vicinity of Debarech but at the still higher altitude of 3,700 m.

The inhabitants of the villages belong to two principal ethnic groups: the Amharans and the Tigrinyans. Haematological tests were made only on native males and females whose ages ranged from 18 to 50 years. The values obtained from the haematological samples from the three villages are shown in Table 7.1. The results are separated by sex.

The tests of statistical comparisons applied to this data show that the values of red blood cell count are significantly higher in the men from Debarech ($P = 0.02$) than in the men from Adi-Arkai. As for the women,

Table 7.1. *Haematological findings in the Ethiopian study*

|  | Adi-Arkai (1,500 m) | | | Debarech (3,000 m) | | | Geech (3,700 m) | | |
|---|---|---|---|---|---|---|---|---|---|
|  | $x$ | $N$ | S.E. | $x$ | $N$ | S.E. | $x$ | $N$ | S.E. |
|  | | | | Males | | | | | |
| RBC | 3,982,127 | 47 | 59,090 | 4,274,324 | 37 | 92,340 | 4,078,662 | 15 | 88,770 |
| WCC | 6,007 | 46 | 265 | 6,073 | 37 | 291 | 6,845 | 15 | 495 |
| PCV | 54.6 | 46 | 0.73 | 58.9 | 45 | 1.09 | 55.0 | 15 | 0.82 |
| Hb | 107.8 | 47 | 0.70 | 101.2 | 56 | 1.94 | 102.9 | 17 | 1.21 |
|  | | | | Females | | | | | |
| RBC | 3,597,368 | 19 | 85,320 | 3,622,500 | 12 | 102,600 | — | — | — |
| WCC | 5,497 | 19 | 331 | 6,097 | 17 | 268 | — | — | — |
| PCV | 50.2 | 17 | 0.93 | 53.3 | 20 | 1.07 | — | — | — |
| Hb | 97.3 | 19 | 1.17 | 91.5 | 17 | 1.40 | — | — | — |

RBC, red blood cells; WCC, wet cell count; PCV, packed cell volume; Hb, haemoglobin concentration.
From Harrison *et al.* (1969).

195

Table 7.2. *Mean corpuscular volume and haemoglobin concentration for the Ethiopian study*

| | Adi-Arkai (1,500 m) | | Debarech (3,000 m) | | Geech (3,700 m) |
|---|---|---|---|---|---|
| | Male | Female | Male | Female | Male |
| Mean corpuscular volume | 117.4 | 117.5 | 119.5 | 125.5 | 115.5 |
| Mean haemoglobin concentration (%) | 33.4 | 33.3 | 28.7 | 29.2 | 31.6 |

From Harrison *et al.* (1969).

there is no significant difference between the two villages in the observed values. On the other hand, the haematocrit concentrations in the subjects of both sexes are higher in Debarech than in Adi-Arkai ($P < 0.001$). The results of this study appear quite contrary to the expected findings since the high-altitude villages show low red blood cell and haemoglobin concentration values. Indeed, the values found for these populations are lower than those of the inhabitants of Adi-Arkai, who live at the lowest altitude. Harrison *et al.* (1969) are of the opinion that the lower values for haemoglobin content are very difficult to explain.

Ruling out any technical error (the determinations were made by the same team, with the same batches of reagents), as well as the presence of sick individuals among the subjects studied, we must first consider the methodology used. The haemoglobin assay, during that investigation, was by means of the optical method of Sahli. The values thus obtained are certainly less precise than measurements made with photo-electric colorimeters.

However, the estimated mean corpuscular volumes and the average amount of haemoglobin suggest non-methodological explanations for the low RBC values and haemoglobin concentrations. Table 7.2 summarises the observed data for the various groups examined. Because of possible technical error, one must be careful about interpreting the generalised macrocytosis. Nevertheless, the low values of mean corpuscular content in haemoglobin, which is significant in the inhabitants of Debarech, suggests a possible macrocytic anaemia. These results take on particular significance when we bear in mind the high frequency of such anaemias among populations living in tropical zones (Bernard & Ruffie, 1966). Many forms of these anaemias have their origin in deficiencies, such as nutritional ones.

For the individuals of the three localities, there is little variation among the values for white blood cell counts and the absence of hyperleucocytosis must be noted. However, a differential count of the white blood cells

196

reveals an interesting phenomenon. Thirty-three individuals out of seventy from Adi-Arkai, and seventy-two out of 116 from Debarech have a greater proportion of lymphocytes than granulocytes. The authors, however, could not provide any direct evidence on the relationship between the lymphocytosis and the pathological state of the populations.

Finally, the existence of high eosinophil counts, particularly in the group of people from Adi-Arkai, was significant. These abnormal levels, in some cases considerable (the level may reach 40% in certain individuals), is certainly due to a high degree of parasitic infestation by Nematoda, which is widespread over that zone.

Harrison *et al.* (1969), however, do not specify whether the values for eosinophilia, which are lower at Debarech, can be explained completely by variations in Nematoda infestation or why there should be such an altitude-related variability in eosinophil counts.

*Studies reported from the Himalaya region*

Numerous research schemes under the sponsorship of IBP were concerned with the study of Himalayan populations. They are summarised in Chapter 18 of *The Biology of Human Adaptability* (Pugh, 1966). Until recently very few of the results observed in the course of those various programmes have, in fact, been published. In 1972, Morpurgo *et al.* (1972) made a study of the variations of the haemoglobin values in the Nepalese Sherpas and the Europeans who were taking part in an Italian expedition to the Himalayas. The results of their study will be discussed in the part of this chapter concerned with the Bohr effect. Therefore in this section we shall only report the data gathered during the first British expedition in the Himalayas, as reported by Pugh (1966) and give a summary of the recent works carried out in India by Bharadwaj, Singh & Malhotra (1973).

Pugh compared the haemoglobin level of the circulating blood of the Nepalese Sherpas and the members of the British teams who took part in the Himalayan expeditions between 1952 and 1961. The values he found are entered in Table 7.3.

During the same expedition Pugh noted the haemoglobin variations in the Sherpas at two different altitudes: 3,800–4,000 m and 5,800 m. For twenty-two Sherpas living between 3,800 and 4,300 m, the average value is 17.1 g/100 ml. For ten Sherpas living at 5,800 m the determinations gave a value of 17.9 g/100 ml. At the same altitude, the values measured for the Europeans gave an average of 19.6 g/100 ml. As a rule, the haemoglobin level in the Sherpas born and living on the Himalayan plateau tends to be lower than that of the Europeans, even when the latter have spent a relatively long time at the altitude.

In India, Bharadwag *et al.* (1973) analysed the influence of high altitude

Table 7.3. *Haemoglobin levels in Sherpa and English mountaineers*

|  | Subjects | N | Hb rate (g/100 ml) | |
|---|---|---|---|---|
|  |  |  | During ascent | Before expedition |
| Mount Cho O Yu, 1952 | Europeans | 8 | 20.3(18.3–22.0) | 14.9 |
|  | Sherpas | 6 | 19(17.6–20.0) | 13.6 |
| Mount Everest, 1953 | Europeans | 12 | 20.9(19.3–23.3) | — |
|  | Sherpas | 18 | 19.3(15.1–23.6) | — |

From Pugh (1966).

Table 7.4. *Haematological findings on three population samples from India*

|  | Tamil people from South India | | Natives |
|---|---|---|---|
|  |  | After continuous stay of 10 months at 3,692 m | Residing at Ladakh |
| Parameter | At sea level |  |  |
| Red blood count | 5.32±0.39 | 6.16[a]±0.68 | 6.36[a]±1.17 |
| (million/ml) | N = 39 | N = 23 | N = 24 |
| Haemoglobin | 15.10±1.28 | 18±1.45 | 18.30±2.40 |
| (g/100 ml) | N = 39 | N = 23 | N = 24 |
| Haematocrit (%) | 47.90±3.18 | 56.10[a]±3.91 | 54.30±5.60 |
|  | N = 40 | N = 80 | N = 24 |

[a] Significant difference.

on two groups of subjects: one group of people born and living at an altitude of about 3,900 m in the province of Ladakh, and the other a Tamil ethnic group who had been recently moved to 3,900 m from their habitual environment near sea level. A sample of people from the Tamil ethnic background living in Madras were used as the control group.

In their study, the authors considered the effect of high altitude on the water and mineral composition of the body and measured various parameters (weight, hydrated mass, percentage of fat, minerals). They also carried out anthropometric and physiological determinations (oxygen consumption, vital lung capacity, red cell traits and plasma proteins). The haematological data gathered from that study will be the only ones considered in this section (Table 7.4).

The observed values for red blood cell count were significantly higher for the newcomers to altitude and the natives born at altitude than they were for the individuals living at sea level ($P < 0.05$). The haematocrit per cent was significantly higher in the Tamil who had stayed at 3,692 m for ten months ($P < 0.05$) than in subjects of the same ethnic group living at sea level. On the other hand, the haematocrit per cent values of Ladakh natives, although higher, were not significantly different from those found for the groups living in Madras.

The authors stress the modifications of the haematological equilibrium in the subjects transferred to high altitude noting that rises in the erythrocyte concentration in haemoglobin and increased circulating red cell counts are part of the human organism's adaptive response to chronic hypoxia.

### Studies reported from Latin America

The two American continents are transected from north to south by a mountainous backbone which has had a basic effect on the physical environment and provided unusual conditions of human settlement. In North America the very high mountainous zones have remained practically uninhabited, whereas in Central and South America the greatest density of population is to be found in higher altitude zones. It is, therefore, not surprising to find that the largest quantity of research on the effects of altitude has been carried out in this part of the world.

Garruto (1973) has recently made an exhaustive bibliographical review and a thorough analysis on which we have heavily relied for this presentation.

### 'Altiplanides' – natives of altitude

The data included under this title refer to populations from the Andean plateau. These are groups where the historical and cultural data indicate long-term residence at constant altitude (Department of Puno in Peru, Departments of La Paz and Oruro in Bolivia). The findings (Quilici, 1968; Hurtado, 1971; Moulin, 1971; Gourdin *et al.*, 1972; Whittembury & Monge, 1972; Garruto, 1973) lead us to suggest that significant polycythaemia exists in these groups only under pathological conditions, i.e. when disease conditions which affect pulmonary ventilation are also present. In support of this conclusion, it is interesting to note that the increase in the number of red cells as the subjects get older is clearly less noticeable in the individuals from Puno than in those of the other two groups studied by Whittembury & Monge (1972) (Fig. 7.6). Although erythropoiesis has not been extensively studied among these 'Altiplanides', it appears that the results are much the same as those observed at sea level (Reynafarje, Ramos, Faura & Villaviencio, 1964; Ruffie, Larrouy & Vergnes, 1966; Gourdin *et al.*, 1972). Such an observation agrees very well with a number of others made in the various aspects of haematology which will be presented in the present chapter.

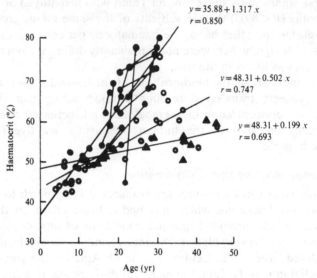

Fig. 7.6. Mass plot of haematocrits as a function of age. ●, Morococha 4,500 m; ○, Cerro de Pasco 4,200 m; ▲, Puno 3,800 m.

In our conclusion, we shall see the importance of the 'Altiplanides' populations, the only ones, in our opinion, genetically adapted to altitude.

### 'Andides' – recent natives or ones living at varying altitudes

These groups of people have been much more extensively studied owing to the high percentage of such individuals living at high altitude. The results of these studies universally (Garruto, 1973) indicate high values for all the criteria of erythaemia: high counts in the number of red cells, high values for haemoglobin, large percentages for haematocrit, etc.

As Reynafarje (1966a) stresses, these alterations are proportional to altitude. However, there may be two caveats to this observation.

(1) Ancient human settlements are rarely situated above 4,200 m and the great majority of the populations living at high altitude is found around 4,000 m or below.

(2) The comparative data on populations living on lowlands or transferred from altitude to lowlands both show much lower red cell counts. At least some authors (Moulin, 1971; Garruto, 1973) suggest that much of this difference reflects the anaemia often present in tropical areas and the below-normal values which occur in highlanders moving to the lowlands.

Whittembury, Lozana, Torres & Monge (1968) demonstrated that blood viscosity is proportional to haematocrit but suggested that this finding

200

Table 7.5. *Methaemoglobin levels in normal, anaemic and polyglobulic subjects in two high-altitude samples*

| | | Polyglobulics Hb > 21 g/100 | t-test | Normal subjects 21 < Hb < 14 | t-test | Anaemic subjects Hb < 14 g/100 |
|---|---|---|---|---|---|---|
| Quechua (208) | N | 1 | | 185 | | 22 |
| (Peru) | m | | | 5.31% | $P < 0.0005$ | 10.92% |
| 3,500 m | S.E. | | | 0.15 | | 1.36 |
| t-test | | | | $P < 0.05$ | | $P < 0.05$ |
| Metis-Aymara (71) | N | 12 | | 19 | | 40 |
| (La Paz) | m | 2.07% | $P < 0.0005$ | 3.66% | $P < 0.0005$ | 9.31% |
| 3,500 m | S.E. | 0.29 | | 0.46 | | 0.81 |

*m*, mean of haemoglobin rate (%).

has limited haemodynamic consequences. In general, at altitude, the mechanism producing polycythaemia is an increase in the secretion of erythropoietin stimulated by anoxia at the tissue level (Reynafarje, 1966a; Carmena & Rivas, 1971; Garruto, 1976). The very few studies on haematopoiesis corroborate that erythropoietin acts on the bone marrow and thereby stimulates erythropoiesis (Merino & Reynafarje, 1949; Huff *et al.*, 1951; Reynafarje, 1966b; Sanchez, Crosby & Merino, 1966; Faura *et al.*, 1969). The red cell formation thus initiated also acts to increase iron absorption by the intestine (Reynafarje, Lozano & Valdiviezo, 1959; Villaviciencio & Reynafarje, 1966).

When 'Andides' are moved to the lowlands a decrease in the section of erythropoietin and in the absorption of iron has been found (Carmena *et al.*, 1968). Another factor of possible consequence at altitude is the frequent increase in methaemoglobin. Values for methaemoglobin are elevated in anaemic individuals but are normal in polycythaemic subjects (Ruffie, Vergnes & Hobbe, 1966; Gourdin *et al.*, 1972) (Table 7.5). Here again, in spite of rather poor samples, there appears to be a marked difference between the true natives and the others. The 'Andides' show a greater variance in the levels of residual methaemoglobin, a process which, according to the authors, reveals the existence of an adaptation at the molecular level. Such an adaptation would result from the presence of a more active system of diaphorasis in the 'Altiplanides'.

Finally, it has been shown that the oxyhaemoglobin dissociation curve is shifted to the right owing to an increase in the erythrocyte 2,3-diphosphoglycerate (2,3-DPG) (Eaton, Brewer & Grover, 1969; Lenfant, Torrance & Finch, 1969; Morpurgo *et al.*, 1970). This increase of the Bohr effect enables more oxygen to be freed at the tissue level. To our knowledge, it has never been studied in the 'Altiplanides'.

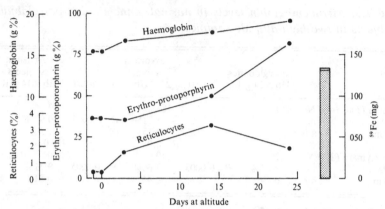

Fig. 7.7. Haemoglobin, reticulocytes, erythro-protoporphrin studies and iron utilization in lowland subjects, during the first days of exposure in altitude. Hatched column, haemoglobin synthesis; open column, other needs. From Reynafarje (1966*b*).

No modification of the haemoglobin structure has been reported in either the 'Andides' or the 'Altiplanides' (Cabannes & Schmidt-Beurrier, 1966; Morpurgo *et al.*, 1970; Arends, 1971; Sendrail & Quilici, 1972).

### Lowland natives living at high altitude

For this group reported results appear different from both previous groups. A haematological response appears in the first hours of exposure to decreased partial oxygen pressure. An initial response of plasma loss is observed which together with a peripheral vasoconstriction appears to spare the oxygen reserves of the more vital tissues (Buskirk *et al.*, 1966). Ringenbach (1973) previously stressed that under these conditions there were differences between the values obtained from the venous blood and the capillary blood.

There is a stimulation of erythropoietin secretion which occurs rapidly and the number of red cells increases in the first hours (Reynafarje, 1966*b*) (Fig. 7.7). This reaction period lasts about a fortnight, after which a state of balance, always characterised by erythaemia, seems to occur (Albrecht & Albrecht, 1970). Finally, a shift to the right of the haemoglobin dissociation curve has been found (Lenfant *et al.*, 1969; Morpurgo *et al.*, 1970; Lenfant, Torrance & Reynafarje, 1971), but this shift is less marked than in the natives of altitude. For the lowlanders moving to altitude no study of 2,3-DPG values have been reported.

A more thorough analysis of the reported shifts in the oxygen dissociation curve will be presented later in this chapter.

Table 7.6. *Haematological values for three North American samples*

| Group | N | Whole blood haemoglobin (g% ±1 s.d.) | Haematocrit (±1 s.d.) |
|---|---|---|---|
| **Men** | | | |
| Ann Arbor | 100 | 15.22 (±1.14) | 46.67 (±2.81) |
| Leadville | 119 | 18.02 (±1.35) | 51.74 (±3.54) |
| Leadville polycythaemic | 44 | 19.79 (±1.79) | 60.03 (±4.84) |
| **Women** | | | |
| Ann Arbor | 59 | 13.56 (±0.94) | 42.40 (±2.44) |
| Leadville | 34 | 15.79 (±1.61) | 46.24 (±3.24) |
| Leadville polycythaemic | 5 | 19.32 (±1.57) | 57.80 (±5.58) |

From Eaton *et al.* (1969, p. 605).

### Studies reported for North America

The observations of North American authors on the inhabitants of some villages in the Rocky mountains will be reviewed in this section. In 1969, Eaton *et al.* published haematological data on the inhabitants of a Colorado village (Leadville), altitude approximately 3,000 m. The aim of the study was to analyse the relationship between intra-erythrocyte 2,3-DPG level and altitude polycythaemia.

The subjects were adults of both sexes (150 men and 59 women) who had been living in the Leadville region for several months. They were divided into a polycythaemic group and a non-polycythaemic group according to their haematocrit level. Haematocrit values of 58% for the men and 53% for the women were considered as the normal upper limits for that altitude.

Table 7.6 summarises the observed variations of the rates of haemoglobin and the haematocrit in the inhabitants of Leadville compared with a control group living near Detroit (Ann Arbor).

Clearly the haemoglobin and haematocrit values of the polycythaemic subjects were significantly higher than those of the normal subjects living at high altitude or at low altitude.

The differences between normal subjects at altitude and the inhabitants of Ann Arbor were not significant. The authors specify that all subjects were whites.

If we examine the distribution of polycythaemic individuals according to their ages, we may notice that, as Brewer & Eaton (1970) point out, the phenomenon rarely appears in persons under 30 years of age but shows an increasing frequency with age. In Leadville 5% of the men over 35 years were polycythaemic. Their haematocrit level was 56%, and in some instances even higher and reaching remarkable values of 70% or even 80%.

Working from these data, Eaton & Brewer suggest a tentative explanation for the polycythaemia found at Leadville. They postulate that a significant factor in the origin of the exaggerated increase in the number of circulating red cells is related to the proportion of intra-erythrocyte 2,3-DPG. In polycythaemic subjects, the level of 2,3-DPG is significantly higher than in the normal subjects living at the same altitude. The correlations between rise in 2,3-DPG and polycythaemic reaction will be discussed later.

Eaton, Weil & Grover (1970) studied the haemoglobin dissociation curve in a few polycythaemic individuals, and to their surprise found that in spite of a high 2,3-DPG concentration, there was no perceptible shift of the curve to the right. To explain this unexpected finding the authors suggested that there were other significant factors contributing to polycythaemia, including respiratory disorders. In fact, out of twenty-four polycythaemic subjects in the group of Leadville residents, four individuals were affected with silicosis and emphysema. Of the remaining twenty, for whom no pathological cause could be identified, Eaton *et al.* (1970) suggested the possible influence of the carbon monoxide contained in cigarette smoke. All of the residual polycythaemics were habitual smokers.

It is indeed the case for smokers that the carbon monoxide works on the respiratory function of haemoglobin through three series of mechanisms: (1) it transforms a certain fraction of haemoglobin into non-functional carboxy-haemoglobin: (2) it makes the haemoglobin dissociation curve shift to the left: (3) it reduces the oxygen saturation of the arterial blood. These elements put together will result in an erythropoietic stimulation. Chronic poisoning by carbon monoxide must therefore be considered as being one of the factors that may start and aggravate polycythaemia in high-altitude residents.

### Other haematological data

Red-cell modifications are not the only haematological observations made in relation to altitude although they are the ones most often emphasised.

#### White cells

We have already noted that most studies of white-cell changes in relation to altitude have not produced reproducible results (Chiodi, 1950; Anduaga & Ramos, 1967; Ramos, Anduaga, Reynafarje & Lopez, 1971; Garruto, 1973, 1976).

For the native of altitude living in the lowlands, which, it must be borne in mind, are tropical lowlands, major modifications are observed (Ramos *et al.*, 1971; Moulin, 1971), particularly a significant rise in eosinophilic counts. Anduaga & Ramos (1967) reported an increase in the white cell

204

count of lowland natives when they are exposed to altitude. This elevation was only for the first 8 days after which a slight monocytosis remained.

### Platelets, haemostasis

At the level of the platelets, on the other hand, and contrary to some assertions (Reynafarje, 1966b), highly significant differences in relation to altitude have been observed.

Thrombo-embolic affections are known to be quite exceptional in the natives of altitude (Caen *et al.*, 1973). Starting from this observation, the coagulating factors have been systematically examined. The first unusual aspect found at high altitude was the existence of a thrombo-cythaemia (Quilici, Ruffie, Moulin & Berthier, 1969; Caen, Feingold & Michel, 1971b). Moreover, the exploration of the various stages of haemostasis has revealed an equilibrium quite noteworthy for its normality (Figallo, Vidurrizaga & Merino, 1964; Ergueta & Vidaurre, 1967; Ergueta & Rodriguez, 1972).

The study of platelet aggregation under the influence of certain agents (ADP, collagen, bovine fibrinogen) has shown a normal degree of aggregation with ADP (Caen *et al.*, 1971a); an increased degree of aggregation with collagen and a decreased one with bovine fibrinogen. On the other hand, if thrombo-cythaemia is associated with the altitude native the functional responses of the platelets seem to be due to the living conditions, particularly to nutritional factors, instead of hypoxia as such. Caen *et al.* (1973) draw attention to the fact that most lipids and proteins in the native came from plants and that the caloric intake proved normal.

Although no epidemiological investigation has been made on natives from lowlands living at altitude, experience suggests that for these individuals thrombo-embolic accidents are frequent. This may be related to the circulatory effects of the polycythaemia found among the newcomers and the reported thrombo-cythaemia (Figallo, Merino, Yuen & Sanchez, 1966; Caen *et al.*, 1973).

Concerning fibrinolysis, a few authors have described a certain tendency to elevation in the fibrinolytic process but the results remain difficult to interpret (Ergueta & Rodriguez, 1972; Caen *et al.*, 1973; Chohan, Singh & Balakrishnan, 1974).

### Plasma modifications and the blood ionogram

We have already stressed that the altitude and length of exposure appear to affect the plasma volume (see also Dill *et al.*, 1969; Sanchez, Merino & Figallo, 1970). A few recent studies have also suggested that there may be other altitude associations with blood components which require further exploration (Cazorla, 1962; Picón-Reátegui, 1966; Picón-Reátegui,

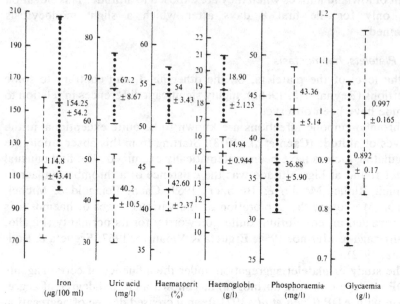

Fig. 7.8. Variations of various blood parameters in subjects native to altitude (La Paz, .....) and in 100 subjects transferred to the lowlands (Santa Cruz, ---).

Buskirk & Baker, 1970; Garmendia, Arroyo & Muro, 1970; Constans, Freminet & Constans, 1972; Henrotte *et al.*, 1972). Constans *et al.* (1972) studied two samples of 100 individuals each of the same sex and the same age group. One group lived in La Paz (3,600 m) while the other resided in Santa Cruz (400 m). They observed a number of differences as summarised in Fig. 7.8.

The variations in the haemoglobin, haematocrit and plasma serum iron have been previously discussed. But it should also be noted that glycaemia and phosphoraemia levels were lower in the subjects living at high altitude. The differences may well be connected with the modifications in energy metabolism, and further study of the topic is desirable (Picón-Reátegui, 1966).

In Fig. 7.8 it may also be noted that the uric acid shows a very high value in the altitude subjects. The authors believe that these elevated levels may be evidence of a more rapid degradation of nucleic substances during accelerated erythroformation or alternatively the results of a quicker turnover in the circulating erythrocytes. The latter might be the case since in their normal state the erythrocytes contain appreciable quantities of uric acid resulting from the degradation of purine nucleotides.

With regard to the variations in the blood ionogram, an hypokaliaemia

206

and an hyponatraemia have been observed in the native of altitude as well as in the native of the lowlands when transferred to altitude, but whereas the latter very quickly recovers normal rates when returned to low altitude, the native of altitude keeps those values after moving to the lowlands (Henrotte & Ruffie, 1969; Lozano *et al.*, 1969; Henrotte, Depraitere & Ruffie, 1970; Zeballos, Galdos & Quintanilla, 1973).

At altitude erythrocytes show a very high magnesium concentration, whereas their plasma concentration does not vary with altitude (Henrotte *et al.*, 1972). The authors consider that the high level of magnesium in the cells is a 'reinforcing of the respiratory enzymatic equipment of the cells' and believe it has an adaptive value against anoxia. Such an interpretation may also pertain to the findings made in the field of haemostasis, where the bivalent ions play a well-known activity role.

### Immune defences

There is a rather obvious decline in infectious agents with altitude (Clegg, Harrison & Baker, 1970). It therefore seems fundamental from a theoretical as well as practical point of view to study the relations between immunological defences and altitude. In high-altitude populations high levels of $\gamma$-globulins have been reported, even for newborn babies (Ruffie *et al.*, 1967; Bocanegra *et al.*, 1971). However, Ruffie & Larrouy (1966), in the group of subjects they studied, showed that the $\gamma$-globulin level in the 'Altiplanides' did not vary when they were transferred to the tropical

Table 7.7. *Proteinogram of two ethnic groups from high altitude before and after transfer to the lowlands*

| | N | Protein level (g/l) | Albumin | $\alpha1$ (%) | $\alpha2$ (%) | $\beta$ (%) | $\gamma$ (%) |
|---|---|---|---|---|---|---|---|
| Quechua (Andides) living on *altiplano* | 60 | 64.04 | 38.52 | 4.22 | 10.92 | 17.75 | 25.63 |
| S.E. | | | 6.32 | 1.23 | 2.69 | 3.26 | 5.41 |
| Quechua (Andides) transferred | 60 | 70.52 | 38.77 | 3.20 | 9.20 | 20.38 | 28.45 |
| S.E. | | | 3.55 | 0.88 | 2.03 | 4.92 | 4.20 |
| Aymara (Altiplanides) living on *altiplano* | 60 | 70.48 | 48.83 | 2.11 | 10.08 | 13.79 | 25.18 |
| S.E. | | | 6.45 | 0.69 | 3.33 | 4.00 | 6.34 |
| Aymara (Altiplanides) transferred | 60 | 70.36 | 48.02 | 4.09 | 9.47 | 12.90 | 25.51 |
| S.E. | | | 5.92 | 1.36 | 1.53 | 3.86 | 4.18 |
| Mocetenes (lowland natives) | 60 | 70.65 | 40.13 | 4.81 | 10.71 | 17.34 | 27.01 |
| S.E. | | | 7.75 | 1.50 | 2.83 | 7.35 | 7.06 |

lowlands (Table 7.7), whereas it increased appreciably in the 'Andides' undergoing the same change. It seems then that permanent living at altitude has caused a loss in the capacity for immunological adaptation when 'Altiplanides' face an environment richer in microbial agents.

### Cytogenetics

Some sixty karyotypes were studied in individuals from Puno and La Paz (Malaspina, Quilici & Ergueta, 1970). No chromosomic anomalies were found. Hypoxia caused a slowing of mitosis when cell growth was studied by means of in-vitro cell cultures at high altitude. However, the observed slowing down of the mitoses does not seem to take place *in vivo*, at least in healthy subjects.

### Bohr effect and altitude

Numerous studies have been concerned with the blood pH and the analysis of the blood gases, but these are outside the scope of this chapter. However we would like to try to make a synthesis of the present state of our knowledge concerning a problem which undoubtedly represents one of the most fundamental aspects of adaptation to altitude: namely the Bohr effect and its relationship to altitude.

By a process known as the Bohr effect, the position of the oxygen dissociation curve (ODC) is known to depend on the intra-erythrocyte pH (Poyart, Bursaux & Freminet, 1972; Rørth, 1972). Thus there is a variation in the affinity of the haemoglobin for oxygen when the pH in the environment changes. At a pH greater than 6 the loss of an oxygen molecule results in the $H^+$ ions linking with the haemoglobin molecule. The molecule of deoxygenated haemoglobin in its normal state thus bonds the ester phosphate of the 2,3-DPG ($B_{141}$) which is strongly dissociated in the intra-cellular environment. The total process is properly called the alkaline Bohr effect. Below pH 6, an opposite reaction has been observed experimentally, but such a change probably does not occur naturally.

Studies concerning the Bohr effect in man living in altitude are relatively limited. The fundamental mechanisms have been known since the standard works of Lenfant *et al.* (1968). Studies carried out on normal adults transferred to 4,500 m (at Morococha, in central Peru) showed the existence of a constant relation between three parameters: the values of the $P_{50}$ (the partial pressure of oxygen required for 50% saturation of haemoglobin), the pH and the 2,3-DPG rate. During the first hours following their arrival the subjects showed an elevation of the $P_{50}$ of 50% above the usual value at sea level, and a tendency for the blood pH to increase as a consequence of respiratory alkalosis. Along with the change, the haematometric constants were altered (cell count, haemoglobin concentration) and the red cells showed a significantly greater

208

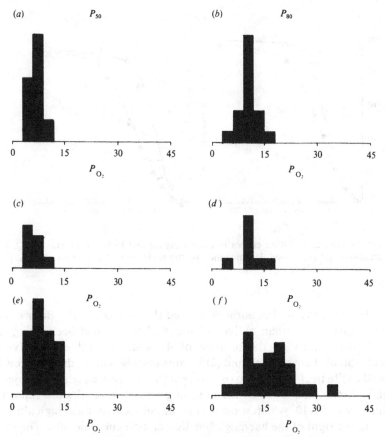

Fig. 7.9. Distribution of Bohr effect in European and Peruvian Indians. (*a*), (*b*), Europeans at sea level; (*c*), (*d*), Europeans above 3,500 m; (*e*), (*f*), Peruvian Indians above 3,500 m. From Morpurgo *et al.* (1970).

amount of DPG. According to Lenfant, the 2,3-DPG rate altered quickly (during the first 36 h of exposure to altitude hypoxia) and the change was of significant proportions (an increase from 27 to 38%). The changes in this metabolite are the product of the combined effects of the elevation of the blood pH resulting from the desaturation of the haemoglobin and from the respiratory alkalosis. During the experiments carried out at Morococha, Lenfant and others also created an experimental acidosis in two subjects of the same group by administering acetazolamide. This procedure prevented the elevation of the blood pH. In these two individuals, Lenfant *et al.* did not observe any increase of the 2,3-DPG level.

Studies of the Bohr effect at altitude have also been carried out in the Andes (at Cerro de Pasco, in Peru) and in Nepal by Morpurgo *et al.* (1970,

209

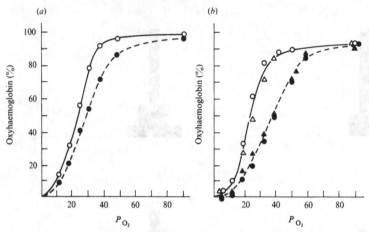

Fig. 7.10. Oxygen dissociation curves in Europeans (*a*) and Peruvian Indians (*b*). ○, first determinations and △, second determination on the same haemolysate. From Morpurgo *et al.* (1970).

1972). In Peru, the Italian authors studied the changes in the Bohr effect on: (1) a group of Indian natives of the Andes who had been living for several generations at the altitude of 4,350 m (a total of twenty-six subjects, all of them adults), and (2) Europeans (eight individuals) recently adapted to life in chronic hypoxia. Morpurgo *et al.* chose eighteen subjects living in Rome (sea level) as the control group. The results are reported in Figs. 7.9 and 7.10, which show that in the subjects living at high altitude a shift to the right of the haemoglobin dissociation curve occurs. The shift occurred at pH values of 7.4 and 6.7; at the latter pH, however, the mean values of $P_{50}$ and $P_{80}$ of the subjects of European origin adapted to altitude are lower than those of the Indian natives of the Andes. Thus the Amerindians had a higher Bohr effect than the Europeans.

This physiological difference, perhaps due to a change in the affinity of the haemoglobin molecule for its 'ligands' (including 2,3-DPG) would result in an easier release of oxygen to the tissues. Such a modification would constitute a highly interesting adaptive mechanism. However, when a similar study was conducted by the same team in Nepal they did not encounter the same results (Morpurgo *et al.*, 1972). The Italian authors studied groups of Sherpas (twenty-four porters) living in the region of Solo Khumbu at an altitude of 4,950 m. The members of the Italian expedition (twelve subjects) were studied at the same altitude. At low altitude (i.e. 1,200 m) Sherpas from Kathmandu were measured as a control group. The principal measurements analysed in this study are summarised in Table 7.8.

Table 7.8. The results of Bohr effect studies in Nepal

| Altitude | P50 pH 7.4 | | P50 pH 6.8 | | Bohr effect | | g Hb/100 ml of blood | | 2,3- DPG ($\mu$mol/ml of blood) | |
|---|---|---|---|---|---|---|---|---|---|---|
| | Low | High | Low | High | Low | High | Low | High | Low | High |
| *Means* | | | | | | | | | | |
| Europeans | 22.23 | 21.67 | 28.69** | 28.92 | 6.46 | 7.25 | 16.87**** | 16.10** | 2.16**** | 3.52**** |
| Sherpas | 21.62 | 21.25 | 27.38** | 27.42 | 5.77 | 6.17 | 14.16**** | 16.76**** | 2.97**** | 3.63 |
| *Variances* | | | | | | | | | | |
| Europeans | 1.19** | 2.97** | 2.56** | 8.27** | 2.10** | 6.02** | 0.28**** | 1.50** | 0.041**** | 0.441**** |
| Sherpas | 3.59** | 2.93 | 1.92 | 4.63 | 1.03 | 1.97** | 4.06**** | 2.45 | 0.092** | 0.441*** |

Levels of significances of the comparisons:

   *$P \leqslant 0.1$;
   **$P \leqslant 0.05$;
   ***$P \leqslant 0.01$;
   ****$P \leqslant 0.001$. The probabilities have been reported only for the comparisons with statistically significant (or almost so) differences.

*The biology of high-altitude peoples*

The mean values of $P_{50}$ at pH 6.8 do not differ significantly. The authors also observed that the erythrocyte 2,3-DPG levels were normal. The discrepancy between the two series of experiments performed by the same laboratory group, with absolutely identical techniques, is difficult to explain.

Finally, the results published by Weikopf & Severinghaus (1972) at the Barcroft Laboratory (3,790–4,300 m) in California should be noted. The authors studied sixteen normal adults at the altitude of the laboratory for 10 days in succession. Tests made at very short intervals (6, 12, 24, 36, 48, 68 h and so on) showed no significant changes in the values of $P_{50}$, and few variations in the pH. Weikopf does not provide any explanations for these rather unexpected results.

## Conclusion

This review of the studies carried out to the present day in the field of haematology at high altitude enables us to draw a certain number of conclusions but, above all, opens up new prospects for future research.

The Andes, more than any other region in the world, has made it possible to show that populations exist that are genetically adapted to altitude. Their high biological specialisation, very probably irreversible, may account for the 'Altiplanides' failing in their attempts to establish settlements in the tropical lowlands. The confusion too frequently made between those we have called 'Altiplanides' and 'Andides' is still the principal cause of the difficulties encountered when interpreting the published results. The short-term effects of hypoxia seem to be clearly defined, but the longer-term reactions to hypoxia are still not well measured and we thus have only a very fragmentary knowledge of what the effects may be.

The multiple variables associated with altitude, the different environments, and above all the different historical and cultural situations, have conditioned distinct adaptive responses. In the true resident native at altitude, whose genetic stock has been able, through the generations, to retain the mutations most favourable to life at high altitude, the metabolic chains, those of the red cells in particular, have adapted to a more efficient use of the limited amount of ambient oxygen.

In contrast, the natives of low altitude who find themselves at altitude avail themselves of faster but less elaborate and more 'expensive' processes to ensure sufficient oxygenation of their tissues. These processes expose them to acute pathology, often serious and sudden.

Finally, in individuals who have been living at altitude for moderate time periods, but long enough for them to have modified their genetic stocks, various reactions appear. The price of these often seems to be some chronic pathology. These pathologies are very probably the factor that

212

makes possible, in future generations, the selection of genes most favourable to life at high altitude.

### References

Albrecht, E. & Albrecht, H. (1970). Altérations métaboliques et études hématologiques à des altitudes allant jusqu'à 6,200 mètres. Possibilités de traitement. *Agressologie*, **11**, 163.

Albrecht, H. P. & Littell, K. J. (1972). Plasma erythropoietin in men and mice during acclimatization to different altitudes. *Journal of applied Physiology*, **32**, 54–8.

Anduaga, G. & Ramos, J. (1967). Variaciones leucocitarias en la exposición aguda à las grandes alturas. *Archivos del Instituto de Biologia Andina*, **2**, 75–85.

Arends, T. (1971). Hemoglobinopathies and enzyme deficiencies in Latin American populations. In *The ongoing evolution of Latin American populations*, ed. F. M. Salzano, pp. 509–59. Springfield, Ill.: C. C. Thomas.

Barcroft, J. (1925). *The respiratory function of the blood*, Part I, 'Lessons from high altitudes'. London: Cambridge University Press.

Bernard, J. & Ruffie, J. (1966). *Hématologie Géographique*, p. 13 and pp. 22–9. Paris: Masson et Cie.

Bharadwaj, N., Singh, A. P. & Malhotra, M. S. (1973). Body composition of the high altitude natives of Ladakh. A comparison with sea level residents. *Human Biology*, **45**, 423–34.

Bocanegra, M., Botto, O., Ortiz, E., Medina, A., Carpio, M., Izaguirre, C., Laos, C., Velarde, I., Marticorena, E. & Paredes, A. (1971). Proteínas séricas. Estudio en madres y recién nacidos de la altura. *Archivos del Instituto de Biologia Andina*, **4**, 44–53.

Buskirk, E. R., Kollias, J., Picón-Reátegui, E., Akers, R., Prokop, E. & Baker, P. T. (1966). Physiology and performance of track athletes at various altitudes in the United States and Peru. In *The effects of altitude on physical performance*, ed. R. F. Goddard, pp. 65–72. Albuquerque, N.M.: Lovelace Foundation.

Cabannes, M. R. & Schmidt-Beurrier, A. (1966). Recherches sur les hémoglobines des populations indiennes de l'Amérique du Sud. *L'Anthropologie, Paris*, **70**, Parts 3–4.

Caen, J. P., Drouet, L., Bellanger, R., Michel, H. & Henon, P. (1973). Etude sur les hauts plateaux andins de la corrélation entre thromboses artérielles et veineuses, fonctions plaquettaires, fibrinolyse à l'anoxie et régime alimentaire.

Caen, J. P., Ergueta, J., Michel, H., Daufresne, A., Poupart, C. & Dhuime, G. (1971a). Modifications de l'agrégation plaquettaire a l'adénosine diphosphate (ADP) chez les Boliviens et les Peruviens de l'altiplano. Etude statistique. *Comptes rendus des séances de l'Académie des sciences, Paris*, **272D**, 505–8.

Caen, J. P., Feingold, N. & Michel, H. (1971b). Plaquettes et environnement. In *Pre-adaptation et adaptation genetique*, ed. C. Bouloux & J. Ruffie, pp. 133–40. Paris: Institut National de la Santé et de la Recherche Médicale.

Carmena, A. O. & Rivas, E. (1971). Urinary excretion of delta-aminolaevulinic acid and phosphobilinogen in young natives of 13,000 feet brought to sea level. *Haematologia, Budapest*, **5**, 3.

Carmena, A. O., Segade, A., Cavagnaro, F. J. & DeTesta, N. G. (1968). Changes

213

in iron metabolism in natives of 13,000 feet brought down to sea level. *Nature, London*, **217**, 70–1.

Cazorla, A. (1962). Some biochemical aspects of the haematie of high altitude adapted man. Report, National Institutes of Health Grant No. 08576.

Chiodi, H. (1950). Blood picture at high altitude. *Journal of applied Physiology*, **2**, 431–6.

Chohan, I. S., Singh, I. & Balakrishnan, K. (1974). Fibrinolytic activity at high altitude and sodium acetate buffer. *Thrombosis and Diathesis haemorrhagica*, **32**, 65.

Clegg, E. J., Harrison, G. A. & Baker, P. T. (1970). The impact of high altitudes on human populations. *Human Biology*, **42**, 486–517.

Constans, J., Freminet, A. & Constans, H. (1972). Etude comparative de divers paramètres biologiques chez les populations Amérindiennes vivant en altitude puis transplantées en basses terres. *Comptes rendus des séances de l'Académie des sciences, Paris*, **274D**, 1832–4.

Dill, D. B., Horvath, S. M., Dahms, E. T., Parker, E. R. & Lynch, R. J. (1969). Hemoconcentration at altitude. *Journal of applied Physiology*, **27**, 514–18.

Dollfus, O. (1973). La cordillère des Andes. Présentation des problèmes géomorphologiques. *Revue de Geógraphie physique et de Géologie dynamique*, **15**, 157–76.

Dollfus, O. & Lavallee, D. (1974). *Ecologie et occupation de l'espace dans les Andes*. Lima: L'Institut Français d'Etudes Andines.

Eaton, J. W., Brewer, G. J. & Grover, R. F. (1969). Role of red cell, 2-3 diphosphoglycerate in the adaptation of man to altitude. *Journal of laboratory and clinical Medicine*, **4**, 603–9.

Eaton, J. W., Weil, J. & Grover, R. (1970). Studies of red cell glycosis and interactions with carbon monoxide, smoking and altitude. In *Red cell metabolism and function*, ed. G. J. Brewer, pp. 95–114. New York: Plenum Press.

Ergueta, C. J. & Rodriguez, A. (1972). Estudio de la fibrinolysis en habitantes à 3,700 m de altura. *Bolletin del Institutio Boliviano de Biologia de Altura*, **16**, vol. 4, No. 1.

Ergueta, C. J. & Vidaurre, N. J. (1967). Estudio de la coagulacion de la sangre en habitantes à 3,600 m y 4,000 m de altura sobre el ni vel del mar. *Cuadernos del Hospital General*, La Paz.

Faura, J., Ramos, J., Reynafarje, C., English, E., Finne, P. & Finch, C. A. (1969). Effect of altitude on erythropoiesis. *Blood*, **33**, 668–76.

Figallo, M. A., Merino, C. F., Yuen, C. A & Sanchez, C. (1966). Blood platelets studies in high altitude hypoxia. *Clinical Research*, **14**, 435.

Figallo, M. A., Vidurrizaga, C. M. & Merino, C. F. (1964). Studies of human blood coagulation in high altitude. Presence of fibrinolytic activity in some cases. *Xth Congress of the International Society of Haematology*, Stockholm.

Garmendia, F., Arroyo, J. & Muro, M. (1970). Glicemia del nativo normal de altura. *Archivos del Instituto de Biologia Andina*, **3**, 209–16.

Garruto, R. M. (1973). Polycythemia as an adaptive response to chronic hypoxic stress. Doctoral Dissertation in Anthropology, The Pennsylvania State University.

Garruto, R. M. (1976). Hematology. In *Man in the Andes: A Multidisciplinary Study of High Altitude Quechua*, ed. P. T. Baker & M. A. Little, pp. 261–82. Stroudsburg, Pa.: Dowden, Hutchinson & Ross.

Gourdin, D., Vergnes, H., Constans, J., Quilici, J. C. & Guttierez, N. (1972). La ferrihemoglobina en el hombre que vive à grandes alturas. *Bolletin del Instituto Boliviano de Biologia de Altura*, **19**, vol. IV, No. 4.

Haldane, J. (1926). Acclimatization to high altitude. *Nature, London,* **118,** 702–3.

Harrison, G. A., Küchemann, C. F., Moore, M. A. S., Boyce, A. J., Baju, T., Mourant, A. E., Godber, M. J., Glasgow, B. G., Kopeć, A. C., Tills, D. & Clegg, E. J. (1969). The effects of altitudinal variation in Ethiopian populations. *Philosophical Transactions of the Royal Society of London,* **256B,** 147–82.

Harrop, G. (1922). The relation of the diffusion constant to mountain sickness. *Proceedings of the Society of Biology and Medicine.*

Henrotte, J. G., Constans, J., Constans, H., Bisseliches, F. & Coudert, J. (1972). Le magnésium érythrocyataire et plasmatique des populations amérindiennes du corridor interandin. *Archives Internationales de Physiologie et de Biochimie,* **80,** 941.

Henrotte, J. G., Depraitere, R. & Ruffie, J. (1970). L'ionogramme sanguin des populations amérindiennes du corridor interandin et les processus adaptatifs. *Compte Rendus des Séances de la Société de Biologie,* **164,** No. 3.

Henrotte, J. G. & Ruffie, J. (1969). Sur les variations de la kaliémie des populations amérindiennes du corridor interandin. *Compte Rendus des Séances de l'Académie des Sciences, Paris,* **269D,** 339–41.

Huff, R. L., Lawrence, J. H., Siri, N. E., Wassermann, L. R. & Hermessy, T. G. (1951). Effects of changes in altitude on hematopoietic activity. *Medicine,* **30,** 197.

Hurtado, A. (1971). Acclimatacion à la altura. El hombre andino. *Primeras Jornadas de Medecina Aerospacial, E.M.G.,* Lima.

Lenfant, C., Torrance, J., English, E., Finch, C. A., Reynafarje, C., Ramas, J. & Faura, J. (1968). Effect of altitude on oxygen binding by hemoglobin and on organic phosphate levels. *Journal of clinical Investigation,* **47,** 2652–6.

Lenfant, C., Torrance, J. D. & Finch, C. A. (1969). The regulation of hemoglobin affinity for oxygen in man. *Transactions of the Association of American Physicians,* 82.

Lenfant, C., Torrance, J. D. & Reynafarje, C. (1971). Shift of the $O_2$ Hb dissociation curve at altitude. Mechanism and effect. *Journal of applied Physiology,* **30,** 625–31.

Lord, B. I. & Murphy, M. J. (1973). Hematopoietic stem cell regulation. II. Chronic effects of hypoxic-hypoxia on C.F.U. kinetics. *Blood,* **42,** 89–98.

Lozano, R., Torres, C., Marchena, C., Whittembury, J. & Monge, C. (1969). Response to metabolic (ammonium chloride) acidosis at sea level and at high altitude. *Nephron,* **6,** 102.

McDonald, T. P., Lange, R. D., Congdon, C. C. & Toya, R. E. (1970). Effect of hypoxia, irradiation and bone marrow transplantation on erythropoietin levels in mice. *Radiation Research,* **42,** 151.

Malaspina, L., Quilici, J. C. & Ergueta, C. J. (1970). Modifications des conditions de cultures cellulaires en haute altitude. *Compte Rendus des Séances de la Société de Biologie,* **164,** No. 5.

Merino, C. F. & Reynafarje, C. (1949). Bone marrow studies in the polycythemia of high altitude. *Journal of laboratory and clinical Medicine,* **34,** 637–47.

Monge, C. (1929). *Les Erythrémies de l'Altitude.* Paris: Masson et Cie.

Monge, C., Lozano, R. & Whittembury, J. (1965). Effect of blood-letting on chronic mountain sickness. *Nature, London,* **207,** 770.

Monge, M. C. & Monge, C. C. (1966). *Altitude diseases.* Springfield, Ill.: C. C. Thomas.

Morpurgo, G., Battaglia, P., Bernini, L., Paolucci, A. M. & Modiano, G. (1970).

215

Higher Bohr effect in Indian natives of Peruvian highlands as compared with Europeans. *Nature, London,* **227**, 387–8.

Morpurgo, G., Battaglia, P., Carter, N. D., Modiano, G. & Passi, S. (1972). The Bohr effect and the red cell 2-3 DPG and Hb content in Sherpas and Europeans at low and high altitude. *Experientia,* **28**, 1280–3.

Moulin, J. (1971). *Hematimetrie et cytologie en milieu tropical de l'Amérique du Sud.* Centre d'Hémotypologie CNRS.

Picón-Reátegui, E. (1966). Efecto de la exposición crónica a la altura sobre el metabolismo de los hidratos de carbono. *Archivos del Instituto de Biologia Andina,* **1**, 255–85.

Picón-Reátegui, E., Buskirk, E. R. & Baker, P. T. (1970). Blood glucose in high altitude natives and during acclimatization to altitude. *Journal of applied Physiology,* **29**, 560–3.

Poyart, C. F., Bursaux, E. & Freminet, A. (1972). Effet Bohr et affinité de l'hémoglobine pour l'oxygène. *Annales Biologie clinique,* **30**, 213–17.

Pugh, L. G. C. (1966). A programme for physiological studies of high-altitude peoples. In *The biology of human adaptability,* ed. P. T. Baker & J. S. Weiner, pp. 521–32. Oxford: Clarendon Press.

Quilici, J. C. (1968). *Les Altiplanides du corridor interandin.* Centre d'Hémotypologie du CNRS.

Quilici, J. C., Ruffie, J., Moulin, J. & Berthier, J. (1969). Variations de la thrombocythémie avec l'altitude dans les populations humaines de la région Andine. *Compte Rendus des Séances de l'Académie des Sciences, Paris,* **268D**, 2423–5.

Ramos, J., Anduaga, G., Reynafarje, C. & Lopez, F. (1971). Variaciones leucocytarias en sujetos nativos de las grandes alturas que descienden a nivel del mar. *Archivos del Instituto de Biologia Andina,* **4**, 38–43.

Reynafarje, C. (1966a). Physiological patterns: hematological aspects. In *Life at high altitudes,* pp. 32–9. Scientific Publication No. 140. Washington, DC: Pan American Health Organization.

Reynafarje, C. (1966b). La policitemia de las grandes altura. *Archivos del Instituto de Biologia Andina,* **1**, 142–50.

Reynafarje, C., Lozano, R. & Valdiviezo, J. (1959). The polycythemia of high altitudes: iron metabolisms and related aspects. *Blood,* **14**, 433.

Reynafarje, C., Ramos, J., Faura, J. & Villaviciencio, D. (1964). Humoral control of erythropoiesis activity in man during and after altitude exposure. *Proceedings of the Society for experimental Biology and Medicine,* **116**, 649.

Ringenbach, M. G. (1973). L'acclimatation hématologique à l'altitude. *Bordeaux medical,* **17**, 2769.

Rørth, M. (1972). Hemoglobin interactions and red cell metabolism. *Series Haematologica,* Vol. V, Munksgaard.

Ruffie, J. (1972). Leçon inaugurale. Chaire Anthropologie Physique. Collège de France, Paris, 59.

Ruffie, J., Ducos, J., Larrouy, G., Marty, Y. & Ohayon, E. (1967). Sur la fréquence élevée des anticorps anti-A et anti-B de haute titre dans les tribus amérindiennes. Rôle possible comme facteur sélectif. *Comptes Rendus des Séances de l'Académie des Sciences, Paris,* **264D**, 1792–5.

Ruffie, J. & Larrouy, G. (1966). Le problème de l'adaptation des populations de l'altiplano Bolovie dans les basses terres. *Journal de la Société des Américanistes,* Tome LV–LVI.

Ruffie, J., Larrouy, G. & Vergnes, H. (1966). Hématologie comparée des populations amérindiennes de Bolivie et phénomènes adaptatifs. *Nouvelle Revue d'Hématologie,* **6**.

Ruffie, J., Vergnes, H. & Hobbe, T. (1966). Sur la réversibilité de la méthémo-globinisation des hématies chez les populations indigènes du corridor inter-andin. Essai d'interprétation. *Compte Rendus des Séances de l'Académie des Sciences, Paris*, **262D**, 1956–8.

Sanchez, C., Crosby, E. & Merino, C. F. (1966). Bilirubinemia in the polycythemia of high altitude. *Proceedings of the Society for experimental Biology and Medicine*, **123**, 478–81.

Sanchez, C., Merino, C. & Figallo, M. (1970). Simultaneous measurement of plasma volume and cell mass in polycythemia of high altitude. *Journal of applied Physiology*, **28**, 775–8.

Schooley, J. C., Garcia, J. F., Cantor, L. & Havens, V. W. (1968). A summary of some studies on erythropoiesis using anti-erythropoietin immune serum. *Annals of the New York Academy of Sciences*, **149**, 266.

Sendrail, A. & Quilici, J. C. (1972). Etude des hémoglobines des habitants du corridor interandin. *L'Anthropologie, Paris*, **74**, No. 3–4.

Stohlmann, F. (1959). Observations of the physiology of erythropoietin and its role in the regulation of red cell production. *Annals of the New York Academy of Sciences*, **77**, 710.

Vellard, J. (1965). Principaux types raciaux des Andes de la Bolivie. *Compte Rendus des Séances de l'Académie des Sciences, Paris*, **261**, 227.

Viault, F. (1890). Sur l'augmentation considérable des globules rouges dans le sang chez les habitants des Hauts Plateaux de l'Amérique du Sud. *Compte Rendus des Séances de l'Académie des Sciences, Paris*, **112**, 295.

Villaviciencio, D. & Reynafarje, C. (1966). Determinacion del hierro y transferrina no saturada en sujetos nativos de altura y del nivel del mar. *Archivos del Instituto de Biologia Andina*, **1**, 319–23.

Weikopf, R. B. & Severinghaus, J. W. (1972). Lack of effect of high altitude on hemoglobin oxygen affinity. *Journal of applied Physiology*, **33**, 276–7.

Whittembury, J., Lozano, R., Torres, C. & Monge, C. (1968). Blood viscosity in high altitude polycythemia. *Acta Physiologica Latino-America*, **18**, 355–9.

Whittembury, J. & Monge, C. C. (1972). High altitude, haematocrit and age. *Nature, London*, **238**, 278–9.

Zeballos, J., Galdos, B. & Quintanilla, A. (1973). Plasma osmolality in subjects acclimatized at high altitude. *Lancet*, **7797**, 230–1.

**Additional bibliography on high-altitude haematology**

Bernard, J. & Ruffie, J. (1966). *Hématologie Géographique*, pp. 327 and 329. Paris: Masson et Cie.

Bouloux, C. J. (1968). *Contribution à l'étude des phenomènes pubertaires en très haute altitude*. Centre d'Hémotypologie du CNRS.

Delrue, G., Vischer, A. & Bouckart, J. P. (1933). Modifications du taux de glutathion sanguin durant le séjour en haute altitude. *Compte Rendus des Séances de la Société de Biologie*, **113**, 942.

Figallo, M. A., Bustamente, E. L., Guevara, E. H. & Aste-Salazar, H. (1969). Human hemoglobin in studies at high altitude. *Clinical Research*, **17**, 31.

Hurtado, A. (1932). Studies at high altitude: blood observations on the Indian natives of the Peruvian Andes. *American Journal of Physiology*, **100**, 487–505.

Hurtado, A., Rotta, A., Merino, C. & Pons, J. (1937). Studies of myohemoglobin at high altitudes. *American Journal of the Medical Sciences*, pp. 194–708.

Katsumoto, K. & Watanabe, S. (1970). Changes of blood cells at high altitude. *Japanese Circulation Journal*, **34**, 77–81.

Lawrence, J. H., Huff, R. L., Siri, W., Wasserman, L. R. & Hennessy, T. G. (1952). A physiological study in the Peruvian Andes. *Acta medica Scandinavica*, p. 142.

Lenfant, C. & Sullivan, K. (1971). Adaptation to high altitude. *New England Journal of Medicine*, **284**, 1305.

Mandelbaum, I. M. *et al.* (1973). Erythrocytes, enzymes and altitude. *Biomedicine*, **19**, 517–20.

Marticorena, E., Torrance, J. D., Lenfant, C. & Cruz, J. (1970–1). Oxygen transport mechanisms in residents at high altitude. *Respiratory Physiology*, **1181**, 15.

Mazess, R. B. (1970). Cardio-respiratory characteristics and adaptation to high altitude. *American Journal of physical Anthropology*, **32**, 267.

Merino, C. F. (1950). Studies on blood formation and destruction in the polycythemia of high altitudes, *Blood*, **5**.

Moncloa, F., Gomez, M. & Hurtado, A. (1965). Plasma catecholamines at high altitude. *Journal of applied Physiology*, **20**, 1329–31.

Moore, L. G., Brewer, G. J. & Oelshegel, F. J. (1972). Red cell metabolic changes in acute and chronic exposure to high altitude. *Advances in Experimental Medicine and Biology*, **28**, 397–413.

Palacios, *et al.* (1967). Hierro sérico en el recien nacido de la altura.

Peterson, R. F. & Peterson, W. G. (1935). The differential count at high altitude. *Journal of laboratory and clinical Medicine*, **20**, 723–6.

Ramos, J., Reynafarje, C. & Villavicencio, D. (1967). Protoporfirina libre de los hematies del recien nacido, al nivel del mar y en la altura. *Annales del Instituto de Biologia Andina*, **2**, No. 1.

Reynafarje, C. (1970). Control humoral de la eritropoiesis en la altura. *Archivos del Instituto de Biologia Andina*, Vol. 3, No. 5–6.

Reynafarje, C. (1971). Regulacion de la erythropoiesis durante la hypoxia y la hyperoxia. *Primeras Jornadas de Medecina Aerospacial*, E.M.G., Lima.

Rørth, M., Nygaard, S. F. & Parving, H. H. (1972). Effect of exposure to simulated high altitude on human red cell phosphates and oxygen affinity of hemoglobin. *Scandinavian Journal of clinical and laboratory Investigation*, **29**, 329.

Rørth, M., Nygaard, S. F., Parving, H. H., Hansen, V. & Kamsig, T. (1972). Effect of 2 hours exposure to simulated high altitude (4,500 m) on human red cell metabolism. *Scandinavian Journal of clinical and laboratory Investigation*, **29**, 321–7.

Scaro, J. L. & Guidi, E. E. (1970). Relationship between plasma erythropoietin activity and hematocrit ratio in high altitude residents. *Acta Physiologica Latino-Americana*, **20**, 281–3.

Singh, I., *et al.* (1972). Abnormalities of blood coagulation at high altitude. *International Journal of Biometeorology*, **16**, 283–97.

Smith, E. E. & Crowell, J. W. (1967). Role of an increased hematocrit in altitude acclimatization. *Aerospace Medicine*, **38**, 39–42.

Stohlmann, F., Rath, C. E. & Rose, J. C. (1954). Evidence for a humoral regulation of erythropoiesis. Studies on a patient with polycythemia secondary to regional hypoxia. *Blood*, **9**, 721–8.

Westergaard, H., Jarnum, S., Preisig, R., Ramsoe, K., Tauber, J. & Tygshup, N. (1970). Degradation of albumin and IgG at high altitude. *Journal of applied Physiology*, **28**, 728–32.

# 8. The food and nutrition of high-altitude populations

E. PICÓN-REÁTEGUI

A substantial portion of the world's population lives at high altitude. However, there is surprisingly little information on the effects, if any, of altitude on the nutritional status of human beings residing permanently at altitudes ranging from 2,500 to 5,000 m and more above sea level. Theoretically altitude, through its effects on factors such as soil and weather conditions, isolation, diminished atmospheric pressure, diminished partial pressure of oxygen and changes in the proportion of radiations of different wave length in the sun's rays could affect food and nutrient availability, food habits, food requirements, and, of course, nutritional status in these populations.

I will attempt to provide an overall picture of the subject. However, difficulties in reaching the world's literature dealing with problems of nutrition in high-altitude peoples compelled me to present views and observations based primarily on the Peruvian Andes.

## Food production

Food production is intimately tied to altitude, mediated by weather, watering and soil conditions. Environmental temperature varies with altitude. Little (1968) has reported a decrease of approximately 2.2 °C for each 500 m increase in altitude in the southern Peruvian Andes. The altitude of the permanent snow-line varies with latitude; thus the highest altitude at which food can be produced and at which human settlements can be established, varies also. In the Peruvian Andes, the largest settlements are located at altitudes ranging from 2,500 to 3,500 m above sea level. In this altitudinal zone, the soil is relatively fertile and weather conditions make possible the cultivation of a wide variety of cereals, tubers, legumes, vegetables and fruits. In the valleys sugar, coffee, cocoa and bananas are also cultivated. Besides agricultural products, the residents raise poultry and several types of livestock. Of these, cattle are the most important.

Above 3,500 m the soil is poor, thin and stony. From 4,000 to 4,800 m above sea level, the environmental temperature varies from 22 °C outdoors during the daylight hours, to below freezing during the early hours of the morning (Little, 1968). Only those plants which are resistant to frost, such as tubers and cereal-like goosefoot plants of the genus

219

*The biology of high-altitude peoples*

*Chenopodium*, are able to grow at these altitudes. The main activities at these highest altitudes are mining and pastoral herding of American camelids and sheep, agriculture being much less important.

Vegetation is almost absent above 4,800 m, though a few individuals are found working as shepherds up to 5,200 m in the southern Peruvian Andes, and, as temporary mine workers, up to 5,800 m in the Chilean Andes.

**Food availability**

The geographical and environmental conditions which characterize the mountainous regions of the world appear to mediate against an optimum diet for high-altitude communities. The availability of supplementary foods from other geographical zones is limited by the very low socio-economic status of the majority of these communities and the harshness of the terrain, which contributes to a lack of roads for a transportation network. Moreover, food production is meagre in quantity as well as in variety due to the fact that most of these communities continue to follow the same traditional peasant agriculture system for subsistence as their ancestors practiced several thousand years ago. Finally, soil and weather conditions tend to be more and more inadequate for agriculture, as altitude increases.

In fact, as can be seen in Table 8.1, twenty-one different foodstuffs were reported for a community at about 3,416 m above sea level (Collazos *et al.*, 1960), but only twelve for a community at about 4,000 m (Mazess & Baker, 1964). In other studies, Gursky (1969) reported that the natives of two towns situated at about 4,000 m above sea level, consumed, respectively, twenty-five and twenty-three different kinds of food products. In contrast only twelve were available in another town of the same zone, but situated at an altitude of about 4,300 m above sea level (see Table 8.2). Collazos *et al.* (1954) also remarked on the small number of foodstuffs available in those communities at high altitude in contrast to the forty or fifty items consumed by sea-level populations. However, it must be noted that the isolation of the settlement and the socio-economic level of the small communities studied play a role in food availability. This effect is obvious even in communities which are situated close to each other. For example, the three communities studied by Gursky (1969) belong to the District of Nuñoa, Province of Melgar, Department of Puno in the southern Peruvian Andes. The largest variety of foodstuffs available for consumption was recorded in the town of Nuñoa. This town is the capital of the district and there is transport by trucks to the railroad which communicates with the tourist center of Cusco and the commercial center of Juliaca. Most of the owners of the nearby haciendas and the political and ecclesiastical leaders of the district live in the town. Most of the town's population

220

Table 8.1. *Foods utilized (g/day) at four different levels of altitude*

| Products | Vicos[a] (2,747 m) (g/day) | Chacan[a] (3,416 m) (g/day) | Puno[a] (3,870 m) (g/day) | Nuñoa[b] (4,000 m) (g/day) |
|---|---|---|---|---|
| Barley | 51.9 | 40.0 | — | 52.6 |
| Corn | 151.4 | 250.9 | — | 24.8 |
| Wheat | 106.6 | 44.4 | — | 13.9 |
| Wheat (flour) | — | — | 7.7 | — |
| Wheat (kernels) | — | — | 4.3 | — |
| Bread | 23.6 | 5.6 | 98.3 | — |
| Spaghetti | — | — | 25.8 | — |
| Semola | — | — | 8.0 | — |
| Rice | — | — | 68.7 | — |
| Oats | — | — | 8.8 | — |
| *Quinoa* | — | 24.0 | 24.3 | 45.4 |
| *Cañihua* | — | — | — | 44.9 |
| Broad beans | 69.8 | 60.4 | 7.0 | — |
| Chick peas | — | — | 3.3 | — |
| Potatoes | 346.7 | 202.2 | 184.8 | 741.0 |
| *Chuño* | — | 31.9 | 30.3 | 470.2 |
| Other tubers | 216.9 | 115.1 | 19.2 | 27.8 |
| Meat | 41.0 | 27.5 | 204.0 | 92.8 |
| Fish | — | — | 14.5 | — |
| Milk | — | 2.6 | 200.8 | — |
| Cheese | — | — | 12.2 | — |
| Eggs | — | — | 10.7 | — |
| Butter | — | — | 11.0 | — |
| Lard | 4.7 | 2.8 | 12.3 | 4.6 |
| Oil | — | — | 20.5 | — |
| Sweet lemon | — | — | 3.2 | — |
| Orange | — | — | 42.8 | — |
| Apple | — | — | 29.5 | — |
| Banana | — | — | 33.3 | — |
| Onions | 5.2 | 15.0 | 41.5 | — |
| Rocoto | 4.2 | 2.8 | — | — |
| Pepper | 0.9 | — | 2.5 | — |
| Cabbage | 12.2 | 41.9 | 14.7 | — |
| Coleus | 0.7 | 0.8 | — | — |
| Coriander | 0.8 | — | — | — |
| *Huacatay* | 1.0 | — | — | — |
| Garlic | — | 2.6 | 2.0 | — |
| Turnip greens | — | 43.8 | — | — |
| *Chijchipa* | — | 0.2 | — | — |
| Parsley | — | 0.2 | 0.7 | — |
| Salt wort | — | 0.1 | 10.3 | — |
| Rue | — | 0.8 | — | — |
| Mint | — | 0.4 | — | — |
| Lettuce | — | — | 13.3 | — |
| Tomato | — | — | 16.0 | — |
| Carrot | — | — | 21.5 | — |
| Pumpkin | — | — | 23.0 | — |
| Sugar | 20.3 | 19.9 | 82.8 | — |

[a] Data from Collazos *et al.* (1960).
[b] Data from Mazess & Baker (1964).

Table 8.2. *Foods utilized (g/day) in three communities of the district of Nuñoa in the southern Peruvian Andes*

| Products | Sincata (4,000 m) | Nuñoa (4,000 m) | Chillihua (4,300 m) |
|---|---|---|---|
| Wheat (flour) | 6.5 | 51.7 | 56.2 |
| Wheat (kernels) | 4.4 | 24.7 | 77.5 |
| Bread | — | 26.4 | 10.6 |
| Barley | 226.3 | 40.4 | 15.2 |
| Corn | 2.4 | 11.9 | 23.5 |
| Cornmeal | 2.3 | 101.6 | — |
| Rice | — | 3.9 | — |
| Quinoa | 208.4 | 65.0 | 41.1 |
| Cañihua | 51.0 | 12.1 | 66.0 |
| Lima beans | 1.6 | — | — |
| Potatoes | 94.0 | 10.6 | 94.4 |
| Chuño | 141.4 | 40.7 | 17.6 |
| Oca | 37.6 | — | — |
| Año | 43.0 | — | — |
| Meat | 27.7 | 67.1 | 84.1 |
| Cheese | — | 15.1 | — |
| Lard | 33.1 | 3.9 | 2.8 |
| Oil | 4.5 | 4.0 | — |
| Pineapple | — | 8.7 | — |
| Onion | 5.1 | 7.2 | — |
| Sugar, white | 4.7 | 46.0 | — |
| Cane sugar | — | 4.0 | — |

From Gursky (1969).

are engaged in commercial wool transactions or are owners of stores where foodstuffs, soft drinks and alcoholic beverages are sold. The native population lives around the town, engaged in agricultural and pastoral activities.

The rural Indian community of Sincata ranked second in the number and variety of foodstuffs available for consumption. This community is situated 20 km southeast of the town of Nuñoa. It has a small store where only carbonated beverages, alcohol, candies, and bread are sold. Men and women are involved in activities connected with agriculture and herding of sheep, llamas and alpacas. Their relative isolation and their low socio-economic level compels them to subsist, almost entirely, on what they produce.

The community where the variety of food products available for consumption was least, was the hacienda of Chillihua, situated 25 km northwest of the district capital. The economy of this community is entirely based on herding of sheep, llamas and alpacas. The daily morning frosts at this location make cultivation of any plants difficult. In contrast to the local production noted in the community of Sincata, the bulk of the food consumed in Chillihua was obtained in the nearby communities, especially the town of Nuñoa.

While it was found that the diet in the town of Nuñoa contained the largest variety of foods, the rural community of Sincata, according to Gursky's survey, consumed the largest amount of food. He reported a daily individual consumption of 894 g in Sincata, 545 g in the town of Nuñoa, and 489 g in Chillihua. It is also important to note that 96.5% by weight of the food consumed in Sincata came from community production, about 65.7% consumed in Chillihua was local, but only 46.5% of that consumed in the town of Nuñoa was produced in the district.

Despite the fact that the main occupation in Chillihua is herding of sheep, llamas and alpacas, the consumption of animal products represents only 17.7% by weight of the diet. This may be explained by the fact that agricultural products are used as wages (Little, 1968), as well as the fact that as with other populations of low economic level, foods of animal origin are generally bartered for cheaper high-energy foods.

The influence of reducing isolation and an improved economic level on food availability is readily seen in cities such as Puno, capital of the department of the same name, situated at about 4,000 m above sea level, in the southern Peruvian Andes. It is linked by road, rail and air to Cusco and to the Peruvian coast, and by road, rail and sea to Bolivia. In this city, most of the residents are engaged in commerce. There is a university and the residents include the highest dignitaries of the political, ecclesiastical and military power of the department. In a survey on food consumption, carried out by the Instituto de Nutrición, Ministerio de Salud Pública (Collazos *et al.*, 1960), it was recorded that forty different foods were commonly utilized (see Table 8.1).

**Food habits**

In isolated communities of low economic level, man has to adapt his food habits to what he produces, as well as to what he can barter with the surrounding communities. As we have seen before, within comparatively short distances, there are very dramatic environmental changes as altitude increases. This causes vegetable and, of course, animal species to vary from one altitude to another. This explains why, at least in the Peruvian Andes, cereals constitute the staple food up to 3,500 m of altitude. Rice is the most common food up to 2,000 m and maize from 2,000 to 3,500 m. However, in this altitudinal zone, there is also considerable use of barley, wheat, potatoes and broad beans. From 3,500 m upwards tubers, mainly potatoes, substitute for cereals as the staple food. It is worth noting that potatoes are primarily used fresh from 3,000 to 3,800 m, while

223

dehydrated potatoes (*chuño negro* and *moraya*) are preferred at higher altitudes. From 4,000 m up to the snow line, the chenopodias (*quinoa* and *cañihua*) play an important role in the daily menu of the native population. They are an important source of protein of high biological value.

The consumption of pigs and cattle is general up to 2,500 m and sheep up to 3,000 m. At higher altitudes, even in those places where herding constitutes the basis of the economy, the native diet is largely vegetarian. The studies of Gursky (1969) showed that in a herding community situated at about 4,300 m, where the daily morning frost makes vegetable life almost impossible, only 17.8% of the diet's bulk was of animal origin.

In all communities, fowls and guinea-pigs are raised; however, they are rarely reported as forming part of the daily native diet. The same may be said about eggs, fish, cheese or milk. The explanation of this phenomenon might possibly be found in the fact that, as in other communities of low economic level, these products are bartered for higher energy foods, or are sold in order to get tools for general use at home. Collazos *et al.* (1954) reported that significant meat consumption occurred in only one of their four surveys (the second in Chacan, altitude 3,416 m). In this survey, 41% of the families consumed meat, in some form, during the week and pork, beef, mutton, tripe, pork skin, pig's feet, organ meats and guinea-pigs were all being eaten. These authors also recorded the consumption of small quantities of cheese, fresh milk and eggs. Neither Gursky (1969) nor Mazess & Baker (1964) recorded the consumption of these items in communities at about 4,000 m above sea level, although the latter investigators reported that milk was usually consumed from December to May. Both the surveys of Gursky and those of Mazess & Baker were carried out in the month of July.

There is little variety of diet in any given week, though seasonal variation may be substantial. Collazos *et al.* (1954) reported that in Vicos (altitude 2,747 m) there was a high consumption of *ocas* (*Oxalis crenata*) during the first survey in July 1953, and a complete absence of this foodstuff during their second survey in February 1954. The same investigators (1954) reported similar differences with regard to the seasonal consumption of *ollucos* (*Ullucus tuberosus*) and turnip greens in Chacan (altitude 3,416 m). Mazess & Baker (1964) also stated that 'there appears to be a substantial seasonal variation' in the types of food eaten in the District of Nuñoa (altitude, 4,000 m). It must be noted that there is a seasonal variation not only in the kind of food consumed, but also in the amount eaten of a specific food. This is especially seen in the consumption of tubers. For example, Collazos *et al.* (1954) remarked that the potato harvest had just been completed at the time of their second Chacan survey (July 1953) and at this time 100% of the families ate an average of 1,230 g of this tuber per day. At the time of the first survey (December 1951), however, 93 % of the families had an average potato consumption of only 390 g per day. This trend was much the same in Vicos, where 65 % of the families consumed an average of 775 g of potatoes per day during the first survey carried out in July 1952, but during the second study (February 1953) 100 % of the families were consuming an average of 2,900 g per day. In the Nuñoa area, where part of the potato harvest is dehydrated to make *chuño negro* or *moraya*, a high consumption of fresh potatoes was found during the harvest months (May to July). The dehydrated forms gradually replaced the fresh forms during the following months of the year (Mazess & Baker, 1964; Gursky, 1969). Other highland tubers, *año* and *olluco*, were most frequently consumed from April to June (Mazess & Baker, 1964). The seasonality of food uses presumably has implications for nutrition, particularly in relation to vitamin C. The seasonality of food consumption is obviously tied to the relative isolation and low socio-economic level of these high-altitude communities, as well as to a complete lack of a scientific technology for food production, storage and conservation.

In general, the high-altitude natives have only two meals a day, 'breakfast' and 'supper'. The preparation of 'breakfast' begins at about 5.00 a.m. in the morning and the meal is generally served between 6.00 and 8.00 a.m. 'Supper' is taken between 4.00 and 6.00 p.m. Some families have also a third meal, which is served from 12.00 to 2.00 p.m. Occasionally, an herbal infusion sweetened with brown sugar is taken early, before breakfast.

When there are only two meals a day, toasted corn, toasted barley or boiled potatoes are usually taken as snacks at noon. At middle altitudes boiled broad beans are also used as snacks.

The two main meals, 'breakfast' and 'supper', have much the same composition. They generally start with a soup, which is prepared using the staple food of the locality to which herbs, generally onion leaves, are added. For example, at middle altitudes, soups are prepared with a base of cereals or broad-bean flour, to which potatoes, *ocas* or *ollucos* are added, as well as herbs of the season. As a second course, '*mote*', is served. This is boiled corn from which the hulls have been previously removed by placing either calcium oxide or ash in the water. Often the course is a combination of beans cooked with *mote*. In the latter dish, the high content of tryptophan in beans supplements the low content of tryptophan in corn. Boiled tubers are also served separately, seasoned with hot pepper.

At higher altitudes, soups are prepared with a base of *chuño negro* or its leached form, *moraya*, to which *quinoa*, *cañihua*, barley or wheat flour and leaves of onions are generally added, and, frequently, alpaca, llama, or sheep's meat. Since fresh meat is often not available a dried form called *charqui* is usually used. Mazess & Baker (1964) remarked that at times,

# The biology of high-altitude peoples

when meat is not available, rendered animal fat, *sebo*, is substituted. This statement was confirmed by Gursky (1969), whose survey on food consumption showed that *sebo* consumption increased as meat availability decreased. At the highest habitable altitudes it is also common to serve boiled potatoes seasoned with hot pepper as a second course.

## Food requirements

Although there is a large high-altitude population, it has not yet been established whether altitude affects food requirements. On this topic, Mitchell & Edman (1951) stated that 'there seems to be little reason to expect, and scanty evidence to support, a belief that altitude and its associated low oxygen tension will disturb the nutrient requirement of man. The low oxygen supply at altitude will depress, rather than stimulate metabolic processes, and, even more surely than a hot humid environment, will discourage the performance of muscular work and lower the capacity for work'.

The studies of Collazos *et al.* (1960) which show a low calorie consumption in populations residing at altitudes ranging from 2,500 to 3,800 m appear to support the previous quotation. However, these studies were not confirmed by later investigations conducted at higher altitudes (Picón-Reátegui, 1963, 1976; Mazess & Baker, 1964), which suggest that if the high-altitude native deviates from the FAOs (1957) model with regard to calorie requirements, it is in the plus direction.

High altitude and hypoxia are frequently used indiscriminately as synonyms, perhaps because the effects of altitude on animal physiology are due predominantly to a diminished partial pressure of oxygen in the inspired air as a function of barometric pressure. This relationship is exponentially related to altitude, as can readily be seen in Fig. 8.1. However, concerning calorie and/or nutrient requirements, components of the high-altitude environment other than barometric pressure, such as temperature, irregular terrain, increased intensity of ultraviolet light, and reduced water-vapor saturation pressure, may play an important role.

As an effect of the inverse relationship between altitude and barometric pressure, the boiling point of water diminishes as altitude increases. The lower boiling point, because it reduces the destruction or denaturation of nutrients labile to heat, may reduce the raw-food requirement for such nutrients. It may also maintain the availability of other nutrients which are destroyed or denatured by the action of those enzymes contained in some foodstuffs whose activity is stopped by heat.

In some respects hypobaria may increase calorie requirements because of its associated low oxygen tension. As can be seen in Fig. 8.1, the partial pressure of oxygen in the inspired air decreases as altitude in-

226

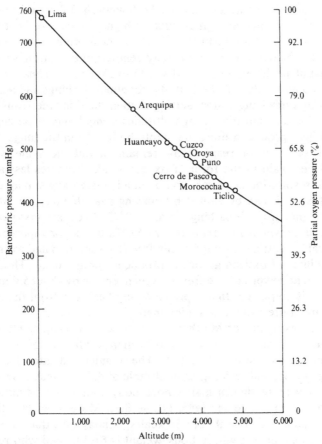

Fig. 8.1. The gas pressures and altitudes of selected Peruvian cities. Taken from Velasquez (1972).

creases. To rectify this change the high-altitude resident has to hyperventilate in order to secure enough oxygen for metabolic processes. Hyperventilation may result in an increase in insensible water loss. The data of Marshall & Specht (1949) obtained on resting subjects indicated that each liter of dry gas carries away from the lungs slightly more moisture at a simulated altitude of 9,150 m than at sea level. Calculations of the correlation coefficients, in this experiment, indicated that whereas water loss from the lungs per unit time is directly dependent upon the ventilation, the aqueous vapor pressure is more closely allied in an inverse manner with the rate of respiration than with tidal or minute volumes.

It is well known that vaporization of water is one of the mechanisms by which the body loses heat. This loss of heat occurs at the rate of

227

## The biology of high-altitude peoples

0.58 kcal per gram of water vaporized (Newburgh, Johnston, Lashmet & Sheldon, 1937). Thus, at high altitudes a higher amount of heat is lost in the process of respiration. In this situation, heat production is increased in order to maintain a constant body temperature. Matthews (1932–3) stated that at sea level, in dry cold air, 15 to 25% of total heat production from the human body was dissipated throughout the lungs and respiratory passages in vaporizing water, setting carbon dioxide free from solution and warming air, and that at high altitudes a much higher percentage of the total heat produced must inevitably be lost from the lungs, because heat loss per liter of respired air remains about the same while the potential heat gain in the form of oxygen intake becomes less and less. The rapid respiration necessary to obtain adequate oxygen may dissipate heat as fast as it can be gained by burning available oxygen. Matthews calculated that when breathing air at $-10\,°C$, 30 cal are lost per liter of air respired, of which approximately 27.5 cal are used for vaporizing water and 2.5 cal for setting carbon dioxide free from solution and warming air. Each milliliter of oxygen absorbed produces about 5.0 cal. Thus, unless the inspired and expired air differ in oxygen tension by about 5.0 mm, even if clothing is so perfect that it prevents any heat loss from the skin, the body temperature cannot be maintained.

Using the estimates of Matthews and those of Velásquez (1947, 1972) the amount of heat lost and produced by men residing at 150 and 4,540 m above sea level were calculated. The results indicate that a man weighing 54.6 kg and residing at an altitude of 4,540 m may lose about 30 kcal per day more than a man whose body weight is 61.4 kg and who resides at 150 m above sea level (Table 8.3). More precisely, the high-altitude resident may lose about 1.30 kcal/kg per day more than the sea-level one, i.e. a man at rest whose body weight is 61.4 kg and who resides at 4,540 m above sea level may lose by respiratory pathways about 78 kcal per day more than the same man living at sea level. When the parameter of reference is body surface area, it shows that the high-altitude resident may lose by respiratory paths about 219 kcal/m² per day instead of only 184 kcal/m² per day at sea level, a difference of 35 kcal/m² per day.

Calculations based on studies by Velásquez indicate that a man living at 4,540 m above sea level, with a body weight of 52.2 kg, may have a heat production about 43 kcal per day less than a sea-level student weighing 58.8 kg, and 302 kcal per day less than a sea-level athlete weighing 66.9 kg (Table 8.4). However, since the man at altitude has a smaller body size than the other two, his heat production may actually be as much as 2.7 kcal/kg per day greater than the student and 2.3 kcal/kg per day greater than the athlete living at sea level. This indicates that if the high-altitude resident had the same body weight as the sea-level student or the sea-level athlete, his heat production might be about 160

228

Table 8.3. *Heat loss at sea level (150 m) and at high altitude (4,540 m) due to respiration*

| | Sea level (SL) | High altitude (HA) | |
| | Mean±s.e. | Mean±s.e. | ΔSL −HA |
| --- | --- | --- | --- |
| No. of subjects | 64 | 96 | |
| Body weight (kg)[a] | 61.4±0.89 | 54.6±0.84 | +6.80 |
| Surface area (m²)[a] | 1.69±0.02 | 1.56±0.05 | +0.13 |
| $V_E$ (l)[a] | 7.19±0.21 | 7.89±0.12 | −0.70 |
| $V_E$/m² (l) | 4.25 | 5.06 | −0.81 |
| Heat loss: kcal per day | 311 | 341 | −30.00 |
| Heat loss: kcal/kg per day | 5.04 | 6.34 | −1.30 |
| Heat loss: kcal/m² per day | 184 | 219 | −35.00 |

[a] Data from Velásquez (1972).

Table 8.4. *Estimated heat production at sea level (150 m) and at high altitude (4,540 m)*

| | Sea-level | | High-altitude workers |
| | Students | Athletes | |
| | Mean±s.e. | Mean±s.e. | Mean±s.e. |
| --- | --- | --- | --- |
| No. of subjects | 8 | 7 | 23 |
| Age (yr)[a] | 25.4±0.8 | 24.3±1.2 | 19.1±0.3 |
| Body weight (kg)[a] | 58.8±4.0 | 66.9±2.3 | 52.2±1.0 |
| Height (cm)[a] | 163.9±3.5 | 169.4±0.9 | 157.3±1.1 |
| Surface area (m²)[a] | 1.63±0.1 | 1.76±0.1 | 1.51±1.7 |
| $O_2$ consumption (ml/min)[a] | 231±14 | 267±10 | 225±4 |
| $O_2$ consumption (ml/min²) | 140±3 | 152±6 | 151±3 |
| Heat production: kcal per day | 1663 | 1922 | 1620 |
| Heat production: kcal/kg per day | 28.3 | 28.7 | 31.0 |
| Heat production: kcal/m² per day | 1008 | 1094 | 1087 |

[a] Data from Velásquez (1947).

kcal per day more than the student and 152 kcal per day more than the athlete at sea level. Table 8.4 also shows that on the basis of surface area, the high-altitude resident and the sea-level athlete produce almost the same amount of heat, each living in their respective environments.

Based on the previous calculations, about 18.3% of the heat produced by a man living at sea level, and about 21.0% of that produced by a man living at 4,540 m above sea level, may be given off through respiration.

Basal metabolic rate may be influenced by barometric pressure, quite independently of oxygen tension (Cook, 1945). This function, necessary to maintain all vital processes not under voluntary control as well as to maintain constant body temperature, consumes about 53% of the total energy of the body. The data used to construct Table 8.4 fail to indicate an effect of altitude on basal oxygen consumption. Reports by European and American authors, before and during World War II, agreed with this result. Lewis, Iliff & Duval (1943) concluded that altitudes varying from sea level to 2,180 m were without appreciable effect on the basic metabolic rate of men and women, but they stated that 'where an effect has been presumed, the indications point to an increased metabolism at altitude'. More recent investigations (Gill & Pugh, 1964; Grover, 1963; Mazess, Picón-Reátegui, Thomas & Little, 1969; Picón-Reátegui, 1961) revealed a higher oxygen consumption in resting high-altitude men than in sea-level ones. The differences are even higher when fat-free body mass or one of its components is taken as the standard of reference (Picón-Reátegui, 1961). Gill & Pugh (1964), after analyzing the roles of (1) the extra oxygen cost of increased ventilation, (2) the influence of changes in body composition and (3) environmental factors other than low atmospheric oxygen tension on the physiological significance of their results, concluded that none of the influences considered above, either alone or in combination, would seem to account for the apparent increase in metabolism found in the Himalayan Scientific and Mountaineering Expedition, 1960–1. In conclusion, they stated that there was some support in their data for the hypothesis of Picón-Reátegui that elevated metabolism in man is part of the adaptive process to a low oxygen tension but that the changes observed by them are relatively small in comparison with the error in the method, and more evidence is needed before this hypothesis can be finally accepted.

The ingestion of food calls for an increase in heat production because of the 'specific dynamic effect of food'. The heat expended after the ingestion of a mixed meal is moderate, amounting to about 6% of the heat produced by the body. Giaja (1938) stated that the specific dynamic action of protein was affected by a decrease in oxygen tension. However, Alekseev *et al.* (quoted by Mitchell & Edman, 1951) reported that in four subjects tested at an altitude of 4,200 m, the increase in basal metabolic rate was not substantial after a high protein meal, and the specific dynamic action was said to be normal. However, the experiments of Giaja (1938) suggested that the specific dynamic action was considerably reduced in rats exposed to low barometric pressure. Giaja's results seem to have been confirmed by Miller & Stock (1969) in experiments conducted with human beings at altitudes ranging from 1,700 to 3,000 m.

The studies of Elsner, Bolstad & Forno (1964), Lahiri *et al.* (1967), Mazess (1967), Baker (1969) and Velásquez (1972) showed that the cost

230

of work is not affected at altitudes up to 4,540 m above sea level. Baker (1969) stated that 'the Nuñoa native in his high-altitude habitat (4,000 m above sea level) has a maximum oxygen consumption equal to or above that of the sea level dweller in their oxygen-rich environment'.

Environmental temperature is one of the many factors considered to influence caloric requirements (FAO, 1957; NAS/NRC, 1974). In this respect, the studies of Johnson & Kark (1947) showed that calorie intake increases as environmental temperature decreases. These authors pointed out that the effect on caloric intake cannot be explained in terms of changing basal metabolic rate nor can it be explained in terms of difference in body size or in terms of different activities, although they had no crucial evidence to decide this latter point. Later studies conducted in various environments by Welch, Buskirk & Iampietro (1958) in men performing moderate work, indicate that calorie intake was the same in all climates when calculated on the basis of body weight. The men in this experiment were properly clothed for each environment and uniformly clothed within each given environment. Johnson & Kark (1947) believed that in their experiments calorie expenditure for a given task is greater in cold than in warm climates because of the hobbling effect of arctic clothing and equipment. Buskirk & Mendez (1967) remarked that since the Quartermaster Corps had markedly improved their protective clothing between the early 1940s and the late 1950s, the soldier in a cold environment was actually working in a microclimate within his clothing, comparable to that normally experienced in temperate areas and that in the early 1940s, when clothing insulation was inadequate, cold exposure and associated shivering supplemented muscular activity to increase calorie needs.

In a previous section it has been seen that the environmental temperature decreases as altitude increases. At the highest habitable altitudes, the effect of wind may reduce even more the effective environmental temperature. However, environmental temperature is not the same at a given altitude in every point of the earth, as is shown by the permanent snow-line whose altitude diminishes from the equator to the poles. This fact may point to cold as the limiting factor for human settlement at places close to the poles and to hypoxia in those close to the equator.

In tropical high-altitude regions where diurnal variations in environmental temperature are more important than seasonal changes, environmental temperature is not very low. However, the residents of these altitudes are exposed to cold more than those of developed countries where, during winter, most individuals are protected against the effect of cold by warm clothes, central heating and heated means of transportation. At high altitudes the native communities perform their main occupations outdoors (herding and/or agriculture) without using devices for protection from cold, rain, snow and wind. To this situation it must be added that high-altitude housing, while providing a shelter against rain and wind, does not significantly alter environmental temperature. In this respect, Baker (1966) reported a difference of only 3° or 4° C, and sometimes no difference between indoor and outdoor temperatures. It must be pointed out that where Baker (1966) conducted the experiment, the environmental temperature drops below freezing during the early hours of the morning. The temperature in the huts, during the cold season, is below that encountered in Eskimo dwellings (Mazess & Larsen, 1972), which suggests that the high-altitude inhabitants may well be exposed to a significant cold stress. However, the habit of using clothing and the adoption of some behavior patterns when sleeping, such as early retiring, adoption of fetal position, and having as many as four or five individuals sleeping in the same bed (Baker, 1969; Mazess & Larsen, 1972) may help to maintain good insulation. If, besides this, one realizes that clothing provides good insulation except for hands, feet and face during outdoor activities (Hanna, 1970), one has to conclude that at least during some hours of the day, the high-altitude native is not too cold stressed. The studies of Baker (1969) indicate that not all the Quechua high-altitude population is cold stressed at the same level. Children suffer the greatest cold stress during inactive periods. During the day, women are slightly more cold stressed than men because they are less active, while during the night, men evidence more cold stress than women.

In addition to the behavioral responses to cold, some physiological responses such as an increase in peripheral blood flow contribute to the maintenance of high peripheral temperatures. This physiological response to cold brings about an increase in heat loss throughout the skin, which in turn calls for an increase in heat production. All of these physiological processes are higher in the Quechua high-altitude population than in subjects of European extraction (Baker *et al.*, 1966).

Since more heat is required in cold than in warm environments to maintain thermal equilibrium, it has been recommended that calorie allowances be higher during cold weather. The increase suggested is about 3–5 % for each 10 °C in mean annual environmental temperature below that used for the reference standard of 10 °C (FAO, 1957; NAS/NRC, 1974).

*Irregular terrain*

In a previous section, it was noted that the cost of work is not directly affected by the high-altitude environment. However, the very rough terrain common in high-altitude areas increases the metabolic costs of many activities. This fact probably also calls for an increase in calorie allowances. Terrain effects may be of great importance during the periodic trips

232

to the nearby communities. These trips are made by walking barefoot irrespective of weather. Lactating women transport their babies on their backs while other family members carry the traded merchandise.

## Ultraviolet radiation

Because atmospheric density is significantly less at high altitude total solar radiation is higher than at sea level. Ultraviolet light is of particular importance in relation to vitamin D in man as the exposure of 7-dehydrocholesterol under the skin to the action of ultraviolet light produces natural vitamin D, i.e. vitamin $D_3$.

Provitamin D, or 7-dehydrocholesterol, is omnipresent in our skin. Since the sun's light is richer in ultraviolet radiation at the higher habitable altitudes, theoretically there should be an absence of rickets in these areas. However, the continuous action of a high amount of ultraviolet radiation on the skin may increase the burning effect of sunlight which may diminish the action of ultraviolet light on the conversion of 7-dehydrocholesterol to vitamin $D_3$.

## Sanitation levels

The interaction of infection and nutrition has been stressed by Scrimshaw & Wilson (1961). Infection and/or parasitism may affect nutrition by increasing food requirements, by diminishing food intake, by diminishing absorption and by increasing specific nutrient destruction.

By its association with fever, infection increases metabolism which in turn may increase protein requirements. Since infection also produces anorexia, calorie intake may be inadequate for energy requirements. In such situations the organism uses first its reserves of energy and then utilizes its active body component for energy purposes. A negative balance of nitrogen has been reported in patients suffering from typhoid fever (Shaffer & Coleman, 1909) which suggests that body proteins are being used as a source of fuel. Peters & Van Slyke (1946) suggested that rather than the increase in the metabolism associated with fever, toxic destruction of cellular protein may affect nitrogen balance in infection.

Another factor which may increase protein requirements is the production of antibodies. Infection may also increase food requirements because both the micro-organisms and the host compete for available nutrients.

In a bacteriological study conducted on three communities situated at altitudes ranging from 3,100 to 3,500 m above sea level, Solis *et al.* (1964) showed a fecal contamination of the drinking water. Contamination was of such a magnitude that they considered these populations at high risk

233

from typhoid fever, bacillary dysentery, hepatitis, etc. Moreover, Castro *et al.* (1964), in a parasitological study carried out in the same communities, reported a high incidence of *Balantidium coli* and *Entamoeba histolytica* in its vegetative form. They were not able to find trophozoites. The incidence of *Balantidium coli*, the highest reported in the literature for similar studies of populations, may be related to the custom of the high-altitude natives of sharing even their bedrooms with domestic animals. The incidence of *Balantidium coli* was found to be 100% in the pigs of these localities.

Among the helminths high incidences of *Ascaris lumbricoides, Trichocephalus dispar* and *Enterobius vermicularis* were encountered. These authors also reported a high percentage of persons infested by parasites of tropical origin such as *Anquilostoma duodenalis* and *Strongyloides stercoralis*.

In short, the interaction of all the factors which form the environment at high altitudes may act to elevate the food requirements of the people who inhabit these areas. In balance these factors seem to point toward an increase rather than to a decrease in food requirements, especially in calorie requirements. This is partly confirmed by the data of Picón-Reátegui (unpublished) shown in Table 8.5. By metabolic balance studies on six subjects residing at 4,000 m he found that maintenance of body weight without body composition modification would require a calorie intake 11.5% above FAO (1957) formula predictions. These findings applied to active young men. This result agrees with previous studies on food consumption reported by Picón-Reátegui (1963) and Mazess & Baker (1964).

The data of Picón-Reátegui on metabolic balances also suggested that the high-altitude environment has no effect on protein requirements. Table 8.6 shows the nitrogen balance found in this study where the subjects had a native diet with the following composition: carbohydrates, 583.0 g; protein, 53.0 g; fat, 16.8 g. Protein consumption was in the proportion of 1.0 g per kg body weight. Although only 17.5% of the protein intake was of animal origin, the subjects maintained nitrogen balance. This study also shows that the high-altitude environment does not affect the absorption of proteins, since the apparent digestibility coefficient of the protein mixture was similar to that reported for vegetable proteins (Swaminathan, 1967). However, the results do not necessarily mean that the amino acid mixture of the diet was adequate in cases of pathological or physiological stress. On the other hand, Gursky (1969) reported that children below 3 years old and girls from 9 to 12 years of age, had protein intakes below recommended allowances for their respective age groups. Moreover, Huenemann *et al.* (1955) showed in their surveys at Chacan and Vicos, that except for breast milk, no other form of animal protein appeared in

234

Table 8.5. *Energy balance (kcal per person per day) in six high-altitude natives residing at 4,000 m*

|  | Mean±s.e. |
|---|---|
| Energy predicted | 2687±52 |
| Energy intake | 2792±3 |
| Lost in urine and stools | 250±48 |
| Energy available | 2547±18 |
| Energy dissipated | 3062±143 |
| Balance | −514±132 |

Table 8.6. *Nitrogen balance in six high-altitude natives residing at 4,000 m*

| Source of nitrogen (g/person/24 h) | Mean±s.e. |
|---|---|
| In food intake | 8.6±0.02 |
| Lost in stool | 2.0±0.6 |
| Absorbed | 6.6±0.2 |
| Lost in urine | 5.3±0.8 |
| Balance | +1.3±0.8 |
| Apparent absorption (%) | 76.7±2.3 |

Table 8.7. *Fat balance in six high-altitude natives residing at 4,000 m*

| Source of fat (g/person/24 h) | Mean±s.e. |
|---|---|
| In food intake | 16.8±0.02 |
| Lost in stools | 7.4±0.9 |
| Absorbed | 9.4±0.4 |
| Apparent absorption (%) | 56.0±5.5 |

the diets of children aged 1 to 3 years old. Thus, the diets were virtually devoid of animal protein.

Table 8.7 shows the results of the fat-balance study. The data indicate a low fat intake and a rather low fat absorption. It is not known whether this constitutes a response to some of the factors which form the high-altitude environment or if it is due to the physical characteristics of the shortening used for cooking. The fat used by these communities is rendered sheep or alpaca fat. The author is not aware of any other study on the subject conducted on natives of high altitude. Some contradictory observations on fat tolerance during acclimatization to altitude do exist (Pugh, 1962; Siri, quoted by Consolazio, Matoush, Johnston & Daws, 1968; Rai, Malhotra, Dimri & Sampathkumar, 1975).

Table 8.8. *Composition of the diet at high altitudes*

| Author | Place | Altitude (m) | Total intake (kcal) | Protein (g) | Fat (g) | Carbohy-drates (g) |
|---|---|---|---|---|---|---|
| Ferro-Luzzi *et al.* (1975) | New Guinea | 2,000 | 2,051 | 44.6 | 23.0 | 427.0 |
| Collazos *et al.* (1954) | Vicos | 2,747 | 1,481 | 36.2 | 11.2 | 308.8 |
| Collazos *et al.* (1954) | Chacan | 3,416 | 1,404 | 33.0 | 11.3 | 292.5 |
| Arteaga *et al.* (1968) | Chile | 3,000–4,000 | 2,400 | 60.0 | 64.0 | 400.0 |
| Mazess & Baker (1964) | Nuñoa | 4,000 | 3,204 | 69.0 | 16.0 | 696.0 |
| Gursky (1969) | Nuñoa | 4,000 | 1,479 | 46.4 | 17.7 | 283.6 |
| Gursky (1969 | Sincata | 4,000 | 2,447 | 65.5 | 24.3 | 491.5 |
| Gursky (1969) | Chillihua | 4,300 | 1,494 | 59.9 | 21.5 | 265.1 |
| Picón-Reátegui (1963) | Morococha | 4,540 | 3,638 | 71.0 | 34.0 | 762.0 |

**Composition of the diet**

Table 8.8 shows the composition of the diet consumed by various high-altitude native populations. It must be pointed out that not all of the populations were engaged in the same type of occupation, nor were other populations' characteristics necessarily similar. For example, Ferro-Luzzi, Norgan & Durnin's (1975) data were from boys 15 to 17 years old; those of Collazos *et al.* (1954) were family studies in agricultural communities; those of Arteaga, Lacassie & Castro (1968), Mazess & Baker (1964) and Gursky (1969) were family studies in pastoral communities which had agriculture as a subsidiary occupation. Mazess & Baker excluded from their data individuals under 3 years of age. Finally, those of Picón-Reátegui were individual studies on men in their twenties engaged in mine labor. These variations may account for many of the differences in reported calorie intake.

Table 8.9 shows that with the exception of the data of Arteaga *et al.* (1968) the diet pattern of the high-altitude populations compared to European and American populations is high in carbohydrates and low in fats. It is not known whether this diet pattern is due to socio-economic factors or if it is an adaptive response to life at high altitudes. From a theoretical point of view, it can be postulated that a carbohydrate-rich meal may be the most adequate for an hypoxic environment. Because the molecule of carbohydrates is already oxidized, it uses less oxygen in its metabolism than either proteins or fats. Moreover, Velásquez (1972) pointed out the necessity of a high carbohydrate consumption at high altitudes in order to maintain a high level of carbon dioxide. High carbon dioxide production helps avoid an acid-base disequilibrium.

Table 8.9. *Percentage distribution of calorie sources in the diet in high-altitude groups*

| Author | Place | Altitude (m) | Protein (%) | Fat (%) | Carbo-hydrates (%) |
|---|---|---|---|---|---|
| Ferro-Luzzi *et al.* (1975) | New Guinea | 2,000 | 8.5 | 9.9 | 81.6 |
| Collazos *et al.* 1954) | Vicos | 2,747 | 9.8 | 6.8 | 83.4 |
| Collazos *et al.* (1954) | Chacan | 3,416 | 9.4 | 7.3 | 83.3 |
| Arteaga *et al.* (1968) | Chile | 3,000–4,000 | 9.9 | 23.8 | 66.3 |
| Mazess & Baker (1964) | Nuñoa | 4,000 | 8.6 | 4.5 | 86.9 |
| Gursky (1969) | Nuñoa | 4,000 | 12.5 | 10.8 | 76.7 |
| Gursky (1969) | Sincata | 4,000 | 10.7 | 9.0 | 80.3 |
| Gursky (1969) | Chillihua | 4,300 | 16.0 | 13.0 | 71.0 |
| Picón-Reátegui (1963) | Morococha | 4,540 | 7.8 | 8.4 | 83.8 |

*Protein*

Protein intake, according to various reports on food consumption (see Table 8.8), varies from 33.0 to 71.0 g per person per day. Both Tables 8.8 and 8.9 suggest that protein intake increases with the increase of the calories of the diet. This suggests that the people in communities with low calorie intake may not be receiving an adequate protein allowance. Moreover, it should be noted that the protein ingested is mostly of vegetable origin, as indicated in the reports of Collazos *et al.* (1954), Picón-Reátegui (1963), Mazess & Baker (1964), Arteaga *et al.* (1968), Gursky (1969) and Ferro-Luzzi *et al.* (1975). However, all these investigators reported an apparent good diet in the adult population. This suggests that the adults are getting enough food at the expense of the unproductive section of the community, i.e. children and old people. This fact may have implications for the growth and development of children.

As the results of Picón-Reátegui (Table 8.6) suggest, the adult population, with the exception of pregnant women and those in lactation, may be receiving an adequate mixture of amino acids for maintenance. It appears that these populations, as was shown by the experiments of Alvistur, Picón-Reátegui & Collazos (1966), are able through their native foods to obtain a protein mixture of high biological value.

*Fats*

As can be seen in Tables 8.8 and 8.9, there is an almost general consensus among the investigators concerning the low fat content of the native diet in high-altitude communities. Although a low fat/high carbohydrate diet may be more sound for individuals living in an hypoxic environment, it is not known whether this diet pattern has its origin in an adaptive response to the environmental conditions, or if it is merely the result of the socio-economic conditions of the communities who live at high altitudes.

Besides the observations of Picón-Reátegui (Table 8.7) of a low fat absorption, the author is not aware of other studies on the subject conducted on natives at high altitude. If fat absorption is impaired at high altitudes, as the results of Picón-Reátegui may suggest, it must constitute a factor to be taken account of when formulating recommended allowances for calories, fat-soluble vitamins and essential fatty acids for populations living in the high-altitude parts of the world.

### Vitamin A

There is general agreement that vitamin A intake is very low in high-altitude Andean populations. As can be seen in Table 8.10 it ranges from the very low mean figure of 1.4 retinol equivalents in the agricultural community of Sincata (Gursky, 1969) to 2,203 retinol equivalents in the pastoralist community of Chillihua (Gursky, 1969). The figures of Collazos *et al.* (1954) ranged from 980 retinol equivalents in their December 1951 survey in Chacan, to only 5.1 retinol equivalents found in their June 1953 survey carried out in the same locality. The reason for this variation was the high consumption of turnip greens during December 1951 and the absence of them in June 1953. Arteaga *et al.* (1968), in their studies in three communities of the northern Chilean Andes, found a mean intake of 493 retinol equivalents. Gursky (1969) remarked that the intake pattern of vitamin A was quite variable. The variability was intimately tied to liver consumption which explains why in the district of Nuñoa, Chillihua had the highest and Sincata the lowest intake of this vitamin. An examination of Gursky's data shows that even in Chillihua some individuals were receiving no vitamin A, while others had a daily intake as high as 13,225 retinol equivalents of vitamin A. It is, however, surprising that no indications of hypo- or hyper-vitaminosis A have been reported. Some possible symptoms such as goose flesh, crinkled skin or scaling which are common in high-altitude residents, may in fact be the result of the cold and dry environment and/or personal hygiene. It has also been suggested by Hensen *et al.* (quoted by Grupo Técnico de la OPS, 1970) that these characteristics occur in people with low fat intakes deficient in essential fatty acids. It will be recalled that the diet pattern of the native high-altitude communities is very low in fat. This may also contribute to a low carotene and vitamin A absorption.

The low vitamin A or carotene intake by high-altitude populations with a lack of anatomic lesions is very intriguing and deserves a careful study.

Table 8.10. *Mineral and vitamin contents of the diet in high-altitude communities*

| Author | Place | Calcium (mg) | Phosphorus (mg) | Iron (mg) | Vitamin A[a] | Thiamin (mg) | Riboflavin (mg) | Niacin[b] | Ascorbic acid (mg) |
|---|---|---|---|---|---|---|---|---|---|
| Collazos et al. (1954) | Chacan I | 426 | — | 16.6 | 979.5 | 1.56 | 1.05 | 14.99 | 43.0 |
| Collazos et al. (1954) | Chacan II | 76 | — | 12.8 | 5.1 | 1.71 | 0.92 | 14.32 | 68.0 |
| Collazos et al. (1954) | Vicos I | 125 | — | 17.3 | 25.2 | 1.81 | 0.92 | 15.22 | 70.6 |
| Collazos et al. (1954) | Vicos II | 138 | — | 12.1 | 15.0 | 1.98 | 0.69 | 17.09 | 76.9 |
| Mazess & Baker (1964) | Nuñoa (district) | 441 | 2,119 | 22.0 | 30.9 | 1.80 | 2.10 | 40.70 | 113.6 |
| Arteaga et al. (1968) | Chile | 390 | — | 18.9 | 493.2 | 1.60 | 1.30 | 14.50 | 40.0 |
| Gursky (1969) | Chillihua | 289 | 1,106 | 19.8 | 2,203.5 | 2.40 | 1.74 | 13.06 | 24.9 |
| Gursky (1969) | Nuñoa (town) | 465 | 761 | 12.4 | 1,339.0 | 1.44 | 1.07 | 13.26 | 10.9 |
| Gursky (1969) | Sincata | 870 | 1,706 | 31.8 | 1.4 | 4.16 | 1.92 | 29.62 | 54.7 |

Chacan I, Chacan II and Vicos I, Vicos II refer to the first and second surveys carried out in these locations.
[a] Retinol equivalent.
[b] Niacin equivalent.

## The biology of high-altitude peoples

Such studies should consider the vitamin A content of the dietetic mixture and its concentration in the plasma and the liver of natives at high altitude.

### Thiamin

Since the recommended dietary allowance for thiamin is 0.5 mg per 1,000 kcal (NAS/NRC, 1974), the thiamin content of high-altitude populations may meet such recommended allowances. The fact that water boils at lower temperatures than at sea level may be a factor in preserving the content of thiamin in native foods, during the process of cooking.

As can be seen in Table 8.10 mean thiamin content in the diet of high-altitude Andean populations varies from 1.56 mg (Collazos *et al.*, 1954) to 4.16 mg (Gursky, 1969).

### Riboflavin

Since recommended dietary allowances for riboflavin do not differ significantly when calculated either on the basis of protein allowances, energy intake, or metabolic body size, the National Academy of Sciences (NAS/NRC, 1974) recommends an allowance of 0.6 mg of riboflavin per 1,000 kcal for people of all ages. Based on this standard a native of the Peruvian Andes, with a standard body weight of 52.6 kg and a calorie requirement of 2,743 kcal a day, should have a riboflavin intake of about 1.65 mg a day.

As seen in Table 8.10 only the populations reported by Mazess & Baker (1964) and those of Chillihua and Sincata reported by Gursky (1969) were meeting this recommended dietary allowance. Deficiency may also be common during pregnancy and lactation, when the recommended dietary allowances are increased by 0.3 and 0.5 mg per day, respectively (NAS/NRC, 1974).

### Niacin

The recommended dietary allowance for adults, expressed as niacin equivalents, was set at 6.6 mg per 1,000 kcal and not less than 13 mg at calorie intakes of less than 2,000 kcal (NAS/NRC, 1974). Thus, for the standard man of the Peruvian Andes, whose calorie requirements are about 2,743 kcal, the recommended dietary allowances for niacin would be about 18.1 niacin equivalents. Again only those groups surveyed in the district of Nuñoa by Mazess & Baker (1964) and those of the towns of Nuñoa and Sincata (Gursky, 1969) were meeting their recommended dietary allowances for this vitamin. However, the town of Nuñoa (Gursky, 1969) may not meet requirements in cases of pregnancy and lactation.

240

*Ascorbic acid*

The National Academy of Sciences (NAS/NRC, 1974) has set up the following recommended dietary allowances fo vitamin C: for children up to 11 years of age, 40 mg; for the adult population, 45 mg; during pregnancy, 60 mg; for lactating women, 80 mg. Table 8.10 shows that those populations studied by Arteaga *et al.* (1968) in the Chilean Andes and those studied by Gursky (1969) at Chillihua and the town of Nuñoa may not meet the recommended dietary allowances for ascorbic acid. The intake in Chacan II (Collazos *et al.*, 1954), Vicos (Collazos *et al.*, 1954), the District of Nuñoa (Mazess & Baker, 1964) and perhaps that of Sincata (Gursky, 1969), may meet the recommended dietary allowances for pregnant women. Only the District of Nuñoa (Mazess & Baker, 1964) and possibly Vicos II (Collazos *et al.*, 1954) meet the recommended dietary allowances for lactating women.

Ascorbic acid intake at the highest Peruvian Andes is intimately related to the availability of fresh potatoes. Tubers are a seasonal food and fresh potatoes are consumed in large quantities from April to August. Other tubers such as *año*, *ollucos* and *ocas* are eaten from April to June. After the seasonal harvest the consumption of fresh tubers diminishes slowly and is replaced, as a staple food, by the dehydrated form of potatoes, *chuño*. The ascorbic acid content decreases during the storage of potatoes and it is almost nil in its dehydrated form. According to Collazos *et al.* (1957), ascorbic acid content in fresh potatoes is about 20.5 mg/100 g of edible portion, and only 1.7 mg/100 g in *chuño*.

*Calcium*

On the basis that 320 mg of calcium are lost daily and that 40% of the dietary calcium is absorbed, the National Academy of Sciences (NAS/NRC, 1974) has stated that 800 mg of calcium should be consumed daily by adults and by children 1 to 10 years old, 1,200 mg by pregnant and lactating women and during the rapid growth period which characterizes pre-adolescence and puberty. According to this, only the adult population under no physiological stress and children 1 to 10 years old may have met recommended dietary allowances for calcium in the community of Sincata (Gursky, 1969).

A perusal of Gursky's (1969) data shows a marked difference in calcium intake even among individuals of the same locality. Very low calcium intake was recorded in lactating mothers and growing children. However, it must be noted that Gursky did not record the contribution of breast milk to the total amount of calcium intake. In this respect it should be noted that many children up to two years old and sometimes up to 3 years old, are breast fed. This may add significant calcium to their daily diet.

## The biology of high-altitude peoples

According to Huenemann *et al.* (1965) some mothers of the Peruvian Andes' communities begin to give solid foods to their babies at the age of 5 months; however, others feel that babies should be older than this, somewhere between 6 and 12 months of age before being given solid foods. They receive no special food but share the family diet. Some of these diets contain sources of calcium not commonly recorded in dietary surveys (Baker & Mazess, 1963). For example, in the district of Nuñoa, the native population served once or twice a week a porridge called *catahui lahua*, which according to Baker & Mazess (1963) may add from 300 to 1,200 mg of calcium to the usual diet.

The native diet of these populations probably supplies enough calcium for maintenance in healthy adults under no physiological stress, as shown by the balance studies of Picón-Reátegui (unpublished data). In this study it was found that the native diet supplied 632 mg of calcium daily of which 527 mg, or 83.4%, were absorbed. There was a retention of about 419 mg. These results agree with the data of Hegsted, Moscoso & Collazos (1952) which showed calcium balances with intakes as low as 100 to 200 mg a day. The high absorption of calcium shown by the data of Picón-Reátegui (unpublished data) may be due to the low amount of fat in the native diet and, possibly of greater importance, the high ultraviolet radiation at high altitudes which may increase the availability of vitamin D. As is well-known vitamin D is required for efficient calcium absorption. However, children up to two years of age may not benefit greatly from the high ultraviolet radiation at high altitudes since when not indoors they are heavily clothed and wrapped in cloth. Even the faces of infants are kept covered.

Until recently there was a general agreement that for an efficient absorption of calcium, there must be in the diet a calcium:phosphate proportion of between 2:1 and 1:2. However, it has been found that ratios outside these limits are satisfactory if the intake of vitamin D is adequate (Mitchell, 1964). In high-altitude Peruvian diets, it has been found that calcium:phosphate ratios ranged from 1:4.8 (Mazess & Baker, 1964) to 1:1.6 (Gursky, 1969). In the balance studies of Picón-Reátegui (unpublished data) the calcium:phosphate ratio of the native diet was 1:2.2 and calcium absorption was shown to be excellent.

### Phosphorus

Phosphorus is an almost universal constituent of foodstuffs. This may be the reason why phosphorus deficiency has not been reported in man. The National Academy of Sciences (NAS/NRC, 1974) has set 800 mg as the recommended dietary allowances for phosphorus. As can be seen in Table 8.10 the phosphorus content reported in the diet of high-altitude Peruvian populations ranged from 761 mg (Gursky, 1969) to 2,119 mg (Mazess & Baker, 1964). In the native diet used in the balance studies of Picón-Reátegui (unpublished data) phosphorus content was 1,366 mg.

*Iron*

Table 8.10 shows that the diet of the high-altitude Andean populations generally meets the recommended dietary allowance of iron for adult men and post-menopausal women, which was set at 10 mg per day. The intakes reported by Gursky (1969) at Sincata and Chillihua, Arteaga *et al.* (1968), Mazess & Baker (1964) and perhaps that of Collazos *et al.* (1954) at Vicos I appear to meet the recommended dietary allowances of iron for women in child-bearing age, which was set at 18 mg per day (NAS/NRC, 1974).

An increase in circulating hemoglobin has been associated with the elevation of the place of residence (Hurtado, Merino & Delgado, 1945). These authors reported a circulating hemoglobin of 12.6 g/kg body weight in adult men living at an altitude of 150 m above sea level, and 20.7 g at an altitude of 4,540 m. Thus, a man with a body weight of 58 kg would have a total circulating hemoglobin of 731 g at sea level and 1,201 g at an altitude of 4,540 m. In general, the human body has 45 mg of iron per kg of body weight, 75% of which is found in the circulating hemoglobin. This means that a man of body weight 58 kg increases the content of iron in his circulating hemoglobin from 1,958 g at sea level to 3,216 mg at an altitude of 4,540 m above sea level. However, this fact may not constitute a factor for increasing the recommended dietary allowances of iron for the adult native population, since there is a balance between blood formation and destruction at every altitude (Reynafarje, Lozano & Valdivieso, 1959). Moreover, the permanent residents of high altitude in whom the equilibrium of erythropoietic activity was demonstrated did not show a significant difference in iron absorption as compared with people born and living at sea level (Reynafarje, 1963). However, the same may not hold true for children who must increase their blood volume as they grow. Gursky (1969) noted that in his sample, children below 3 years of age did not meet recommended dietary allowances for their age. This finding becomes more significant in light of the fact that the diet of young children is not supplemented with iron in spite of the low iron content of the milk. In general the endowment of iron at birth is usually not sufficient to meet needs beyond 6 months of age.

## The biology of high-altitude peoples

*Iodine*

Kelly & Snedden (1960) suggested that the mountain slopes of the Himalayas, Alps, Pyrenees and Andes are the world's most notorious foci of endemic goiter. The etiology of goiter has been related to nutritional, toxic and infective factors. The studies of Stanbury *et al.* (1954), Pretell *et al.* (1969), Koutras, Papapetrou, Yataganas & Malamos (1970), Karmarkar, Deo, Kochupillai & Ramalingaswami (1974) and others indicated that a lack of iodine in the diet, as well as in the drinking water, is the most probable cause of goiter in high-altitude areas. Fierro-Benites, Degroot, Paredes-Suaréz & Peñafiel (1967) suggested that in spite of a chronic deficiency of iodine, populations living over 3,500 m of altitude do not suffer from endemic goiter. However, Salinas (1972) and Pretell *et al.* (1969) did not find a diminution of the prevalence of goiter above the altitude limit set by Fierro-Benites *et al.* (1967).

### Summary and conclusions

Most of the native communities which have developed in the high mountain areas of the world are formed by individuals from low socio-economic levels who generally live in isolated conditions. This fact makes most high-altitude peoples dependent on their own food production. As environmental and geological conditions do not allow the growth of a large variety of plants, the diets are monotonous, high in carbohydrates and low in fats. Although the protein mixture of the diet is low and sometimes free from animal protein it appears adequate for nitrogen balance in adult populations free from pathological or physiological stress. As the foodstuffs are produced to be consumed by the family only during the period between one harvest and the next, calorie and nutrient content of the diet may meet recommended dietary allowances just after harvesting, but tend to be more and more deficient as the next harvest is approached. The unproductive section of the community, i.e. children and old people, as well as pregnant and lactating women, are probably the most affected by this situation. Deficient diets during pregnancy and during the growth of the future mother (Thomson & Billewicz, 1963) may possibly contribute to the low body weights found in newborn infants by Goodwin (1974) in the Peruvian Andes (see Chapter 4). Babies of low body weight, exposed to the high-altitude environmental aggression, especially to wide fluctuations of environmental temperature, certainly contribute to the high neonatal mortality as reported by Mazess (1965) analyzing mortality figures from the 1958 Peruvian census (see Chapter 11).

As has been mentioned before, protein intake may be quantitatively and qualitatively adequate for maintaining nitrogen balance in adults. However, the amino acid mixture may not be adequate for meeting the

Food and nutrition

requirements of growing children. This problem may be further complicated during periods of calorie shortage, when the productive section of the community is fed at the expense of children and old people, i.e. growing children may receive a diet deficient not only in proteins of high biological value but also in calories. This may be a significant factor in the slow growth observed by Frisancho (1966) in the southern Peruvian Andes.

The problem of vitamin A deserves special consideration in view of the discrepancies between the survey results showing low vitamin A intake and the lack of an overt symptomatology needed to diagnose a vitamin A deficiency. The subject is quite intriguing and deserves careful study with emphasis on the vitamin A content in plasma, in liver biopsies and in the liver of cadavers of the native population. A careful survey of the vitamin A and carotene content of the composite diet consumed by these communities would, of course, also be necessary.

The high ultraviolet radiation at high altitudes may make possible in adults the absorption of a high percentage of the calcium available into the intestines; however, small children who are protected from the sun's rays, either in their homes, or wrapped in blankets, may suffer some level of rickets. The elongation or attenuation of early physical growth (Frisancho, 1966), motor development and deciduous tooth eruption processes (Baker, 1969), may all be related to the nutritional problem.

The fact that water boils at a lower temperature than at sea level may be a significant factor in reducing the destruction or denaturation of some nutrients, such as thiamin and ascorbic acid. However, on the other hand, it may also prevent a complete sterilization of the water which is generally contaminated with micro-organisms from human and other animal feces. The micro-organisms, either by competing with the host, or by producing fever may increase food requirements.

### References

Alvistur, C. E., Picón-Reátegui, E. & Collazos-Chiriboga, C. (1966). Biologic value of the composite diet of two Andean communities. In *Human adaptability to environment and physical fitness*, ed. M. S. Malhotra, pp. 231–6. Madras, India: Defence Institute of Physiology and Allied Sciences.
Arteaga, A., Lacassie, Y. & Castro, N. (1968). Estudio de la alimentación y del estado nutritivo de la población indigena de la precordillera de Arica, Chile. *Revista Chilena de Pediatria*, **39**, 631–44.
Baker, P. T. (1966). Micro-environment cold in a high altitude Peruvian population. In *Human adaptability and its methodology*, ed. H. Yoshimura & J. S. Weiner, pp. 67–77. Tokyo, Japan: Japan Society for the Promotion of Sciences.
Baker, P. T. (1969). Human adaptation to high altitude. *Science, Washington*, **163**, 1149–56.
Baker, P. T., Buskirk, E. R., Picón-Reátegui, E., Kollias, J. & Mazess, R. B. (1966). Regulación de la temperatura corporal en indios Quechuas nativos de la altura. *Archivos del Instituto de Biologia Andina*, **1**, 286–98.

# The biology of high-altitude peoples

Baker, P. T. & Mazess, R. B. (1963). Calcium: Unusual sources in the highland Peruvian diet. *Science, Washington,* **142,** 1466–7.

Buskirk, E. R. & Mendez, J. (1967). Nutrition, environment and work performance with special reference to altitude. *Federation Proceedings,* **26,** 1760–7.

Castro, E., Arana, J., Delgado, L., Lamas, J., Tello, L., Zúñiga, J. & Lumbreras, H. (1964). Estudio parasitológico en cuatro comunidades rurales del Cuzco. In *Estudio Médico Antropológico en Cuatro Comunidades Indígenas,* pp. 31–73. Lima, Peru: Facultad de Medicina 'Cayetano Heredia', Universidad Peruano de Ciencias Medicas Y Biologicas.

Collazos, C., Moscoso, I., Bravo de Rueda, Y., Castellanos, A., Cáceres de Fuentes, C., Roca, A. & Bradfield, R. B. (1960). La alimentación y el estado de nutrición en el Peru. *Anales de la Facultad de Medicina, Lima,* **43,** 7–343.

Collazos, C., White, H. S., Huenemann, R. L., Reh, E., White, P. L., Castellanos, A., Benites, R., Bravo, Y., Loo, A., Moscoso, I., Cáceres, C. & Dieseldorff, A. (1954). Dietary surveys in Peru. III. Chacán and Vicos. Rural communities in the Peruvian Andes. *Journal of the American dietetic Association,* **30,** 1222–30.

Collazos, C., White, P. L., White, H. S., Viñas, E., Alvistur, T. E., Urquieta, R., Vásquez, J., Díaz, C., Quiroz, A., Roca, A., Hegsted, D. M. & Bradfield, R. B. (1957). La composición de los alimentos peruanos. *Anales de la Facultad de Medicina, Lima,* **40,** 232–66.

Cook, S. F. (1945). The inhibition of animal metabolism under decompression. *Journal of aviation Medicine,* **16,** 268–71.

Consolazio, C. F., Matoush, L. O., Johnston, H. L. & Daws, T. A. (1968). Protein and water balances of young adults during prolonged exposure to high altitude (4,300 meters). *American Journal of clinical Nutrition,* **21,** 154–61.

Elsner, R. W., Bolstad, A. & Forno, C. (1964). Maximum oxygen consumption of Peruvian Indians native to altitude. In *Physiological effects of high altitude,* ed. W. H. Weihe, pp. 217–23. New York: Macmillan.

FAO (1957). *Calorie requirements. Second Committee on Calorie Requirements. FAO Nutrition Studies 15.* Rome: Food and Agriculture Organization of the United Nations.

Ferro-Luzzi, A., Norgan, N. G. & Durnin, J. V. G. A. (1975). Food intake, its relationship to body weight and age, and its apparent nutritional adequacy in New Guinea children. *American Journal of clinical Nutrition,* **28,** 1443–53.

Fierro-Benites, R., Degroot, L., Paredes-Suárez, M. & Peñafiel, W. (1967). Yodo, bocio y cretinismo endémicos en la región andina del Ecuador. *Revista Ecuatoriana de Medicina y Ciencias Biológicas,* **5,** 15–30.

Frisancho, A. R. (1966). Human growth in a high altitude Peruvian population. Master's Thesis in Anthropology, The Pennsylvania State University.

Giaja, J. (1938). La dépression barométrique et l'action dynamique spécifique des proteides. *Compte Rendus des Seances de la Société de Biologie,* **128,** 687–8.

Gill, M. B. & Pugh, L. G. C. E. (1964). Basal metabolism and respiration in men living at 5,000 m (19,000 feet). *Journal of applied Physiology,* **19,** 949–54.

Goodwin, J. (1974). Altitude and maternal and infant capabilities. In *Horizons in perinatal research,* ed. N. Kretchmer & E. G. Hasselmeyer, pp. 84–117. New York: John Wiley & Sons.

Grover, R. F. (1963). Basal oxygen uptake of man at high altitude. *Journal of applied Physiology,* **18,** 909–12.

Grupo Técnico de la OPS (1970). *Hipovitaminosis A en las Américas. Publicación Cientófica No. 198.* Washington, DC: Organización Panamericana de la Salud

Oficina Sanitaria Panamericana, Oficina Regional de la Organizacion Mundial de la Salud.

Gursky, M. J. (1969). Dietary survey of three Peruvian highland communities. Master's Thesis in Anthropology, The Pennsylvania State University.

Hanna, J. M. (1970). Comparison of laboratory and field studies of cold response. *American Journal of physical Anthropology*, **32**, 227–32.

Hegsted, D. M., Moscoso, I. & Collazos, C. (1952). A study of the minimum calcium requirement of adult men. *Journal of Nutrition*, **46**, 181–201.

Huenemann, R. L., Collazos, C., Hegsted, D. M., Bravo de Rueda, Y., Castellanos, A., Dieseldorff, A., Escobar, M., Moscoso, I., White, P. L. & White, H. S. (1955). Nutrition and care of young children in Peru. IV. Chacan and Vicos, rural communities in the Andes. *Journal of the American dietetic Association*, **31**, 1121–33.

Hurtado, A., Merino, C. & Delgado, E. (1945). Influence of anoxemia on the hematopoietic activity. *Archives of Internal Medicine*, **75**, 284–323.

Johnson, R. E. & Kark, R. M. (1947). Environment and food intake in man. *Science, Washington*, **105**, 378–9.

Karmarkar, M. G., Deo, M. G., Kochupillai, N. & Ramalingaswami, V. (1974). Pathophysiology of Himalayan endemic goiter. *American Journal of clinical Nutrition*, **27**, 96–103.

Kelly, F. C. & Snedden, W. W. (1960). Prevalence and geographical distribution of endemic goitre. In *Endemic goitre*, p. 27. Geneva: World Health Organization.

Koutras, D. A., Papapetrou, P. D., Yataganas, X. & Malamos, B. (1970). Dietary sources of iodine in areas with and without iodine-deficiency goiter. *American Journal of clinical Nutrition*, **23**, 870–4.

Lahiri, S., Milledge, J. S., Shattopadhyay, A. P., Bhattacharyya, A. K. & Sinha, A. K. (1967). Respiration and heart rate of Sherpa highlanders during exercise. *Journal of applied Physiology*, **23**, 545–54.

Lewis, R. C., Iliff, A. & Duval, A. M. (1943). Further consideration of the effect of altitude on basal metabolism. A study on young women residents of Denver. *Journal of Nutrition*, **26**, 175–85.

Little, M. A. (1968). Racial and developmental factors in foot cooling Quechua Indians and U.S. whites. In *Human adaptation to high altitude*. Final Progress Report, US Army Surgeon General, Contract No. DA-49-193-MD-2260. University Park, Pa.: The Pennsylvania State University.

Marshall, L. H. & Specht, H. (1949). Respiratory water vapor at simulated altitude. *American Journal of Physiology*, **156**, 299–310.

Matthews, B. H. C. (1932–33). Loss of heat at high altitudes. Proceedings of the Physiological Society. *Journal of Physiology*, **77**, 28–9P.

Mazess, R. B. (1965). Neonatal mortality and altitude in Peru. *American Journal of physical Anthropology*, **23**, 209–13.

Mazess, R. B. (1967). Group differences in exercise performance at high altitude. Doctoral Dissertation in Anthropology, University of Wisconsin.

Mazess, R. B. & Baker, P. T. (1964). Diet of the Quechua Indians living at high altitude. *American Journal of clinical Nutrition*, **15**, 341–51.

Mazess, R. B. & Larsen, R. (1972). Response of Andean highlanders to night cold. *International Journal of Biometeorology*, **16**, 181–92.

Mazess, R. B., Picón-Reátegui, E., Thomas, R. B. & Little, M. A. (1969). Oxygen intake and body temperature of basal and sleeping Andean natives at high altitude. *Aerospace Medicine*, **40**, 6–9.

Miller, D. S. & Stock, M. J. (1969). The effect of altitude on thermogenesis in man. *Proceedings of the Nutrition Society*, **28**, 74A.

Mitchell, H. H. (1964). *Comparative nutrition of man and domestic animals*, vol. II. New York & London: Academic Press.

Mitchell, H. H. & Edman, M. (1951). *Nutrition and climatic stress*, pp. 96–167. Springfield, Ill.: Charles C. Thomas.

NAS/NRC (1974). *Recommended dietary allowances*. A Report of the Food and Nutrition Board, National Research Council. Washington, DC: National Academy of Sciences.

Newburgh, L. H., Johnston, M. W., Lashmet, F. H. & Sheldon, J. M. (1937). Further experiences with the measurement of heat production from insensible loss of weight. *Journal of Nutrition*, **13**, 203–21.

Peters, J. P. & Van Slyke, D. D. (1946). *Quantitative clinical chemistry interpretations*, 2nd edn, vol. 1. Baltimore, Md.: Williams & Wilkins.

Picón-Reátegui, E. (1961). Basal metabolic rate and body composition at high altitude. *Journal of applied Physiology*, **16**, 431–4.

Picón-Reátegui, E. (1963). Intravenous glucose tolerance test at sea level and at high altitudes. *Journal of clinical Endocrinology and Metabolism*, **23**, 1256–61.

Picón-Reátegui, R. (1976). Nutrition. In *Man in the Andes: A Multidisciplinary Study of High Altitude Quechua*, ed. P. T. Baker & M. A. Little, pp. 208–36. Stroudsburg, Pa.: Dowden, Hutchinson & Ross.

Pretell, E., Moncloa, F., Salinas, R., Guerra-García, R., Kawano, A., Gutiérrez, L., Pretell, J. & Wan, M. (1969). Endemic goiter in rural Peru: effect of iodized oil on prevalence and size of goiter and on thyroid iodine metabolism in known endemic goitrous populations. In *Endemic goiter*, ed. J. B. Stanbury. Scientific Publication No. 193. Washington, DC: Pan American Health Organization.

Pugh, L. G. C. E. (1962). Physiological and medical aspects of the Himalayan scientific and mountaineering expedition, 1960–61. *British Medical Journal*, **ii**, 621–7.

Rai, R. M., Malhotra, M. S., Dimri, G. P. & Sampathkumar, T. (1975). Utilization of different quantities of fat at high altitude. *American Journal of clinical Nutrition*, **28**, 242–5.

Reynafarje, C. (1963). Hematologic changes during rest and physical activity in man at high altitude. In *The symposium on the physiological effects of high altitude, Interlaken*, pp. 73–85. Oxford: Pergamon Press.

Reynafarje, C., Lozano, R. & Valdivieso, J. (1959). The polycythemia of high altitude: iron metabolism and related aspects. *Blood*, **14**, 433–55.

Salinas, R. (1972). Existencia de bocio en el pais. In *Anales del Primer Congreso Peruano de Nutrición*, pp. 9–19. Lima, Peru: Asociación Peruana de Nutricion.

Scrimshaw, N. S. & Wilson, D. (1961). Nutricion e infeccion. *Guatemala Pediatrica*, **1**, 72–82.

Shaffer, P. A. & Coleman, W. (1909). Protein metabolism in typhoid fever. *Archives of Internal Medicine*, **4**, 538–600.

Solis, G., Azcárate, C., Cabrera, J., Barbe, G. & Carrillo, C. (1964). Estudio bacteriológico de treinta fuentes de agua en cuatro comunidades rurales del Cusco. In *Estudio Médico Antropológico en Cuatro Comunidades Indígenas*, pp. 8–29. Lima, Peru: Facultad de Medicina 'Cayetano Heredia', Universidad Peruana de Ciencias Médicas y Biológicas.

Stanbury, J. B., Brownell, G. L., Riggs, D. S., Perinetti, H., Itoiz, J. & Del Castillo, E. B. (1954). *Endemic goiter.* Cambridge, Mass.: Harvard University Press.

Swaminathan, M. (1967). Availability of plant proteins. In *Newer methods of nutritional biochemistry with applications and interpretations*, ed. A. A. Albanese, vol. III, pp. 197–241. New York & London: Academic Press.

Thomson, A. M. & Billewicz, W. Z. (1963). Nutritional status, maternal physique and reproductive efficiency. *Proceedings of the Nutrition Society*, 22, 55–60.

Velásquez, T. (1947). Metabolismo basal en la altura. Bachelor's Thesis. Lima, Peru: Facultad de Medicina, Universidad Nacional Mayor de San Marcos.

Velásquez, T. (1972). Análisis de la Función Respiratoria en la Adaptación a la Altitud. Tesis de Doctorado. Lima, Peru: Programa Académico de Medicina Humana, Universidad Nacional Mayor de San Marcos.

Welch, B. E., Buskirk, E. R. & Iampietro, P. F. (1958). Relation of climate and temperature to food and water intake in man. *Metabolism*, 7, 141–8.

# 9.  The responses of high-altitude populations to cold and other stresses

M. A. LITTLE & J. M. HANNA

Low atmospheric oxygen pressure is just one of a broad spectrum of climatic features that tend to make high-altitude environments stressful to humans. Throughout the world, increases in terrestrial elevation are accompanied by decreases in ambient temperature and humidity and corresponding increases in air turbulence, wind flow and solar radiation, particularly short-wavelength radiation. These unique climatic features, when combined with variations in relief and vegetational cover, interact to produce complex weather phenomena that show, in turn, marked regional variations according to latitude and altitude.

The material that follows in this chapter focuses upon climatic stresses with an emphasis on cold stress and adaptation of human populations residing in the Andes and the Himalayan mountains. Although these highland regions are not the only regions of the world where human populations are subject to high-altitude climatic stress, they are, nevertheless, the only high-altitude areas where systematic studies of climatic stresses have been conducted. Topics to be discussed include: general climatic characteristics of highland zones, cultural patterns of adaptation to climatic stress, studies of biological modes of cold adaptation, the influence of drugs on cold tolerance, and the multivariant stresses of hypoxia, cold, aridity and high solar radiation.

## Climatic characteristics of high-altitude zones
### General characteristics

The microclimate near the ground at high altitude has several basic thermal features that can be distinguished from sea-level microclimate.

First, at any given latitude, seasonal variation in monthly temperatures is less at high altitude than at sea level. This holds for tropical and temperate highlands, and, hence, there is a trend from 'continental-type' climates at sea level to 'maritime-type' climates at high altitude. Moreover, as the equator is approached, seasonal variations virtually disappear. For example, in the Austrian Alps at latitude 47° N the temperature range between the warmest and coldest months is 21 °C at 200 m elevation and 14 °C at 3,000 m (Geiger, 1965, p. 444). In the Andean city of Cuzco, Peru (latitude, 13° S) at 3,300 m the annual monthly range is 6 °C, while in Quito, Ecuador (latitude, 0°) at 2,800 m, it is less than 1 °C (Nelson, 1968).

Second, whereas seasonal variations are minimized at high altitude, diurnal variations are maximized. As in deserts, diurnal variation, especially during dry, clear seasons, can show a range as great as 30 °C. This results from the high levels of long-wavelength (thermal infrared) radiation that occur under cloudless skies during the day and the escape of long-wavelength radiation to the clear skies at night. Under overcast skies, the diurnal variation in temperature decreases.

Third, there is an inverse relationship between altitude and environmental temperature such that temperatures become colder with increasing elevation. There is, however, no uniform value for decline in °C/m increase in elevation that applies around the globe, since seasonal, latitudinal and regional topographic variations determine the changes. In both temperate and tropical mountains, the steepest temperature declines occur during the warmest months. In the temperate Austrian Alps, there is a 1 °C drop in temperature for each 250 m increase in elevation during January and a 1 °C drop per 160 m in July (Geiger, 1965, p. 444). At higher elevations within the equatorial zone on the Peruvian *altiplano*, temperature declines are more abrupt at 1 °C per 95 m during the cold (dry) season and 1 °C per 80 m during the warm (wet) season (Little, 1968, pp. 359–61).

Local variations in topography influence temperature and humidity, especially with respect to diurnal ranges. When comparing valleys, slopes and peaks, there is a roughly vertical division where, during the day, valley temperatures are warmest and peak temperatures are coldest. At night, there is an inversion where both valley and peak temperatures are low and slope temperatures are relatively high. The same general pattern for the three topographic features holds for humidity, except that humidity peaks at night when temperatures are low. Valley floors, then, despite their lower elevation, tend to display the greatest diurnal temperatures and humidity fluctuations.

Solar radiation increases with altitude because of the reduction in atmospheric oxygen, nitrogen and ozone molecules which serve to absorb and scatter radiation at various wavelengths. Again using the Austrian Alps as an example, the mean global radiation (cal/cm² per day) at this latitude increases about 25% from 200 m to 3,000 m elevation under clear skies. Under partly overcast skies, cloud reflection augments the mean global radiation to produce a 150% rise between the same two elevations although absolute values are lower than under clear skies (Geiger, 1965, p. 443). With solar radiation, temperate highlands show pronounced seasonal variations, while in tropical highlands, seasonal variation is less pro-

252

nounced and there is greater influence of cloud cover, haze and dust associated with dryness. Ultraviolet radiation at above 3,000 m in the tropics is more than double that at sea level because of the reduced amount of ozone present at high altitude (Buettner, 1969).

### The Peruvian altiplano

The *altiplano* of the central Andes is a high plateau that surrounds Lake Titicaca on the border of Peru and Bolivia. The relief of the *altiplano* is not as severe as that found either to the north or to the south, and the

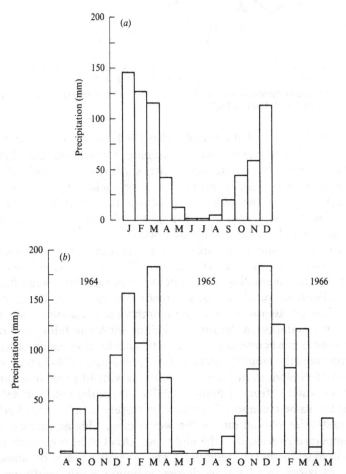

Fig. 9.1. (*a*) Forty-year averages of monthly precipitation at Chuquibambilla, Peru (3,910 m, latitude 14° 45' S). (*b*) Monthly precipitation between August, 1964 and May, 1966 at Nuñoa, Peru (4,000 m, latitude 14° 30' S).

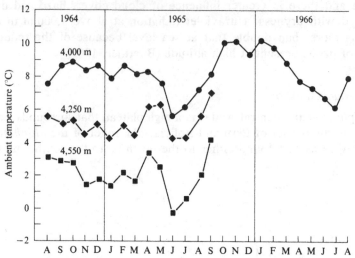

Fig. 9.2. Mean monthly temperatures at three elevations in the region of Nuñoa, Peru, between August, 1964 and August, 1966.

base elevation of most of the broad valleys is between 3,800 and 4,200 m above sea level. The vegetation of the *altiplano* is commonly called *puna* and consists of grasses, herbaceous plants, shrubs, mosses and lichens. Some stunted trees (*Polylepis* sp.) are found on sheltered slopes, but for all practical purposes the *altiplano* is above the timberline, and the countryside is basically natural alpine grassland. Human settlement is between the basal elevation and about 4,800 m.

Seasonality of climate on the *altiplano* results from the typical tropical monsoon pattern of wet and dry seasons. The rainy season begins in September, reaches a monthly peak of between 10 and 20 cm precipitation in January, February or March, and extends through April. The months from May through August are dry with monthly precipitation usually less than 1 cm. Precipitation in the form of hail or snow can fall at any time of the year but is most common in the months that bracket the dry season. Fig. 9.1 compares the monthly precipitation from August 1964 to May 1966 in the town of Nuñoa, Peru (latitude 14° 30′ S) with 40-yr mean monthly precipitation values from Chuquibambilla, Peru (latitude 14° 45′ S), roughly at the same elevation. Precipitation closely parallels cloud cover and overcast skies prevail during the wet season. This moderates wet-season temperatures such that the night-time effect of heavy clouds is to reduce outgoing longwave radiation. The overall pattern of ambient temperatures is one in which the diurnal variation is greater during the dry months because of temperature extremes, whereas the mean daily temperatures are warmer during the wet and cloudy months largely because of the increase in night-time temperatures.

254

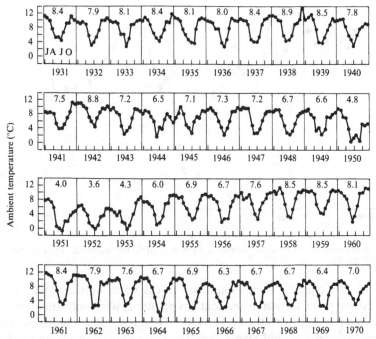

Fig. 9.3. Mean monthly temperatures from 1931 to 1970 at Chuquibambilla, Peru (3,910 m, latitude 14° 45′ S). Values at the top of each 12-month period are mean annual temperatures.

Mean monthly temperatures at three elevations are presented for Nuñoa in Fig. 9.2. These data are not wholly representative of either the *altiplano* or altitude variations because they cover a short time span and stations were located on different land features. For example, the station at 4,000 m was sited in a broad valley, the station at 4,250 m on a mountain slope, and the station at 4,550 m on a mountain ridge. Nevertheless, the general pattern of seasonal trends and temperature declines with altitude can be observed.

Despite the relative constancy in seasonality of temperature and precipitation, there are rather remarkable year-to-year variations in temperature. This is illustrated in Fig. 9.3 with a 40-yr record of mean monthly temperatures from the weather station at Chuquibambilla. The comparatively warm 1930s and early 1940s with mean annual temperatures ranging between 7.5 and 8.9 °C were followed by a cooling trend that reached a low of 3.6 °C in 1952. This 40-yr trend in mean annual temperature is plotted in Fig. 9.4. Such information on long-term variations in ambient temperature is important when drawing inferences about the present and past conditions of cold stress experienced by human populations. For example, the cold tests of Peruvian Indians that are discussed in a later

255

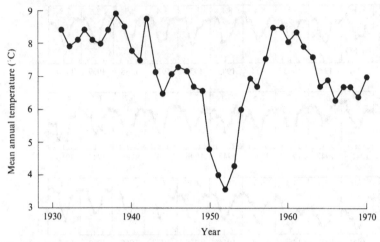

Fig. 9.4. Forty-year trend in mean annual temperature at Chuquibambilla, Peru (3,910 m, latitude 14° 45′ S).

section of this chapter were conducted in the years between 1965 and 1967, a period when *altiplano* ambient temperatures were only moderately cool when compared with the early 1950s. Therefore, any conclusions on the extent of natural cold stress imposed upon human residents of the *altiplano* should be based upon long-term rather than short-term climatic data, when available.

### The Himalayas

The Himalaya and Karakoram mountains are inland and temperate (latitude 28–37° N) in contrast to the coastal and equatorial central Andean *altiplano* (latitude 14° N–18° S). Also, many of the northern slopes and intermountain valleys of the Himalayas are in rain shadows and, hence, quite dry (Clegg, Harrison & Baker, 1970). In general, the climate of the Himalayas is slightly warmer than would be expected at these latitudes and altitudes because the higher northern ranges act as climatic barriers preventing the cold winds of central Asia from passing southward (Robinson, 1967, p. 116). In the eastern Himalayas of Nepal, Sikkim and Bhutan, the timberline is at 4,100 m (Swan, 1967). Sherpa settlements in Nepal range between 3,600 and 4,000 m during the winter months, and up to and even beyond 5,000 m during the summer months (Fürer-Haimendorf, 1963).

Temperature and some precipitation data from three elevations are illustrated in Fig. 9.5. The seasonal range in mean monthly temperatures is greater than on the Peruvian *altiplano* at comparable altitudes. More-

256

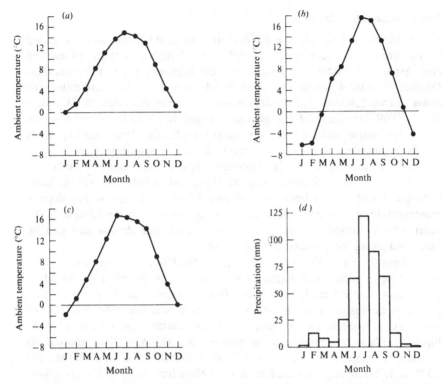

Fig. 9.5. Mean monthly temperatures at (*a*) Yatung, Tibet from 1924 to 1943 (3,000 m, latitude 27° 29′ N); (*b*) Leh, India from 1881 to 1942 (3,500 m, latitude 34° 09′ N); (*c*) Lhasa, Tibet from 1941 to 1948 (3,700 m, latitude 29° 40′ N); (*d*) monthly precipitation at Lhasa. (Data from Meteorological Office, 1966.)

over, summer and winter seasonality is clearly defined, with winter temperatures substantially lower than in Peru even at elevations 1,000 m higher (see Fig. 9.2). Yatung and Lhasa, Tibet, have monsoon rains which coincide with the warm summer months. Leh, a small town in the Karakoram mountains of the Ladakh region of Kashmir, is in an arid zone which receives an average precipitation of less than 100 mm per year.

As a consequence of the higher latitudes and colder winters of the Himalaya and Karakoram mountains, it is likely that highland residents suffer somewhat greater seasonal cold stress than highlanders from the central Andes. In the case of the Sherpa from Nepal, who move their settlements to greater elevations during the warmer months, some cold stress is probably experienced sporadically throughout the year.

# The biology of high-altitude peoples

## The Ethiopian highlands

The Ethiopian highlands are divided into an eastern and western group by the great rift valley which stretches the length of Africa. In marked contrast to the Andes and Himalayas, the highest peak in Ethiopia (Ras Dashan) is only 4,620 m above sea level. However, there are extensive areas above 2,500 m that are inhabited by human populations (Harrison *et al.*, 1969). Because of the abrupt changes in elevation down into the rift valley and a series of western escarpments, the Ethiopian highlands can be characterized as a high plateau. Hence, population movement in and out of the highlands is probably greater than in the Andes and Himalayas. Three highland zones are identified in Ethiopia: (1) the *kolla* is tropical and ranges between 900 and 1,500 m; (2) the *woina dega* is intermediate at between 1,500 and 2,500 m; and (3) above 2,500 m is the *dega* which consists of open short grassland and shrubs and can be identified as alpine pasture (Stamp, 1964).

Well within the northern equatorial zone, the Ethiopian highlands extend about 1,300 km between latitudes 5° N and 17° N. Ambient temperature conditions, consequently, are quite similar to those of the Peruvian *altiplano* at corresponding elevations. Harrison and his colleagues (1969) recorded temperatures at three elevations during the course of their high-altitude investigations in the northwest highlands. They noted daily maximum temperatures of 35–40 °C at 1,500 m, 35 °C at 3,000 m, and 20 °C at 3,700 m. Nights were frequently below freezing at the intermediate elevation, while temperatures below 0 °C were common at the highest elevation. Precipitation patterns are typically monsoon in the highlands with the wet season extending from June through October. Annual precipitation, largely in the form of rain, amounts to about 100 cm.

## Cultural and behavioral responses to high-altitude stress

Cultural mechanisms which provide a more comfortable microclimate under cold ambient conditions have been developed by all high-altitude residents. While the form of the adaptations has been modified by economic and social factors, the commonality of a high-altitude physical environment has given them a certain unity. All cultural adaptations to high-altitude life will thus tend to provide some means for satisfying basic biological needs. These include: (1) increasing individual microclimate temperatures, (2) reducing heat loss through convection and radiation and (3) perhaps increasing relative humidity. In a practical sense these adjustments are made with material culture, and for high-altitude farmers this is generally limited to shelter and clothing. Other institutionalized and idiosyncratic cultural adjustments have not been studied to our knowledge.

258

Some suggestions as to their form and function have been discussed by Baker (1966*a*).

## Houses and shelter

At high altitude a basic consideration in house construction is to provide a microclimate more suitable than the one which exists naturally. Those aspects of the environment which can be modified by use of primitive technology include low ambient temperatures, propensity for heat loss due to convection, and the potential for heat loss through long-wavelength radiation. The ideal house form, then, should be a draft-free structure of tight construction with a low surface-area to volume ratio. The roof should be well insulated and the walls should be constructed of massive materials in order to dampen the daily extremes of temperature which occur in all high-altitude regions (Fitch & Branch, 1960; Rappaport, 1968). Descriptions of houses in high-altitude regions suggest that they fit this model to some degree.

### Andean houses

Andean houses have received a good deal of attention from IBP. Detailed descriptions can be found in Baker (1966*a*, *b*), Hanna (1968), Little (1968) and Larsen (1973). There are two basic house designs. The first uses adobe or sod and is a permanent building. This type is usually found in towns or *haciendas* and represents a major investment. The second design is constructed of piled fieldstone, is semipermanent, and is cheap to construct. It is more characteristic of areas where the population is largely pastoral.

The adobe building has a rectangular floor plan with average dimensions of 5 m by 10 m. The roof is gabled with a peak 4 m to 5 m from the ground. Frequently the first meter of the walls will be made of stone to resist erosion due to rain draining from the roof. The roof is typically constructed of tile, grass or, in more affluent families, corrugated tin. The door is small and its height seldom exceeds 1.3 m. Doors are usually wooden, but in some cases blankets or old *ponchos* may be used to cover the openings. Walls are usually plastered with mud to form an air-tight structure. The roof is tightly fitted, regardless of the material used. In some cases a wooden floor may be added but usually a natural dirt floor is preferred. Rooms may be employed for cooking, sleeping or storage.

Baker (1966*b*) measured interior temperatures of adobe houses during the cold and dry season and found that well-constructed adobe houses could maintain an interior night-time temperature about 7 °C above ambient temperature. Even when ambient temperatures approached freezing, floor level inside the structure remained warmer. Larsen (1973) has also studied adobe house temperatures and found somewhat lower temperature gradients. During the coldest period of the night the indoor temperatures were only 5 °C above ambient values. Again Larsen found indoor temperatures to remain above freezing even if outdoor temperatures did not. He also found that the daytime temperatures in adobe buildings remained *lower* than ambient; thus, the overall thermal effect of these houses was to reduce the diurnal changes in microclimate temperature. That is, indoor values are warmer than ambient at night and cooler by day.

Baker (1966b) and Larsen (1973) have also studied the second type of house made of stone. The floor plan of these houses is circular or rectangular with the upper walls sloping slightly inward. The roof is always constructed of grass and supported by tree limbs. The diameter is quite variable as is the height. The walls are of piled fieldstone. If the house is to be occupied for an extended period of time the stones are carefully piled to eliminate cracks. Those large holes which remain and those at eye level are used as windows; that is, they serve to admit daylight and provide for observation of the surrounding terrain. These houses may have either a wooden door or a piece of old cloth may be used to cover the entrance.

The thermal protection offered by stone houses seems to be less than that found in adobe structures. Baker reports that there is only 3.7 °C difference between indoor and outdoor temperatures and that sometimes interior temperatures fall below freezing. Larsen's observations confirm this, for he too found the floor temperatures in some stone houses to fall below freezing.

Unfortunately the effectiveness of these houses has been evaluated only in terms of thermal stress; other parameters have not been quantified. It can only be speculated that other important differences exist. The most likely is a differential protection against convective heat loss. The simply constructed, loosely mortared walls of stone houses must provide for considerable convective heat loss during windy evenings. On the other hand the hemispherical design of the stone house with its highly insulative grass roof suggests a favorable surface-area to volume relationship which might offset the heat loss due to convection.

Since neither adobe nor stone houses are heated, the contribution of cooking fires is of interest. Both Baker and Larsen have studied their effects, and apparently there is a transitory increase in interior temperature at meal time but this is rapidly dissipated.

Andean houses offer partial protection. Maintenance of a warm microclimate seems less important than protection against convective and radiation heat loss.

### Himalayan houses

The houses of the Sherpa seem to be well constructed and designed to meet the demands of a cold environment. There have been no climatological studies nor are there any quantitative data as to their effectiveness; however, some pictures and descriptions are available (Maron, Rose & Heyman, 1956; Pugh, 1966).

Sherpa houses are constructed of heavy stone and have wooden roofs which are held in place with stones. Most are two-storey structures whose interior dimensions are rather small. Overall dimensions of the rectangular floor plan are 10 or 12 m by 2 or 4 m (Fürer-Haimendorf, 1964). They are apparently tightly constructed because the first floor, which is windowless, is also reported as extremely dark. The first floor is reserved as animal quarters while the second is reserved for human habitation. Cooking is done indoors with the smoke escaping through a small hole in the roof.

The use of the first floor as animal quarters might add to the insulation between the floor of the human section and the ground. If it is tightly constructed the insulation would be still greater. Since animals must be quartered indoors during some periods of the year the construction of a single two-storey house rather than a separate house and barn suggests an attempt to reduce the surface-area to volume relationship. It is unfortunate that quantitative thermal data are not available since the Sherpa house seems to be engineered to withstand a rigorous climate, and some confirmation would be of great interest.

### Ethiopian houses

Quantitative thermal data are also lacking for the houses of the Ethiopian plateau, although fairly complete descriptions are available (Simoons, 1960). The basic design is typical of sub-Saharan Africa. The floor plan is circular, the walls are of wattle and daub and the roof is conical thatch. There is a single door and no windows. The overall dimensions of the structure are variable. A fireplace is located in the center of the floor although in some cases two fireplaces may be built. The second is for added warmth. On occasion some animals may also be brought inside to spend the night, the inhabitants perhaps taking advantage of the greater heat production of the cattle.

Again the cylindrical shape with its low surface-area per unit volume is used. The single door, no windows, the use of mud plaster, the presence of animals, and the occasional maintenance of two fires suggest that protection from the cold must be taken into consideration in Ethiopian plateau housing. Again there is no quantitative evidence available to estimate the efficiency of this rather interesting structure.

## The biology of high-altitude peoples

### Clothing

Because the temperature of all high-altitude regions falls below thermal neutral at some period of the day, and because of constant cooling due to radiation and convection, native peoples must resort to clothing of some sort to maintain thermal equilibrium. The problems surrounding clothing design reflect the same problems encountered in housing design, that is, maintaining a microclimate which is warmer and less exposed than the natural one.

The design of clothing must adhere to some general principles for reduction of heat loss. Several of these have been developed through long experience in the Arctic and should be readily applicable to altitude cold (Forbes, 1949; Kennedy, 1961; Hanna, 1968).

The first principle is to entrap air and utilize its insulative properties to enhance those of the fabric. There is an inherent advantage in the use of this principle at altitude because as air density decreases its insulative qualities increase. At 4,000 m, the insulation of air in Clo units* is about 130% of the sea-level value (Belding, 1949, p. 358). Air can be entrapped within the fabric itself in two ways. First, the fiber selected can be one which contains air spaces and resists compacting. That is, normal use of the fiber should not result in loss of entrapped air. Wool is the most desirable of the natural fibers in this respect; it can provide up to six times as much resistance to heat flow as a cotton material of the same thickness (Fourt & Harris, 1949). The second technique for retaining air within the fabric is the construction technique. A loosely woven fabric will contain a greater amount of entrapped air than a tightly woven one.

The second principle which can be utilized in the design of high-altitude clothing is a reliance upon multiple layers rather than upon a single, bulky garment. Layered clothing has two obvious advantages: (1) there is an air space between layers and (2) the layers can be removed or added to adjust to the immediate needs of the user. The importance of this latter consideration can scarcely be overstressed. The greatest threat to the insulation provided by clothing is that the air space will fill with water. Water with its high conductivity can remove heat at a rate up to 25 times that of air (Bullard & Rapp, 1970). Further, water evaporates from the surface of the garment and removes 0.58 kcal/g. An overclothed individual will perspire and lessen the effectiveness of his clothing. Peeling off the extra insulation and reducing heat buildup is thus desirable.

A third principle used at altitude is the layering of heavy and less permeable material over more coarsely woven material. Thus the outer layer should be of tight weave to reduce dampness and convective heat transfer.

* The Clo is a unit of insulation having a mean thermal resistance of 0.155 °C . m²/W. One Clo is equivalent to ordinary clothing worn indoors (Kerslake, 1972).

Finally there is the principle of color. At altitude there is a high ambient radiation so that solar heat gain is potentially high. Selecting darker colors should enhance the heat gain and offer some thermal advantage. Studies of reflectivity suggest that black color has a several-fold thermal advantage over white (Blum, 1945).

Consideration of these principles suggests a model for high-altitude clothing. The fabric should be wool or a similar fiber which entraps air. It should be constructed with a loose weave for internal clothing and a tighter weave for exterior wear. Clothing should be used in layers which can be removed as the situation dictates and the outer garments should be black or some other dark color to take advantage of the high ambient radiation.

### Andean clothing

In general Andean clothing conforms to the model. There are regional variations in details but all variants revolve about the same pattern.

Men in the Andes wear woolen homespun pants which are mid-calf in length. They are worn over one or more layers of loosely-knit woolen underwear. A knitted, sleeveless undershirt is used under a cotton shirt with long sleeves. The latter is usually purchased in a local store. A colorful jacket, matching the pants, is also used. The outfit is completed with a felt hat and a *poncho*. The *poncho* is made of a tightly woven material, usually wool or cotton and extends to the mid-calf.

Women wear several woolen skirts and a long sleeved jacket of similar material. They also may use knitted underwear but like the men wear a manufactured cotton blouse. Women carry over their shoulders a shawl which is similar in construction to the *poncho* of the men with the exception that it is somewhat shorter and is held about the neck with a pin. They also wear felt hats or on occasion the more traditional 'mortarboard' variety. Skirts are usually black or dark red as are jackets.

In cool weather, several layers of clothing may be used at the same time. Usually the newer garments are used on the outside and the older ones on the inside. Men may use a stocking cap under the felt hat while women pull the shawl over the neck as protection. On occasion a larger, warmer shawl may be utilized over the smaller one. In the cold, gloves are sometimes used by either sex, but are generally rare. Footwear is limited to rubber tire sandals and frequently even they are not used.

In general the Quechua Indian seems well protected in terms of clothing, except for the feet, which are always exposed. Despite the potential for cold stress to the feet being rather high, there is no effort to protect them.

Hanna (1968, 1970) and Larsen (1973) have reported on the effectiveness of Quechua clothing. In one controlled laboratory study, men and women were tested at 10 °C on two different days. One day the men wore athletic shorts and the women brief smocks. On the other day they wore their usual clothing but without hats, shawls or *ponchos*. Over the 2-h exposure in 10 °C air, surface temperatures, rectal temperature and oxygen consumption were measured. The differences between the clothed and unclothed trials provided an estimate of the contribution of clothing to microclimate temperatures.

Under these conditions the microclimate temperature with clothing was 4 °C higher. The caloric savings over the two hours were 139 kcal for men and 109 kcal for women. The surface temperatures which showed the greatest increase with the addition of clothing were those of the hands and feet. Insulation values for men's clothing were 1.21 Clo and for women, 1.43 Clo. These values are substantial and would have been even greater had hats and outer coverings been utilized. Larsen and Hanna found native clothing adequate for most naturally occurring cold situations.

### Himalayan clothing

There have been no detailed studies of Sherpa clothing despite its obvious importance to survival in the Himalayas. From the ethnographic work of Fürer-Haimendorf (1964) some details can be ascertained. There is a description of 'best' clothing, that is clothing reserved for special occasions. For men this includes brocade hats, white shirts, dark woolen coats and colorful boots. Women wear silk blouses of bright colors, long dresses of silk or wool and silk or woolen aprons. Photographs from the same source suggest a different type of clothing for everyday use. Women are shown wearing long-sleeved cotton blouses which are covered by woolen jackets and ankle-length skirts. Men's dress also seems substantial with long jackets, long pants and heavy coats. There seems to be a relationship between altitude and materials used such that cotton is utilized in the lower regions and wool predominates at high elevations (Maron *et al.*, 1956). As among Quechua Indians, most Sherpas seem to walk barefoot.

The protection offered the Sherpa by their clothing has not been documented, but the description of a single well-clothed Sherpa exposed to the cold was given by Pugh (1963). This man wore cotton pants, a cotton shirt, a woolen vest, a sleeveless pullover sweater, an old coat and a large turban. With this clothing and no shoes or gloves he slept behind rocks in the snow at ambient temperatures of −13 to −15 °C. Under laboratory conditions during a 3 to 4-h exposure to 0 °C, he showed no dangerous reduction in any surface temperatures. In the laboratory he also slept well at night, fully clothed at ambient temperatures of −5 to −10 °C. If this is typical of Sherpa clothing, it appears adequate even at very low temperatures.

There are no detailed descriptions of the clothing of Ethiopian highlanders, nor have there been any quantitative estimates of the effect of their clothing on their microclimate. Hence, Ethiopian clothing will not be considered here.

The two varieties of high-altitude clothing which have been studied suggest adequate protection under laboratory conditions. In design they both seem to include the several principles which have been outlined above. Wool is the favored material, at least in the cold regions. There are several layers topped off by a wind-breaking material of dark color. Curiously, neither Quechua Indians nor Sherpas wear boots or thermally protective footwear. Descriptions of foot temperatures during cold exposure suggest that rather than insulation, they rely upon elevated peripheral temperatures to protect their feet (Little, 1969; Pugh, 1963).

## Subsistence activities

In the Andes, Himalayas and the Ethiopian highlands there is a mix of herding and agriculture as a subsistence base. Thus each of the populations has ample exposure to the elements. Herding is usually a daily activity which involves sedentary activities. Under cold conditions this is potentially a stressful condition. Agriculture as practiced in arid upland regions requires planting during the rainy season. Hence the agriculturalist is exposed to dampness as well as cold. In the three areas there is undoubtedly a great variation in type and degree of exposure so that cultural adaptations to exposure are probably quite different.

In the southern Peruvian Andes there are two periods of maximal cold exposure as noted above. Quantitative data have been given by Baker (1968). The most severe period in terms of ambient temperature, radiation cooling and general aridity occurs from May through August. This is the dry season and there is not a great amount of agricultural activity. Exposure is potentially most stressful during the nights of this season when the clear skies promote radiation cooling. During this season night-time temperatures frequently fall below freezing, which could pose a problem for a population with a limited technology and unheated houses. During the days solar radiation is intense.

A second period of cold in the Andes occurs during the rainy season, which begins gradually in September and lasts through to April. This season is characterized by intense thunderstorm activity which produces locally heavy precipitation accompanied by gusty winds. Daily cloud cover limits heat gain from solar radiation, but also reduces long-wavelength radiation exchange with the sky. Perhaps the most critical aspect of the climate during this period is precipitation which may fall in the form of rain, sleet, hail or snow. Heavy winds usually accompany precipitation and enhance heat loss. In addition to cooling the lower extremities, the whole body may become wet and chilled while the individual is outside

265

tending the herds on rainy mornings or afternoons. The planting season, which may begin as early as September and carry over to October, is a time of the year when the individual may suffer an added degree of cold stress. Since the rainy season begins in September, the natives may spend long hours in the fields under cold-wet conditions, thus they cannot escape some exposure.

No clear description of the interaction of climatic stress and subsistence is available for the Himalayas or the Ethiopian highlands, so that it is difficult to assess the kind and degree of exposure in those regions. In Ethiopia ambient temperatures during the day have been reported to reach 20 to 35 °C but they also fall below freezing at night. Solar radiation is intense during all seasons (Harrison *et al.*, 1969). The stresses in this region must be associated with cold nights and intense radiation in the days. In the Himalayas there also appears to be considerable diurnal variation, but in the absence of quantification it is difficult to assess its consequences.

## Cold stress at high altitude

*Field and simulated studies of natural cold exposure*

The only quantitative estimates of cold exposure experienced by high-altitude peoples have been reported from the Andes. As a result of a comprehensive program to determine the degree of cold normally experienced by Quechua Indians, the major dimensions of their cold exposure are now known. In the absence of measurements from other areas these data may serve as a tentative model.

Baker (1966b) measured the microclimate temperatures of sleeping Quechua Indians during the coldest part of the year. Data included rectal, chest, finger, toe, indoor and outdoor temperatures with samples of eighteen men, ten women, and twenty-eight children of both sexes. Microclimate was measured through skin-surface temperatures and body-heat status was estimated through rectal temperature. Measurements were taken periodically as the families slept over an 8-h period. All were measured in their own homes using customary native bedding and under normal behavioral practices.

The results suggest that they were not greatly stressed at this time. While indoor temperatures fell to near freezing levels, rectal temperatures of all groups remained relatively high (see Fig. 9.6). Men and women experienced similar skin and similar rectal temperatures as the night progressed, but children showed a different pattern. There was an initial rapid fall in rectal temperature and then equilibrium for the rest of the night. The final rectal value for children was 36.4 °C, which is not excessively low. There was also a relationship between final rectal temperature and age in children with the younger children falling more rapidly. This suggests a cooling

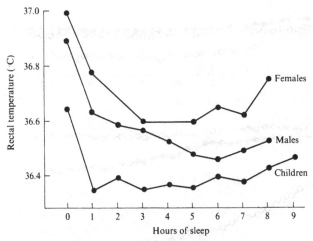

Fig. 9.6. Rectal temperatures of Quechua Indian Men ($N = 18$), women ($N = 10$), and children ($N = 28$) during normal sleeping conditions in native huts. All families were measured in the district of Nuñoa at between 4,000 and 4,600 m. (Adapted from Baker, 1966*b*.)

pattern in which size plays an important part. The younger children were smaller with a greater surface-area-to-mass ratio and hence cooled more rapidly. Skin temperatures did not differ greatly except for toe temperature. Children had the warmest toes while women had the coldest. Except for the first hour of sleep, extremity temperatures were quite warm, so that little real cold stress was experienced.

Baker also reported sleeping behavior for the same subjects and his report suggests attempts to limit heat loss. For example, the family usually slept together in groups of between two and four. All slept in a fetal position even though they were covered with bedding material and were clothed.

The effectiveness of Quechua Indian bedding was assessed in the laboratory and reported by Mazess & Larsen (1972). Six young Quechua Indian men were recruited to sleep overnight under three different conditions. The first condition involved sleeping in a chamber at 23 °C using a light woolen blanket bag. This was a thermal neutral approximation. The second condition involved use of the same blanket bag but at a room temperature of only 4 °C, and the third condition was at 4 °C with native bedding. The latter temperature is slightly warmer than average indoor temperatures reported by Baker (1966*b*). The men would normally have slept with clothing, but this was not permitted in order to derive data that were comparable with other published sources. Similarly, each subject slept alone rather than in family groups as would have occurred under

267

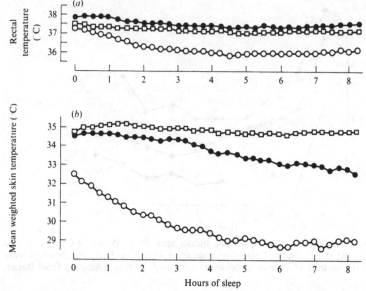

Fig. 9.7. Rectal temperatures (*a*) and mean weighted skin temperatures (*b*) of six Quechua Indian men during three successive night-time sleeping tests in the laboratory. 'Warm blanket' conditions (□——□) were at a room temperature of 23 °C with a single-layer woollen blanket-bag used; 'cold blanket' conditions (O——O) were at a room temperature of 4 °C with the same blanket-bag used; 'native bedding' conditions (●——●) were at a room temperature of 4 °C with subjects employing customary bedclothes of sheepskin undercovers and hand-woven woolen blankets as overcovers. Tests were conducted in Nuñoa, Peru at 4,000 m. (Adapted from Mazess & Larsen, 1972.)

normal sleeping conditions. During the 8-h night, various surface temperatures, rectal temperature and metabolic rate were monitored.

Fig. 9.7 illustrates the night-time changes in rectal and mean weighted skin temperatures. Use of native bedding allowed body temperatures to remain close to thermal neutral control levels. Much of the drop in mean weighted skin temperature with native bedding, however, resulted from temperature declines at the surfaces of the hand and foot. Finger, foot and toe temperatures dropped 8, 14, and 15 °C, respectively, during the native bedding exposure, and toe temperature actually approached the low value achieved during the blanket-bag test at 4 °C. There were also periodic increases in metabolic rate indicating that brief bouts of shivering took place. These cooling patterns might not have occurred had clothing been worn and 'clustered' sleeping been permitted. In general, the investigations of Baker (1966*b*) and Mazess & Larsen (1972) suggest that native bedding, when used in conjunction with clothing and clustered sleeping, is adequate to meet the conditions of cold normally encountered at night.

268

Baker's (1966*b*) study of sleeping temperatures was undertaken during nights of the dry season when ambient temperatures were at their annual minimum. Larsen (1973) presented data collected during the same season which support Baker's conclusions. He found that although children tended toward slightly greater heat loss than adults, the actual cold stress experienced by Quechua Indian men, women and children was not severe.

Larsen (1973) also measured microclimate temperatures throughout the day. Ambient temperatures ranged from $-6$ °C at 6 a.m. to 20 °C at 1 p.m., but despite the great range the mean exterior temperatures on the surface of the clothing fluctuated from about 16 to 30 °C. Beneath the clothing, individual microclimate temperatures at the level of the chest averaged between 30 and 35 °C. Women maintained the warmest microclimate with men slightly cooler and children the coldest. In children, who as a group were coldest, chest surface temperatures never fell below 29.5 °C, although hand and foot surfaces averaged slightly less than 20 °C in the late afternoon–early evening hours. Their microclimate temperatures were coldest early in the morning and just before retiring in the evening. This corresponds to Baker's (1966*b*) observations that children start the night colder and wake up colder.

In the Andes during the dry season, cold stress is night-time stress. The greatest cold during this period is experienced by children. They are colder before retiring, they are colder during sleep, and they wake up colder. If cold acts as a selective force, children are the unit to study (Little & Hochner, 1973).

A second period of potential cold stress occurs in the rainy season. Precipitation falls during the afternoons when the population is outdoors – either in the fields or tending the herds. Rain, snow, sleet and hail are driven by high winds providing potential for evaporative heat loss and convective heat loss.

Several Quechua Indian families were followed through their daily activities during this season by Hanna (1968, 1970). Ten families, including adults and children, were studied throughout the day from early morning to bedtime. Each two hours, rectal temperature, skin temperatures and climatic parameters were measured. A total of 355 temperature observations were taken on sixty-two different individuals whose ages ranged from three to eighty years. The families represented both herding and agricultural subsistence complexes which Thomas (1973) has shown may occasion different types of cold exposure. In the low-altitude regions, where mixed herding and agriculture were practiced, actual involvement in agriculture activities was not observed so that most observations were on herding-related activities. Thomas (1973) has shown that metabolic heat production during herding is much lower than that for most agricultural activities, so agriculturalists were at lower-than-normal levels of heat pro-

269

Table 9.1. *Mean temperature (°C) responses of adults and children*

| Site | N | Boys, 3–15 yr of age | Girls, 3–15 yr of age | Men, 15–80 yr of age | Women, 15–80 yr of age |
|---|---|---|---|---|---|
| Forehead | 13 | 31.6 | 31.3 | 31.7 | 31.7 |
| Chest | 13 | 33.4 | 33.5 | 33.5 | 33.7 |
| Arm | 13 | 30.3 | 29.9 | 32.7 | 31.5 |
| Hand | 13 | 26.4 | 24.9 | 28.7 | 28.3 |
| Finger | 13 | 22.3 | 21.5 | 24.7 | 26.0 |
| Leg | 13 | 29.0 | 26.6 | 30.9 | 28.5 |
| Foot | 13 | 26.4 | 23.5 | 28.4 | 27.3 |
| Toe | 13 | 21.7 | 20.2 | 22.6 | 22.5 |
| Rectal[a] | 10 | 37.4 | 37.3 | 37.2 | 36.8 |
| Ambient[b] temperature | 13 | 11.4 | 11.2 | 10.8 | 11.7 |

Temperatures are means of five measurements taken at 2-h intervals throughout the day.
[a] Some children would not accept the rectal probes, so oral temperature was taken. Simultaneous rectal and oral temperatures on eight young boys and girls showed rectal temperature was 0.3 °C higher. This correction was made.
[b] Sum of daily environmental temperatures corresponding to the times of surface temperature measurements.

duction during the tests. These observations may then approach the coldest conditions to which these people are normally exposed during this season.

Men were more frequently involved in herding while women remained closer to home. Women and small children could then seek refuge from the storms in their houses. When the sky was clear both women and children were outside taking advantage of the warm sun. Older children attended the herds, usually in groups from two to five. They were actively playing and chasing animals most of the time so that their heat production was elevated. Men frequently accompanied the children but also visited neighbors or examined the maturing crops. Thomas (1973) has described these activities in detail.

A summary of the temperature data for the fifty-eight individuals whose measurements were most complete is presented in Table 9.1. The temperatures represented averages for the various sites taken throughout the day. Surface temperatures did not appear to fall very low; indeed, the values are near thermal neutral temperatures for Quechua Indians (Mazess & Larsen, 1972). Even hand and foot temperatures, which fell to the lowest mean values, were always above 20 °C. If these are taken as representative temperatures, then cold stress was not great. None of the rectal temperatures were depressed and the reduced rectal temperature in the group of women probably reflected inactivity while remaining near the house.

Table 9.2. *Skin temperature differences between subjects tested at lower elevation (4,000 m) and higher elevation (4,300 m) field sites*

| Site | N | Lower elevation (4,000 m) | Higher elevation (4,300 m) | Difference |
|------|---|---------------------------|----------------------------|------------|
| Forehead | 48 | 32.0 | 30.3 | 1.7 |
| Chest | 48 | 34.1 | 32.8 | 1.3 |
| Arm | 48 | 31.4 | 30.5 | 0.9 |
| Hand | 48 | 27.7 | 26.1 | 1.6 |
| Finger | 48 | 25.3 | 21.6 | 3.7 |
| Leg | 48 | 29.6 | 27.8 | 1.8 |
| Foot | 48 | 26.6 | 25.8 | 0.8 |
| Toe | 48 | 22.8 | 20.2 | 2.6 |
| Ambient | 48 | 12.4 | 10.1 | 2.3 |

The effects of differential altitude were estimated by dividing the sample into those studied at higher elevations and those studied at lower elevations (4,300 and 4,000 m, respectively). Table 9.2 summarizes these observations and suggests that skin-surface temperatures, which reflect microclimate, do show significant differences with altitude. At the higher elevations, where temperatures are lowest, there is still no evidence of excessively low surface or core temperatures. However, finger and toe temperatures are approaching levels that can produce some discomfort.

To determine the maximum cold which might be anticipated under the conditions of this study, the minimum ambient temperature observed during each day was recorded. The average was 6.2 °C and it usually occurred in mid-afternoon when the ground was damp. The surface and rectal temperatures which corresponded to these lowest ambients are presented in Table 9.3. Toe and finger temperatures were lowest. The mean rectal temperature is normal, 37.15 °C, suggesting that there was no general heat loss. The minimum temperatures recorded at each site are also presented in the table. The low rectal values of several women who were sitting inside houses at this time and who had been inactive over most of the day are included in this average value. Larsen (1973) also found the lowest core temperatures among inactive women during the dry season, so that in general, women might be somewhat cooler than men due to their greater periods of inactivity.

A regression of minimum temperature on body weight was performed. It was anticipated that smaller people would experience greater cooling, but the correlation between core temperature (as estimated by rectal temperature) and ambient temperature was not significant. Thus, during

271

Table 9.3. *Temperature of various sites measured when daily dry bulb temperatures were lowest*

| Site | N | Temperature (°C) | | |
|---|---|---|---|---|
| | | Mean | S.D. | Minimum |
| Head | 54 | 29.6 | 3.3 | 21.0 |
| Chest | 54 | 32.3 | 2.4 | 27.3 |
| Forearm | 54 | 29.1 | 3.2 | 20.5 |
| Hand | 54 | 22.5 | 4.6 | 13.1 |
| Finger | 54 | 18.8 | 6.1 | 10.0 |
| Leg | 54 | 27.1 | 3.6 | 22.0 |
| Foot | 54 | 20.9 | 4.9 | 13.4 |
| Toe | 54 | 16.5 | 5.2 | 7.7 |
| Rectal[a] | 54 | 37.15 | 0.65 | 36.00 |
| Ambient | 9 | 6.5 | 1.9 | 3.3 |

[a] Rectal temperature was measured on 40 subjects. For the remaining 14, oral temperature was measured and 0.3 deg C added. The 14 who refused rectals were all children so that correction is based upon simultaneous measurements of oral and rectal temperatures in eight children who were willing to have both measured.

the day body size appears not to be related to heat loss from the core. A similar regression of body weight against toe temperature was undertaken to determine if body size had any effect on peripheral cooling during the cold period. The correlation was $r = 0.37$ ($P < 0.05$), which is significant for fifty-eight pairs of observations. The relationship is described by: minimum toe temperature = 0.049 (body weight in kg) + 12.5. Hence, body size appears important for peripheral cooling but not for heat loss from the core. The intercept (at zero body weight) is theoretically the minimum temperature to which anyone would be exposed, yet the value of 12 °C is still above the temperature required to produce tissue injury.

The absence of a significant relationship between body weight and rectal temperature and the significant relationship between weight and toe temperature underlines a point made earlier: extremity temperatures may show volume-related cooling but core temperature does not appear to be influenced by size.

A final field study in the Andes involved the simulation of hand and foot cooling in water that is normally experienced by Quechua Indians of both sexes during day-to-day activities (Jones *et al.*, 1976). Contacts of the extremities with cold water are especially stressful, since cold water acts both as a medium for evaporative cooling and as a heat-sink. All highland Indians are routinely exposed to wet-cold conditions by walking barefoot through wet snow, slush and rainpools, and by the need to ford mountain streams. Women wash cooking pots and prepare a form of dried potato

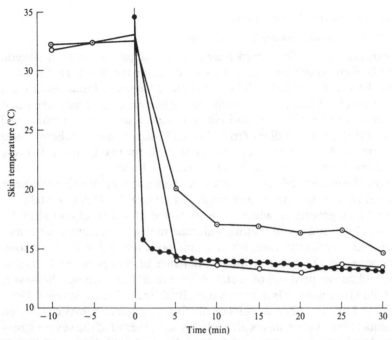

Fig. 9.8. Comparison of foot temperatures of Quechua Indian women (*N* = 9) while washing clothes in the Nuñoa river (⊙——⊙), Quechua Indian men (*N* = 10) while constructing a channel in the same river at a water temperature of 10 °C (O——O), and foot exposure of Quechua Indian men (*N* = 12) in 10 °C water in the laboratory (●——●). Tests were conducted at 4,000 m. (Adapted from Jones *et al.*, 1976.)

in cold water, and men are responsible for drying another form of potato by first allowing the potato to soak for several days in a stream. Men further expose their feet to wet-cold during the preparation of mud used for adobe bricks, the dried earth cooking stove, and caulking for stone huts. Other activities in which both hands and feet are exposed to wet-cold include construction and maintenance of irrigation ditches and channels by Indian men, and the laundering of clothing.

Correspondingly, tests were conducted on women while they washed clothes in the Nuñoa river and men while they constructed a diversion channel in the same river. During these tasks, both men and women had their feet immersed in 10 °C water at from ankle- to mid-calf-depth, and hands immersed intermittently. Foot temperatures dropped rapidly to levels that were comparable to the responses that were elicited during laboratory exposures (see Fig. 9.8), although foot temperatures of women were slightly warmer than men. Clearly, these and similar activities which are carried out weekly produce a high degree of cold stress to the extremities because of the nature of wet-cold exposure.

273

## The biology of high-altitude peoples

#### Whole-body and extremity exposures

The earliest standardized cold tolerance investigations of high-altitude residents were conducted in the early 1960s in the southern Peruvian Andes. Elsner & Bolstad (1963) tested eight Quechua Indian men from the vicinity of La Raya (4,200 m) during night-time cold at between 2 and 5 °C while the men slept in wool blanket-bags. Later, Baker (1966a) and Hanna (1965) studied Indians from Cuzco (3,300 m) and Chinchero (3,800 m) during whole-body exposures to air at 14 °C for two hours and during finger immersion tests in water at 0 °C for one hour.

In both the whole-body cold studies, Andean natives maintained warmer hand and foot temperatures and displayed greater declines in rectal temperature than groups of white subjects tested at sea level and altitude. Hammel (1964) suggested that high-altitude natives represent a pattern of *hypothermic acclimatization*, where stored body heat is lost during whole-body cooling as the result of the maintenance of high peripheral temperatures and consequent fall of rectal temperature. Later studies, however, have failed to confirm this model (Baker, Buskirk, Kollias & Mazess, 1967; Mazess & Larsen, 1972). Results of the finger immersion tests conducted by Hanna (1965) were equivocal, probably as the result of the severe stress imposed by ice-water immersion. Indians did show slightly warmer fingers than whites during the 1-h test, but differences were minimal.

At about the same time these early investigations were being conducted in Peru, similar studies were carried out in the Ladakh area of Kashmir, India (4,000 m). These studies were reported in several papers (Gupta, 1964; Gupta, Karani, Soni & Das Gupta, 1964a; Gupta et al., 1964b; Davis et al., 1966), and involved the testing of low-altitude Jats, high-altitude acclimatized Jats, and native Tibetan highlanders in 60-min whole-body cold tests in 2 °C air. The Jats were Indian soldiers who are members of a cultivator caste of northwest India. Low-altitude Jats were sea-level residents who were tested shortly after arrival at 4,000 m; high-altitude Jats had been resident at 4,000 m for 11 months prior to testing; Tibetans were refugees who had spent a lifetime at elevations greater than 3,500 m.

Results of some of the tests are presented in Fig. 9.9. Since subjects were dressed only in shorts, the test at 2 °C can be identified as an acute cold stress, and is not directly comparable with the whole-body cooling experiments conducted in the Andes. Also, the sample sizes are quite small; hence the results should be interpreted with some degree of caution. Several subjects from the two Jat samples were unable to complete the tests, so the data are incomplete. Tibetans had highest skin temperature, largely resulting from elevated extremity surface temperatures (see the

274

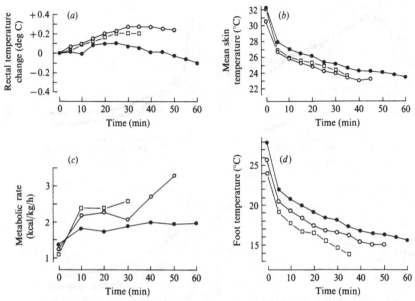

Fig. 9.9. Rectal temperature change (*a*), mean skin temperature (*b*), metabolic rate (*c*), and foot temperature (*d*) of high-altitude Tibetans ($N = 5$, ●——●), Jat soldiers from India who had resided at high altitude for 11 months ('high-altitude Jats', $N = 5$, ○——○), and Jat soldiers who were newcomers to altitude ('low-altitude Jats', $N = 6$, □——□) during whole-body exposure for 1 h at a room temperature of 2 °C. Tests were conducted in the Karakoram mountains of Kashmir at 4,000 m. (Adapted from Davis *et al.*, 1966.)

graph of foot temperature in Fig. 9.9). As shown by the decline in rectal temperature, the Tibetans were probably losing more heat than the two Jat groups. Metabolic rate increase and rate of shivering during the test were very low in the Tibetans. In general, low-altitude Jats shivered more than acclimatized high-altitude Jats, and although rectal and mean skin temperatures were roughly equivalent in the two Jat groups, foot temperature dropped more precipitously in the unacclimatized group.

The two groups of Jat soldiers were also tested at sea level in Delhi, India (not shown). At sea level, both groups shivered less, maintained warmer surface temperatures, and displayed greater declines in rectal temperature.

The Ladakh study can be contrasted with tests of whole-body cooling for two hours at 10 °C in Nuñoa, Peru (4,000 m). Results of these Andean studies are presented in Fig. 9.10. Nuñoa Indians were lifelong residents of high altitude; Mollendo Indians were sea-level residents of the coast who were first or second generation migrants from the highlands to the coast, but had spent a lifetime at sea level; whites were scientists and students who had been lifelong residents at sea level in the United States.

275

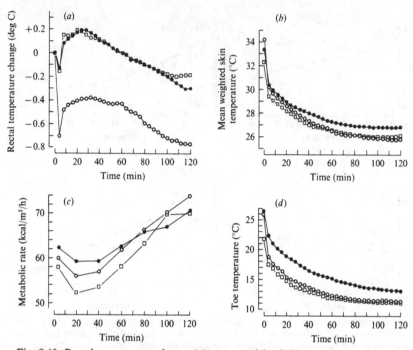

Fig. 9.10. Rectal temperature change (*a*), mean weighted skin temperature (*b*), metabolic rate (*c*) and toe temperature (*d*) of high-altitude Nuñoa Indian men (*N* = 26, ●——●), low-altitude Mollendo Indian men (*N* = 9, ○——○), and US white men (*N* = 15, □——□) during whole-body exposure for 2 h at a room temperature of 10 °C. Tests were conducted in Nuñoa, Peru at 4,000 m. (Adapted from Baker *et al.*, 1967; Weitz, 1969.)

Rectal temperature dropped to the greatest extent in Mollendo Indians and least in whites. However, this may reflect the disparity in initial core temperatures among the three groups. Mollendo Indians began with a relatively high rectal temperature of 37.9 °C, and Nuñoa Indians and whites with rectal temperatures of 37.0 °C and 36.7 °C, respectively. Mean body temperatures were also highest in Mollendo Indians and lowest in whites at the onset of the test, with greatest loss of stored heat in the Mollendo sample. Therefore, it seems likely that group differences in rectal temperature decline over the two hours resulted from the disparity in initial thermal state of the body. Metabolic rate was highest initially in the Nuñoa Indian sample, although during the second hour of the test, there were essentially no differences among the three groups. Trunk surface temperatures of the groups were equal, while extremity temperatures, and hence, mean weighted skin temperatures, of the Nuñoa group were elevated over the other two groups.

During the first hour of exposure, Nuñoa Indians and whites lost

276

roughly equal amounts of heat, but among the Nuñoa Indians this heat was drawn largely from the elevated metabolism rather than stored body heat. Despite the increased metabolic rate of Nuñoa Indians during the second hour of the test, they lost some stored heat, and at a greater rate than whites.

The general pattern of thermal and metabolic responses to whole-body cooling as exemplified by Tibetans (Davis *et al.*, 1966), and Andean Indians during long-duration (Mazess & Larsen, 1972) and short-duration (Baker *et al.*, 1967) exposures to cold is: (1) a lack of a dramatic fall in core temperature (see Figs. 9.7, 9.9 and 9.10), (2) slightly elevated metabolic rate, (3) consistently high extremity surface temperatures and (4) a slightly greater loss of body heat content than whites or sea-level residents tested under comparable conditions.

Highland natives and lowlanders were exposed also to local cold in standardized tests of extremity responses. Tests were conducted both in the southern Peruvian Andes (Nuñoa) and in the Karakoram mountains of Ladakh, India. The Ladakh studies will be discussed in a later section and the studies of Andean natives treated here.

Highland Quechua Indians were tested under a variety of conditions including exposure of the hand and foot to air at 0 °C (Little, Thomas, Mazess & Baker, 1971), exposure of the foot to water at 5, 10 and 15 °C (Little, 1969), and exposure of the hand to water at 4 °C (Little, Thomas & Larrick 1973). In each test, Quechua Indians displayed warmer skin temperatures, and hence, increased blood flow to the surface of the extremities, than white subjects who were variously tested at sea level and at high altitude. Indian–white differences, however, were not uniformly great. Under conditions of moderate cold stress to the foot (air at 0 °C), after one hour of exposure Indian skin temperatures were between 5 and 10 °C warmer than those of whites. Moreover, although there was no distinct pattern of differences between the upper and lower extremities in whites, among the Indian subjects tested, foot surface temperatures were usually warmer than corresponding temperatures of the hand. This may result from differential exposure of the upper and lower extremities and local acclimatization, since among Quechua Indians individuals appear to suffer more cold stress of the feet than the hands during normal activities.

During the more severe tests of hand and foot cooling in water, Indian–white differences were markedly less than those observed during tests in air. For example, the tests of finger cooling in ice water conducted in Chinchero showed minimal group differences (Hanna, 1965); tests of foot cooling at 5 °C (Little, 1969) and hand cooling at 4 °C (Little *et al.*, 1973) showed slight but statistically non-significant differences; while tests of foot cooling at 10 and 15 °C showed somewhat clearer differences (Little,

277

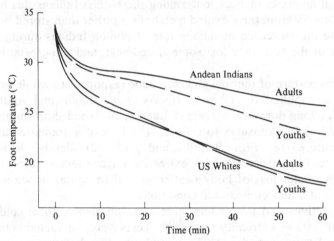

Fig. 9.11. Dorsal foot temperatures of Nuñoa Indian men ($N = 30$, mean age, 32 yr) and boys ($N. = 29$, mean age, 14 yr) tested at 4,000 m and US white men ($N. = 26$, mean age, 27 yr) and boys (N. = 28, mean age, 13 yr) tested at sea level during foot exposure to air at 0 °C for 60 min. (Adapted from Little *et al.*, 1971.)

1969). Apparently, there are limits to the extremity cold tolerance of the Andean native such that his peripheral vasomotor system operates more effectively under conditions of moderate cold stress.

The microcirculatory mechanisms that enable Andean natives to maintain warm hands and feet during both local extremity and whole-body cold exposures have not been fully identified. It does seem that the specialized reaction of cold-induced vasodilation and the more general relaxation of vasoconstrictor control of arterioles act jointly to produce warm extremity surface temperatures during moderate cold stress. Under more severe cold stress, the former response is of greater importance; under mild cold stress, the latter. Whether these mechanisms are based upon developmental or short-term or other acclimatizations is not known. Age differences, however, were observed in tests of foot exposure to air at 0 °C (see Fig. 9.11). Quechua Indian youths were unable to maintain foot surface temperatures at levels that were comparable with Indian adults, whereas both age groups of whites were uniformly lower and equivalent. This suggests some degree of developmental acclimatization in the highland natives. Studies of Athapascan Indians (Williams, Petajan & Lee, 1969) and Eskimos (Miller & Irving, 1962) have shown similar age variations in children and adults.

Sex differences were observed in several studies of Andean natives. Hanna (1970 and unpublished data) found that Quechua Indian women had warmer core temperature and slightly warmer mean weighted skin temperature than men during whole-body exposure to air at 10 °C for two hours. Women also had warmer finger temperatures than men, but the toe temperature of men was warmer during this test. These sex differences were attributed largely to the greater adiposity characteristic of females leading to better insulation against conductive heat transfer to the surface of the body. Differences by sex were also observed in extremity cold tests (Little *et al.*, 1973; Jones *et al.*, 1976; see Fig. 9.8). In hand-immersion experiments in water at 4 °C, Quechua Indian women began with slightly warmer resting hand temperatures than the men, maintained warmer hands throughout the 20-min test, and showed a more rapid temperature recovery following removal of the hand from the cold water (Little *et al.*, 1973). Under field conditions in Nuñoa, women also maintained warmer hand and foot temperatures than men while performing tasks in river water (Jones *et al.*, 1976). Others have also demonstrated a female advantage over males in hand cooling tests of Eskimos (Krog & Wika, 1971–2; Lund-Larsen, Wika & Krog, 1970) and Japanese and Ainus (Kondo, 1969; Suzuki, 1969).

### Basal metabolism

A persistent finding that relates to temperature regulation at high altitude is the elevated basal metabolic rate (BMR) or heat production found in both sojourners and permanent residents at altitude. In the USA Klausen (1966) and Grover (1963) found slightly higher basal metabolism among sojourners at 3,800 m and 4,300 m, respectively. In the Himalayas at 5,800 m, Gill & Pugh (1964) found elevated basal metabolism values in both sojourners (white mountaineers) and permanent altitude residents (Sherpa natives). Further west in the Karakoram mountains, transient increases in BMR were noted in soldiers from lowland India as well as elevated values for an unspecified group of Ladakh permanent residents at 3,300 m (Nair, Malhotra & Gopinath, 1971). Most comparisons have been based upon sea-level standards of Boothby, Berkson & Dunn (1936).

Investigations of Quechua Indians from Peru (Picón-Reátegui, 1961; Mazess, Picón-Reátegui, Thomas & Little, 1969) found that BMR was elevated about 5% above Boothby standards. When lean-body mass as a reference standard was used rather than body surface area, Quechua Indians at 4,500 m showed an even greater basal metabolism than sea-level residents (Picón-Reátegui, 1961). Furthermore, in whole-body cooling tests conducted in Nuñoa, Peru, initial resting metabolic rates were usually higher in the Indians than in whites tested at sea level (Mazess & Larsen, 1972) or high altitude (Baker *et al.*, 1967).

It appears that among sojourners, BMR is initially elevated upon arrival at high altitude, and the increased basal metabolism persists for some time – probably more than a month or two. For example, Burrus *et al.* (1974)

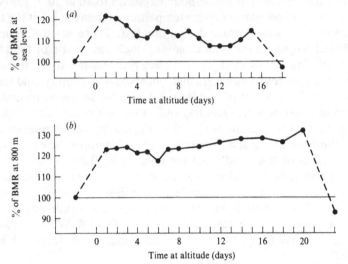

Fig. 9.12. (*a*) Change in basal metabolic rate (BMR) in six white male athletes during a sojourn of 15 days at 3,090 m. (*b*) Change in basal metabolic rate of three white male non-athletes during a sojourn of 19 days at 3,800 m. Tests were conducted in the USA. (Adapted from Burrus *et al.*, 1974.)

recorded BMR values on sea-level residents at altitude for several weeks at White Mountain, California, and found that values were still high after 19 continuous days at 3,800 m (see Fig. 9.12). They concluded that individual variation in responses was great, and although the evidence was not definitive, that the elevated BMR in sojourners was transient. BMR was recorded in eight women during a stay of longer duration at Pike's Peak, Colorado (4,300 m) by Hannon & Sudman (1973). The change in basal metabolism and basal oral temperature during a high-altitude sojourn of 78 days is illustrated in Fig. 9.13. Oral temperature returned to sea-level values after one month, while BMR required about two months before values had returned to normal. The authors attributed almost 50% of the increase in BMR during the initial days of acclimatization to altitude to parallel increases in body temperature and to the greater oxygen require-ments of increased mechanical work of the heart at altitude.

It is likely, then, that the rise in basal metabolism and subsequent fall to sea-level values among sojourners after two or more months at altitude reflects one aspect of the general acclimatizational process that is linked to hypoxia rather than to cold stress. The chronically high BMR of permanent residents of high altitudes is also probably associated with acclimatization to hypoxia, perhaps through increased thyroid activity (Beckwitt, Surks & Chidsey, 1966; Mazess *et al.*, 1969; Hannon & Sudman, 1973). In any

280

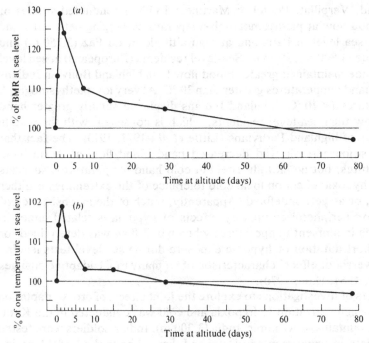

Fig. 9.13. Changes in (*a*) basal metabolic rate (BMR) and (*b*) basal oral temperature (BOT) in eight white women during a sojourn of 78 days at 4,300 m. Tests were conducted in the USA. (Adapted from Hannon & Sudman, 1973.)

case, if the elevated BMR is produced by chronic altitude exposure, then there is a fortuitous advantage gained in combating cold stress as well.

### Cold and hypoxic stresses

Effects of the interaction of cold and hypoxic stresses on humans are not very clear. Negative cross-adaptation to hypoxia and cold was demonstrated in laboratory animals by differential survival (Fregley, 1954; Hale, 1970). However, the research designs necessary to study cross-adaptation in humans are difficult and expensive to apply. For example, laboratory tests of simulated altitude in hypobaric chambers or by use of low-oxygen mixtures are generally of short duration and do not allow time for acclimatization to take place. Several investigations have shown an increase in skin blood flow under laboratory conditions of reduced oxygen pressure (Schneider & Sisco, 1914; Kottke *et al.*, 1948; Newman & Cipriano, 1974). Other investigations have shown inconsistent increases as well as decreases in extremity blood flow under hypoxic conditions (Freeman, Shaw & Snyder, 1936; Gellhorn & Steck, 1938; Abramson, Landt & Benjamin, 1941; Lim & Luft, 1960).

Durand, Verpillat, Pradel & Martineaud (1969) conducted studies of hand blood flow at plethysmograph temperatures ranging between 7 and 43 °C at sea level in Paris and at high altitude in La Paz (3,750 m) and Chorolque (4,800 m), Bolivia. Sea-level residents (Europeans) at sea level and altitude maintained greater blood flow than highland Bolivian Indians at local hand temperatures greater than 30 °C. At very low plethysmograph temperatures (< 10 °C) highland Indians displayed slightly greater hand blood flow than sea-level residents, which is consistent with the results of studies of highland Peruvians (Little *et al.*, 1971, 1973). The fact that altitude sojourners showed decreased hand blood flow at warm hand temperatures, but no altitude effect at cold hand temperatures, suggests that the hypoxic effect on local cold tolerance of the extremities is either minimal, or as yet, undefined. Apparently, much of the inconsistency of studies on peripheral circulatory effects of hypoxia is related either to variations in ambient temperatures when blood flow was determined, or to the short duration of hypoxic exposure during sea-level tests leading to an 'overshoot effect' characteristic of so many rapid adaptive changes (Prosser, 1964; Hale, 1970).

A series of investigations to explore the joint effects of cross-adaptation or cross-acclimatization to hypoxia and cold was conducted in the Karakoram mountains of Kashmir, India (3,300 m). Indian soldiers were tested in a variety of experiments in Delhi and then taken to the Ladakh region at high altitude where sequential tests were continued for six weeks before returning to sea level. A first group of subjects lived under thermal neutral conditions (25–28 °C, ambient) and a second group was exposed to cold conditions (6–11 °C, ambient) for six hours each day during the first three weeks of their sojourn. Conditions were reversed during the second 3-week period for each of the two groups of subjects.

The results of four of the experiments are presented in Fig. 9.14. There were no differences between the two groups for basal oral temperature (Fig. 9.14*a*) which declined during the first four weeks and then began recovering (Nair & George, 1972). This pattern is not wholly at variance with the study of Hannon & Sudman (1973), since the rise in basal oral temperature that they observed peaked during the second day of the high-altitude sojourn (see Fig. 9.13). It is likely that this peak response was missed in the tests of Nair & George (1972). Basal metabolism (Fig. 9.14*b*) increased in both groups during the first week at 3,300 m and remained high in the initially cold-exposed group (Nair, Malhotra & Gopinath, 1971). The decline of BMR in the first group after week 1 of their sojourn was reversed upon imposition of cold exposure. This suggests that both cold and hypoxia were contributing to the elevation of BMR. However, previous attempts to produce an increase in BMR by laboratory cold exposures required several months for any detectable

282

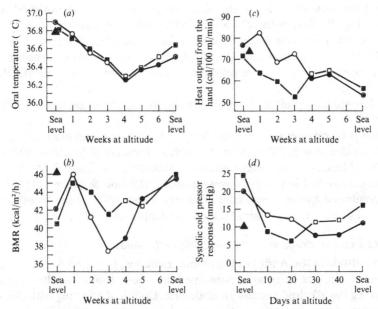

Fig. 9.14. (*a*) Combined effects of hypoxia and cold (■, ●) and hypoxia alone (□, ○) on oral temperature in two groups of soldiers from India (ten subjects in each group) compared with oral temperature in ten high-altitude natives (▲). (Adapted from Nair & George, 1972.)

(*b*) Combined effects of hypoxia and cold and hypoxia alone on BMR in two groups of Indian soldiers (ten subjects in each group) compared with BMR in five high-altitude natives. (Adapted from Nair, Malhotra & Gopinath, 1971.)

(*c*) Combined effects of hypoxia and cold and hypoxia alone on heat output from the hand during hand immersion in water at 4 °C for 30 min in two groups of Indian soldiers (eight subjects in each group) compared with hand heat output in nine high-altitude natives. (Adapted from Nair, Malhotra & Tiwari, 1973.)

(*d*) Combined effects of hypoxia and cold and hypoxia alone on cold pressor response during immersion of the hand in ice water for 2 min in two groups of Indian soldiers (nine subjects in each group) compared with cold pressor responses in ten high-altitude natives. (Adapted from Nair, Malhotra, Tiwari & Gopinath, 1971.)

Tests were conducted in Delhi at sea level and in the Karakoram mountains of Kashmir at 3,300 m.

change following daily, acute exposures (Rennie, Howell & Covino, 1964), and generally have been unsuccessful in simulating the dramatic seasonal changes observed by Hong (1963) in the diving women of Korea.

Heat output from the hand during immersion in water at 4 °C (Fig. 9.14*c*) declined to a greater extent under cold and hypoxic conditions than under only conditions of hypoxia (Nair, Malhotra & Tiwari, 1973). These findings imply that general cold stress and hypoxia independently produce peripheral vasoconstriction during hand cooling in sojourners, whereas highlanders, who were tested in the same experiments, maintained relatively high values for heat output. The cold pressor response, which was induced by hand immersion in ice water for two minutes, was diminished at altitude (Fig. 9.14*d*), with cold and hypoxia jointly producing a slightly greater diminution than hypoxia alone (Nair, Malhotra, Tiwari & Gopinath, 1971). The highland Ladakhis displayed low cold pressor responses, which is compatible with observations on highland Peruvian Indians (Little *et al.*, 1973).

In other studies, Nair and associates employing the same basic design found:

(1) basal blood pressure was unchanged by hypoxia and cold during a 6-week residence at 3,300 m (Nair, Malhotra, Gopinath & Mathew, 1971);

(2) basal heart rate was elevated (Nair, Malhotra, Gopinath & Mathew, 1971) in similar fashion to the observations of Hannon & Sudman (1973);

(3) combined hypoxia and cold produced a reduction in the threshold of *critical flicker fusion* (Nair, Malhotra & Gopinath, 1972).

### Influences of drugs on cold tolerance in the Andes

At high altitude in the Andes of Columbia, Ecuador, Peru and Bolivia, two principal drugs are regularly consumed by most Indian adults – coca (containing the alkaloid cocaine) and alcohol. Coca leaf chewing with *llipta* or calcium dates back to antiquity and is a practice nearly always associated with Indian ethnic groups. Alcohol is drunk in the form of sugarcane alcohol, which is diluted to about 40% by weight before being consumed, and *chicha*, a native corn beer whose manufacture is clearly pre-Hispanic in age. Both drugs are incorporated into ritual functions, and consumption is expected at social gatherings such as fiestas, marriages and funerals. Alcohol consumption is more sporadic than is coca consumption. Dried coca leaves are masticated several times each day with a total consumption of between 20 and 75 g/day (Hanna, 1974). Sugarcane alcohol is only consumed in the evening, on weekends or during fiestas. Among Indians who are pastoralist-farmers in the countryside, rates of alcohol consumption are considerably lower than among those who reside in towns. Basically, then, both drugs are well integrated into Andean Indian life and are important components of the culture of this high-altitude zone.

The effects of these two drugs on cold tolerance and exercise capacity have been explored in a number of studies conducted in the district of Nuñoa, Peru (Picón-Reátegui, 1968; Mazess, Picón-Reátegui, Thomas & Little, 1968; Little, 1970; Hanna, 1971*a*, *b*, 1974; Jones *et al.*, 1976), and in La Raya, Peru (Elsner & Bolstad, 1963).

Alcohol consumption is known to produce peripheral vasodilation and a subjective feeling of warmth (Cook & Brown, 1932; Forney, Hughes, Harger & Richards, 1964; Fewings, Hanna, Walsh & Whelan, 1966). Also, there is some evidence that alcohol metabolism by the body is somewhat

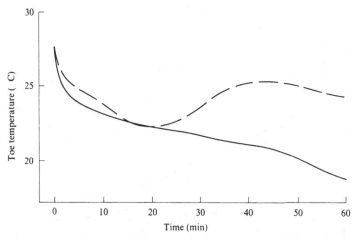

Fig. 9.15. Toe temperature of Nuñoa Indian men (*N.* = 25) during exposure of the foot to air at 0 °C for 60 min with alcohol consumption for 10 min, starting at minute 10 (broken line) and without consumption of alcohol (unbroken line). Tests were conducted at 4,000 m. (Adapted from Little, 1970.)

slower at altitude than at sea level (McFarland & Forbes, 1936), which may permit blood alcohol levels to remain elevated for a slightly longer duration among altitude residents. In light of the known peripheral vaso-dilatory effects of alcohol consumption, tests were conducted on a sample of Nuñoa Indian men to determine the effects of alcohol consumption on temperature responses of the lower extremity (Little, 1970). In this procedure, the foot was exposed to air at 0 °C with and without a 10-min period where the equivalent of 1.1 g of absolute alcohol per kg of body weight was consumed. The results of these tests are illustrated in Fig. 9.15. The short-term effect of the alcohol was almost immediate, producing an elevation of skin temperature that was 5 °C above control values at the termination of the test. In another test of Nuñoa Indians (Jones *et al.*, 1976), subjects walked over a measured course at an ambient temperature of 8 °C while fully clothed. An amount of alcohol equal to that of the foot cooling test was consumed, but had only a minimal effect on skin and rectal temperatures. In this case, activity may have accelerated the clearance of the drug, thus eliminating the thermal response. Alcohol consumption by Andean natives from time to time could contribute to a subjective feeling of thermal comfort, yet aside from this short-term effect, it is unlikely that any unique biological advantages are gained by the drinking patterns at altitude.

Investigations of the biological effects of coca chewing on cold re-sponses at high altitude have presented conflicting results. In two early

285

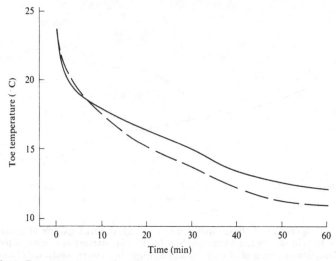

Fig. 9.16. Toe temperature of Nuñoa Indian boys (*N.* = 29) during exposure of the foot to air at 0 °C for 60 min with coca mastication (coca taken *ad libidum* after minute 10, broken line), and without coca mastication (unbroken line). Tests were conducted at 4,000 m. (Adapted from Little, 1970.)

Peruvian studies, coca mastication was found to increase metabolic rate and body temperature (Risemberg, 1944; Zapata Ortíz, 1944). More recently, Picón-Reátegui (1968), during a dietary study of metabolic balance, observed elevated body surface temperatures of Quechua Indians from Nuñoa when a period of coca use was compared with a period of abstention. Each of these studies was carried out under thermal neutral conditions.

Human responses to coca mastication under conditions of moderate cold stress appear to be qualitatively different from those under thermal neutral conditions. For example, Elsner & Bolstad (1963), in tests of four Quechua Indians during whole-body night-time cold exposures, found no influence of coca on rectal temperature, skin temperature or metabolic rate compared with the same subjects during abstention. Local exposure of the foot to cold air at 0 °C, however, resulted in slightly depressed toe temperatures during coca use in a sample of adolescent Quechua Indian boys (Little, 1970). Results of this test are represented in Fig. 9.16. Nearly all of the Indian boys who were tested were only occasional coca chewers and reported that their frequency of use was from twice each year to twice each week. However, 75% of the boys stated that they took the drug more often during cold or rainy weather. Hanna (1971*a*) tested the effects of coca use on whole-body cooling at 15.5 °C in a group of fourteen Quechua

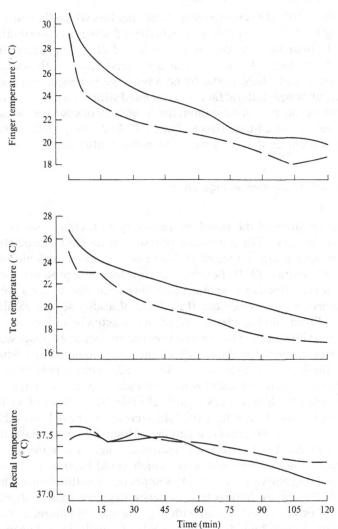

Fig. 9.17. Finger, toe and rectal temperatures of Nuñoa Indian men (*N*. = 14) during whole-body exposure to air at 15.5 °C for 2 h without coca mastication (unbroken line) and with coca mastication (coca taken *ad libidum*, broken line). (Adapted from Hanna, 1971.)

Indian men who were habitual users. Some of the results of Hanna's investigation are given in Fig. 9.17. Subjects were tested during a day of coca use and served as their own controls during a day without chewing coca. Head, arm, chest and thigh surface temperatures showed minimal or no effects of coca use, whereas toe and finger temperatures were depressed by coca use. Rectal temperature declined less during coca use,

287

probably as the result of decreased peripheral heat loss which limited loss of stored body heat. Hanna (1974) interprets these findings as representing an adaption to high-altitude cold by the action of coca to promote the retention of body heat. At sea level in equatorial zones with warmer ambient temperatures, there would be no advantage to conserving body heat. Further, although cultural factors associated with coca use certainly play an important role in its distribution, the frequency of coca use, which is positively related to altitude (Buck, Sasaki & Anderson, 1968, p. 99), supports the contention that coca use is adaptive at altitude.

### Humidity and solar radiation at high altitude

*Cold and aridity*

Many highland regions of the world are relatively arid, either seasonally or throughout the year. The low vapor pressure or humidity associated with aridity is yet another source of environmental stress at high altitude (Petropoulos & Timiras, 1974). For example, low humidity promotes heat loss through more effective evaporative cooling such that under a wide variety of environmental conditions the effects of aridity should parallel those of cold. From another viewpoint, aridity is actually an increase in potential for heat loss. At rest in a thermal neutral environment, insensible perspiration can account for about 16% of the heat produced by resting metabolism (Buskirk, Thomson & Whedon, 1963). With a reduction in humidity it is likely that this value would increase, especially in light of the hyperventilation which occurs at high altitude. A doubling of ventilation at altitude would lead to a twofold increase in heat loss, other conditions being equal. Moreover, a halving of humidity would have the same effect. This combination of low humidity and high ventilation would potentially more than double heat loss, which could become a serious problem in a resting native at altitude. It is noteworthy in this respect that sojourners at altitude are known to experience negative water balance, which although probably associated with hypoxia and cold diuresis, may be due, in part, to low humidity (Pugh, 1964a; Consolazio, Johnson & Krzywicki, 1972).

In addition to modification of heat-loss parameters, aridity might have contributed to present human morphology. Although the data are not available, it can be predicted that populations with a long history of residence at high altitude would have a narrow nose form (Thomson & Buxton, 1923; Weiner, 1954). A general impression of Quechua Indians at high altitude is that they do indeed have narrower noses than other Amerindian groups. Again in a speculative vein, aridity might also influence the morphology of the skin. The external epidermal layer or corneum would certainly be under a constant dehydration process, espe-

288

cially in exposed areas (Harrison & Montagna, 1969, p. 152). Some modification in composition and thickness might be anticipated as the result of constant dehydration.

In brief, there are at least three areas which are susceptible to stress from an environment with low humidity: (1) evaporation from the skin and respiratory tract, (2) nose form and morphology, and (3) the structure and composition of the epidermis. Unfortunately, these and other related problems have not been investigated in native high-altitude peoples.

### Solar radiation

As noted at the beginning of this chapter, solar radiation increases in proportion to altitude above sea level. There is a general increase in total incident radiation as well as in certain wavelengths (Gates, 1966), with a proportional increase in ultraviolet radiation. Specifically there are increases in the 290 to 320 mm wavelengths which are known to produce sunburn tissue damage and catalyze vitamin D synthesis (Blum, 1961; Daniels, 1964). It is this short-wavelength radiation which is known to produce adaptive changes in melanization, keratinization, and a thickening of the stratum corneum (Loomis, 1967; Daniels, 1974; Szabó, 1975).

Protection is afforded the high-altitude resident by clothing, but radiation is intense and must have some consequence. The hands, face and feet which are most frequently exposed to sunlight must also adapt in some manner. It might also be suggested that the eyes make some accommodation to the extra radiant energy they habitually perceive.

In areas of snow fall the solar heat gain may also be quite high. In a snowfield Pugh (1964b) reported a solar heat gain of 360 kcal/m/h, a value nine times that normally produced by the resting body. Under less reflective conditions the gain from soil was 290 kcal/m/h. Unfortunately there has been little interest in quantification of the consequences of solar radiation to high-altitude natives. This should provide a fruitful area for future research.

### Skin color of highlanders

There have been two reflectometric studies of altitude and skin-color relationships. The first was undertaken in the Andes and has been described by Conway & Baker (1972). They compared the skin reflectance of Andean Indian farmers with that of several Amazon basin groups. The Andean Indians were quite similar to the jungle groups in reflectance, although they showed some tendency to be lighter. The authors concluded that the skin reflectance of the Quechua was typical for South American Indians and intermediate between Europeans and Africans.

The other study has been described by Harrison *et al.* (1969). They

289

measured the reflectance of highland versus lowland farmers in Ethiopia. Surprisingly the higher altitude residents were somewhat lighter than the lowland residents.

The significance of these two studies is difficult to ascertain. The higher radiation at altitude has apparently had little effect in producing greater melanization; indeed, the opposite seems to have occurred. If high-altitude peoples are lighter it can be suggested that the need for clothing has led to the same selective pressures that are encountered in more temperate climates (Loomis, 1967). This is, however, a most tenuous suggestion, since there are few data on this topic.

**Overview**

Highland populations from the Andes, the Himalayas and other mountain areas are exposed to a variety of stresses including hypoxia, cold, low humidity and high levels of solar radiation that are imposed by the physical environment. Other stresses such as limited energy and food resources and disease are imposed from other sources, yet interact with climatic stresses requiring specific highland patterns of adaptation to be achieved. All such patterns of adaptation depend upon the interaction of environmental, cultural and human biological parameters for their development. It is therefore necessary, when exploring problems of cold adaptation at altitude, to be aware that cold stress itself is modified by other stresses in its impact on a human population, and that any adaptive patterns that a population develops are compromises to deal with what are often conflicting stresses.

Cultural modes of adaptation to primary stresses from the physical environment appear quite effective in the Andes and Himalayas. In the Andes, the trunk and upper limbs of the body are well insulated by several layers of wool clothing of variable weave. Dark colors are generally employed to increase radiant energy gain, yet brimmed hats are always worn by both sexes to protect the head from intense solar radiation. Conspicuously absent are coverings for the hands and feet so the distal portions of the limbs are exposed to somewhat more cold than other parts of the body. Andean pastoralist bedding of hides and homespun coverings seems equally well designed to combat night-time cold, and although the stone huts of the rural pastoralist are poorly insulated, this may be dictated by the need for seasonal shifts in residence. Investigations of the thermal buffering of Andean clothing and shelter under a variety of conditions have verified their protective function in substantially reducing cold stress.

Despite the appearance of an effective pattern of cultural protection from cold, the highland native, nonetheless, displays two distinct physiological adaptations to cold. These include slightly elevated basal metabolism and greater heat flow to the extremities during generalized and peripheral cold exposure. The transient increase in basal metabolism experienced by sojourners to altitude is stimulated by hypoxia. On the other hand, the chronically high BMR of permanent residents, which is well established (cf. Petropoulos & Timiras, 1974), is likely to be the result of hypoxia and cold acting together. The effects of hypoxia on peripheral blood flow and temperatures are not clear; however, there are indications that among sojourners, blood flow in the hand is depressed under warm conditions and unchanged under cold conditions at altitude (Durand *et al.*, 1969). In permanent residents at altitude, the maintenance of warm hands and feet may facilitate oxygen transport to the tissues, since it is known that lowering blood temperature reduces the dissociation of oxygen by hemoglobin (Comroe, 1965, p. 163).

There is virtually no information on the interrelationships between hypoxia and low humidity or hypoxia and high radiation in humans. Reviews have documented studies of laboratory animals that demonstrated both cross-resistance and cross-sensitivity (Petropoulos & Timiras, 1974). Any suggestions on cross-adaptation in humans have been largely speculative.

Drugs such as alcohol and the coca leaf, which have widespread use in the Andes, influence responses to cold largely via changes in the peripheral circulation. Alcohol acts as a vasodilator and during rest in the cold probably contributes somewhat to a subjective feeling of thermal comfort. Coca, with its more consistent use, acts as a vasoconstrictor and may serve the adaptive function of inhibiting heat loss from the body. Tobacco is not smoked regularly by Andean natives and its widespread use in the Himalayas is unlikely, although it also acts as a vasoconstrictor and could have some effect on temperature regulation (Freund & Ward, 1960). We have no information on either tobacco or marijuana effects at altitude. Studies of the latter would seem warranted in light of its increased use among youths in high-altitude urban centers.

A number of general problem areas associated with cold and other stresses from the physical environment of high altitude remain to be explored. Clearer distinctions must be made between the responses which can be identified as acclimatizations to single stresses and acclimatizations to multiple stresses. This area of cross-adaptation to multiple stresses has been investigated at length on laboratory animals, while information on humans is sorely lacking. In addition, more careful distinctions must be made among the responses of: (1) laboratory subjects during short-term acute exposure to stress(es), (2) sojourners who acclimatize to altitude slowly, (3) sojourners who acclimatize to altitude more abruptly and (4) permanent residents at altitude. Finally, longitudinal studies of the

*The biology of high-altitude peoples*

time-course and general characteristics of the acclimatizational process
are needed as well.

## References

Abramson, D. I., Landt, H. & Benjamin, J. E. (1941). Peripheral vascular responses to general anoxia. *Proceedings of the Society for experimental Biology and Medicine*, **48**, 214–16.

Baker, P. T. (1966a). Ecological and physiological adaptation in indigenous South Americans: with special reference to the physical environment. In *The biology of human adaptability*, ed. P. T. Baker & J. S. Weiner, pp. 275–303. Oxford: Clarendon Press.

Baker, P. T. (1966b). Micro-environment cold in a high altitude Peruvian population. In *Human adaptability and its methodology*, ed. H. Yoshimura & J. S. Weiner, pp. 67–77. Tokyo: Japan Society for the Promotion of Sciences.

Baker, P. T. (1968). Summary of weather data: Nuñoa district, 1964–1966. *Occasional Papers in Anthropology*, **1**, 37–44. University Park, Pa.: The Pennsylvania State University.

Baker, P. T., Buskirk, E. R., Kollias, J. & Mazess, R. B. (1967). Temperature regulation at high altitude: Quechua Indians and US whites during total body cold exposure. *Human Biology*, **39**, 155–69.

Beckwitt, H. J., Surks, M. I. & Chidsey, C. A. (1966). Basal metabolism, thyroid and sympathetic activity in man at high altitude. *Federation Proceedings*, **25**, 399 (Abstract).

Belding, H. S. (1949). Protection against dry cold. In *Physiology of heat regulation and the science of clothing*, ed. L. H. Newburgh, pp. 351–67. Philadelphia, Pa.: W. B. Saunders.

Blum, H. F. (1945). The solar heat load: its relationship to total heat load and its importance in the design of clothing. *Journal of clinical Investigation*, **24**, 712–21.

Blum, H. F. (1961). Does the melanin pigment of human skin have adaptive value? *Quarterly Review of Biology*, **36**, 50–63.

Boothby, W. M., Berkson, J. & Dunn, H. L. (1936). Studies of the energy metabolism of normal individuals: a standard for basal metabolism with a nomogram for clinical application. *American Journal of Physiology*, **116**, 468–84.

Buck, A. A., Sasaki, T. T. & Anderson, R. I. (1968). *Health and disease in four Peruvian villages: contrasts in epidemiology*. Baltimore, Md.: The Johns Hopkins Press.

Buettner, K. J. K. (1969). The effects of natural sunlight on human skin. In *The biologic effects of ultraviolet radiation*, ed. F. Urback, pp. 237–49. Oxford: Pergamon Press.

Bullard, R. W. & Rapp, G. (1970). Problems of body heat loss in water immersion. *Aerospace Medicine*, **41**, 1269–77.

Burrus, S. K., Dill, D. B., Burk, D. L., Freeland, D. V. & Adams, W. C. (1974). Observations at sea level and altitude on basal metabolic rate and related cardio-pulmonary functions. *Human Biology*, **46**, 677–92.

Buskirk, E. R., Thompson, R. & Whedon, G. (1963). Metabolic response to cold air in men and women in relation to total body fat content. *Journal of applied Physiology*, **18**, 603–12.

292

Clegg, E. J., Harrison, G. A. & Baker, P. T. (1970). The impact of high altitudes on human populations. *Human Biology*, **42**, 486–518.

Comroe, J. H. (1965). *Physiology of respiration*. Chicago: Yearbook Medical Publishers.

Consolazio, C. F., Johnson, H. L. & Krzywicki, H. J. (1972). Body fluids, body composition, and metabolic aspects of high-altitude adaptation. In *Physiological adaptations: desert and mountain*, ed. M. K. Yousef, S. M. Horvath & R. W. Bullard, pp. 227–41. New York & London: Academic Press.

Conway, D. L. & Baker, P. T. (1972). Skin reflectance of Quechua Indians: the effects of genetic admixture, sex, and age. *American Journal of Physical Anthropology*, **36**, 267–82.

Cook, E. N. & Brown, G. E. (1932). The vasodilating effects of ethyl alcohol on the peripheral arteries. *Proceedings of the Staff Meetings of the Mayo Clinic*, **7**, 449–52.

Daniels, F. T., Jr (1964). Man and radiant energy: solar radiation. In *Handbook of physiology, Section 4: adaptation to the environment*, ed. D. B. Dill, E. F. Adolph & C. G. Wilber, pp. 969–87. Washington, DC: American Physiological Society.

Daniels, F. T., Jr (1974). Radiant energy. Part A: Solar radiation. In *Environmental physiology*, ed. N. B. Slonim, pp. 276–86. St Louis, Mo.: The C. V. Mosby Company.

Davis, T. R. A., Nayar, H. S., Sinha, K. C., Nishith, S. D. & Rai, R. M. (1966). The effect of altitude on the cold responses of low altitude acclimatized Jats, high altitude acclimatized Jats and Tibetans. *Biometeorology II, Proceedings of the 3rd International Biometeorology Congress*, pp. 191–8. Oxford: Pergamon Press.

Durand, J., Verpillat, J.-M., Pradel, M. & Martineaud, J.-P. (1969). Influence of altitude on the cutaneous circulation of residents and newcomers. *Federation Proceedings*, **28**, 1124–8.

Elsner, R. W. & Bolstad, A. (1963). *Thermal and metabolic responses to cold of Peruvian Indians native to high altitudes*. Technical Report No. AAL-TDR-62-64. Fort Wainwright, Alaska: Arctic Aeromedical Laboratory.

Fewings, J. D., Hanna, M. J. D., Walsh, J. A. & Whelan, R. F. (1966). The effects of ethyl alcohol on the blood vessels of the hand and forearm in man. *British Journal of Pharmacology and Chemotherapy*, **27**, 93–106.

Fitch, J. M. & Branch, D. (1960). Primitive architecture and climate. *Scientific American*, **203** (6), 134–44.

Forbes, W. H. (1949). Laboratory and field studies: general principles. In *Physiology of heat regulation and the science of clothing*, ed. L. H. Newburgh, pp. 320–9. Philadelphia, Pa.: W. B. Saunders.

Forney, R. B., Hughes, F. W., Harger, R. M. & Richards, A. B. (1964). Alcohol distribution in the vascular system: concentration of orally administered alcohol in blood from various points in the vascular system, and in re-breathed air, during absorption. *Quarterly Journal of Studies of Alcohol*, **25**, 205–17.

Fourt, L. & Harris, M. (1949). Physical properties of clothing fabrics. In *Physiology of heat regulation and the science of clothing*, ed. L. H. Newburgh, pp. 291–319. Philadelphia, Pa.: W. B. Saunders.

Freeman, N. E., Shaw, J. L. & Snyder, J. C. (1936). Peripheral blood flow in surgical shock: reduction in circulation through the hand resulting from pain, fear, cold, and asphyxia, with quantitative measurements of the volume flow of blood in clinical cases of surgical shock. *Journal of clinical Investigation*, **15**, 651–64.

Fregly, M. J. (1954). Cross acclimatization between cold and altitude in rats. *American Journal of Physiology*, **176**, 267–74.

Freund, J. & Ward, C. (1960). The acute effect of cigarette smoking on the digital circulation in health and disease. *Annals of the New York Academy of Sciences*, **90**, 85–101.

Fürer-Haimendorf, C. von (1963). The Sherpas of the Khumbu region. In *Mount Everest: formation, population and exploration of the Everest region*, pp. 124–81. London: Oxford University Press.

Fürer-Haimendorf, C. von (1964). *The Sherpas of Nepal: Buddhist highlanders*. Berkeley, California: University of California Press.

Gates, D. M. (1966). Spectral distribution of solar radiation at the earth's surface. *Science, Washington*, **151**, 523–9.

Geiger, R. (1965). *The climate near the ground*. Cambridge, Mass.: Harvard University Press.

Gellhorn, E. & Steck, I. E. (1938). Effect of the inhalation of gases with a low oxygen and increased carbon dioxide tension in the peripheral blood flow in man. *American Journal of Physiology*, **124**, 735–41.

Gill, M. B. & Pugh, L. G. C. E. (1964). Basal metabolism and respiration in men living at 5,800 m (19,000 ft). *Journal of applied Physiology*, **19**, 949–54.

Grover, R. F. (1963). Basal oxygen uptake of man at high altitude. *Journal of applied Physiology*, **18**, 909–12.

Gupta, K. K. (1964). Cold acclimatisation. *Indian Journal of Physiology and Pharmacology*, **8**, 7–14.

Gupta, K. K., Karani, N. D. P., Soni, C. M. & Das Gupta, N. C. (1964a). Effectiveness of cold acclimatisation acquired after 20 exposures. *Indian Journal of Physiology and Pharmacology*, **8**, 243–9.

Gupta, K. K., Nayar, H. S., Nishith, S. D., Rai, R. M. & Karani, N. D. P. (1964b). Cold acclimatisation. *Armed Forces medical Journal (India)*, **20**, 168–79.

Hale, H. B. (1970). Cross-adaptation. In *Physiology, environment and man*, ed. D. H. K. Lee & D. Minard, pp. 158–69. New York & London: Academic Press.

Hammel, H. T. (1964). Terrestrial animals in cold: recent studies of primitive man. In *Handbook of physiology, section 4: adaptation to the environment*, ed. D. B. Dill, E. F. Adolph & C. G. Wilber, pp. 413–34. Washington, DC: American Physiological Society.

Hanna, J. M. (1965). Biological and cultural factors in peripheral blood flow at low temperatures. MA Thesis in Anthropology, The Pennsylvania State University.

Hanna, J. M. (1968). Cold stress and microclimate in the Quechua Indians of southern Peru. *Occasional Papers in Anthropology*, **1**, 196–326. University Park, Pa.: The Pennsylvania State University.

Hanna, J. M. (1970). A comparison of laboratory and field studies of cold response. *American Journal of physical Anthropology*, **32**, 227–31.

Hanna, J. M. (1971a). Responses of Quechua Indians to coca ingestion during cold exposure. *American Journal of physical Anthropology*, **34**, 273–7.

Hanna, J. M. (1971b). Further studies on the effects of coca chewing on exercise. *Human Biology*, **43**, 200–9.

Hanna, J. M. (1974). Coca leaf use in southern Peru: some biosocial aspects. *American Anthropologist*, **76**, 281–96.

Hannon, J. P. & Sudman, D. M. (1973). Basal metabolic and cardiovascular function of women during altitude acclimatization. *Journal of applied Physiology*, **34**, 471–7.

Harrison, G. A., Küchemann, C. F., Moore, M. A. S., Boyce, A. J., Baju, T., Mourant, A. E., Godber, M. J., Glasgow, B. G., Kopeć, A. C., Tills, D. & Clegg, E. J. (1969). The effects of altitudinal variation in Ethiopian populations. *Philosophical Transactions of the Royal Society of London*, **256B**, 147–82.

Harrison, R. J. & Montagna, W. (1969). *Man*. New York: Appleton-Century-Crofts.

Hong, S. K. (1963). Comparison of diving and nondiving women of Korea. *Federation Proceedings*, **22**, 831–3.

Jones, R. E., Little, M. A., Thomas, R. B., Hoff, C. J. & Dufour, D. L. (1976). Local cold exposure of Andean Indians during normal and simulated activities. *American Journal of physical Anthropology*, **44**, 305–13.

Kennedy, S. J. (1961). Clothing and personal protection. In *Man living in the Arctic*, ed. F. Fisher, pp. 56–67. Washington, DC: National Research Council, National Academy of Sciences.

Kerslake, D. McK. (1972). *The stress of hot environments*. London: Cambridge University Press.

Klausen, K. (1966). Cardiac output in man at rest and work during and after acclimatization to 3,800 meters. *Journal of applied Physiology*, **21**, 609–16.

Kondo, S. (1969). A study on acclimatization of the Ainu and the Japanese with reference to hunting temperature reaction. *Journal of the Faculty of Science of Tokyo University*, Section 5, **3**, 253–65.

Kottke, F. J., Phalen, J. S., Taylor, C. B., Visscher, M. B. & Evans, G. T. (1948). Effect of hypoxia upon temperature regulation of mice, dogs, and man. *American Journal of Physiology*, **153**, 10–15.

Krog, J. & Wika, M. (1971–2). Studies of the peripheral circulation in the hand of the Igloolik Eskimo. In *International biological programme, human adaptability project (Igloolik, NWT)*, annual report No. 4, pp. 173–81. Toronto: University of Toronto.

Larsen, R. M. (1973). The thermal microenvironment of a highland Quechua population: biocultural adjustment to the cold. MA Thesis in Anthropology, University of Wisconsin.

Lim, T. P. K. & Luft, U. C. (1960). Body temperature regulation in hypoxia. *The Physiologist*, **3**, 105.

Little, M. A. (1968). Racial and developmental factors in foot cooling: Quechua Indians and US whites. *Occasional Papers in Anthropology*, **1**, 327–537. University Park, Pa.: The Pennsylvania State University.

Little, M. A. (1969). Temperature regulation at high altitude: Quechua Indians and US whites during foot exposure to cold water and cold air. *Human Biology*, **41**, 519–35.

Little, M. A. (1970). Effects of alcohol and coca on foot temperature responses of highland Peruvians during a localized cold exposure. *American Journal of physical Anthropology*, **32**, 233–42.

Little, M. A. & Hochner, D. H. (1973). Human thermoregulation, growth, and mortality. *Module in Anthropology*, No. 36. Reading, Mass.: Addison-Wesley.

Little, M. A., Thomas, R. B. & Larrick, J. W. (1973). Skin temperature and cold pressor responses of Andean Indians during hand immersion in water at 4 °C. *Human Biology*, **45**, 643–62.

Little, M. A., Thomas, R. B., Mazess, R. B. & Baker, P. T. (1971). Population differences and developmental changes in extremity temperature responses to cold among Andean Indians. *Human Biology*, **43**, 70–91.

Loomis, W. F. (1967). Skin-pigment regulation of vitamin D biosynthesis in man. *Science, Washington*, **157**, 501–6.

Lund-Larsen, K., Wika, M. & Krog, J. (1970). Circulatory responses of the hand of Greenlanders to local cold stimulation. *Arctic Anthropology*, **7**, 21–5.

McFarland, R. A. & Forbes, W. H. (1936). The metabolism of alcohol in man at high altitude. *Human Biology*, **8**, 387–98.

Maron, S., Rose, L. & Heyman, J. (1956). *A Survey of Nepal Society. Human Relations Area Files (HRAF), vol. 47*. Berkeley, California: University of California.

Mazess, R. B. & Larsen, R. (1972). Responses of Andean highlanders to nighttime cold. *International Journal of Biometeorology*, **16**, 181–92.

Mazess, R. B., Picón-Reátegui, E., Thomas, R. B. & Little, M. A. (1968). Effects of alcohol and altitude on man during rest and work. *Aerospace Medicine*, **39**, 403–6.

Mazess, R. B., Picón-Reátegui, E., Thomas, R. B. & Little, M. A. (1969). Oxygen intake and body temperature of basal and sleeping Andean natives at high altitude. *Aerospace Medicine*, **40**, 6–9.

Meteorological Office. (1966). *Tables of temperature, relative humidity and precipitation for the world, Part V, Asia*. London: Her Majesty's Stationery Office.

Miller, L. K. & Irving, L. (1962). Local reactions to air cooling in an Eskimo population. *Journal of applied Physiology*, **17**, 449–55.

Nair, C. S. & George, S. (1972). The effect of altitude and cold on body temperature during acclimatization of man at 3,300 m. *International Journal of Biometeorology*, **16**, 79–84.

Nair, C. S., Malhotra, M. S. & Gopinath, P. M. (1971). Effect of altitude and cold acclimatization on the basal metabolism in man. *Aerospace Medicine*, **42**, 1056–9.

Nair, C. S., Malhotra, M. S. & Gopinath, P. M. (1972). Effect of altitude acclimatization and simultaneous acclimatization to altitude and cold on critical flicker frequency at 11,000 ft. Altitude in man. *Aerospace Medicine*, **43**, 1097–1100.

Nair, C. S., Malhotra, M. S., Gopinath, P. M. & Mathew, L. (1971). Effect of acclimatization to altitude and cold on basal heart rate, blood pressure, respiration and breath holding in man. *Aerospace Medicine*, **42**, 851–5.

Nair, C. S., Malhotra, M. S. & Tiwari, O. P. (1973). Heat output from the hand of men during acclimatization to altitude and cold. *International Journal of Biometeorology*, **17**, 95–101.

Nair, C. S., Malhrotra, M. S., Tiwari, O. P. & Gopinath, P. M. (1971). Effect of altitude acclimatization and cold on cold press or response in man. *Aerospace Medicine*, **42**, 991–4.

Nelson, H. L. (1968). *Climatic data for representative stations of the world*. Lincoln, Nebr.: University of Nebraska Press.

Newman, R. W. & Cipriano, L. F. (1974). Effect of hypoxia on temperature regulation of men exposed to 10 °C air temperature. *American Journal of physical Anthropology*, **40**, 146 (Abstract).

Petropoulos, E. A. & Timiras, P. S. (1974). Biological effects of high altitude as related to increased solar radiation, temperature fluctuations and reduced partial pressure of oxygen. In *Progress in biometeorology*, chapter 6, section 7, ed. S. W. Tromp, pp. 295–328. Amsterdam: Swets & Zeittinger.

*Responses to cold and other stresses*

Picón-Reátegui, E. (1961). Basal metabolic rate and body composition at high altitudes. *Journal of applied Physiology*, **16**, 431–4.
Picón-Reátegui, E. (1968). Effect of coca chewing on metabolic balance in Peruvian high altitude natives. *Occasional Papers in Anthropology*, **1**, 556–63. University Park, Pa.: The Pennsylvania State University.
Prosser, C. L. (1964). Perspectives of adaptation: theoretical aspects. In *Handbook of physiology, section 4: adaptation to the environment*, ed. D. B. Dill, E. F. Adolph & C. G. Wilber, pp. 11–25. Washington, DC: American Physiological Society.
Pugh, L. G. C. E. (1963). Tolerance to extreme cold at altitude in a Nepalese pilgrim. *Journal of applied Physiology*, **18**, 1234–8.
Pugh, L. G. C. E. (1964a). Animals at high altitudes: man above 5,000 meters. In *Handbook of physiology, section 4: adaptation to the environment*, ed. D. B. Dill, E. F. Adolph & C. G. Wilber, pp. 861–8. Washington, DC: American Physiological Society.
Pugh, L. G. C. E. (1964b). Solar heat gain by man in the high Himalaya. In *Environmental physiology and psychology in arid conditions*. Liège: UNESCO.
Pugh, L. G. C. E. (1966). A programme for physiological studies of high altitude peoples. In *The biology of human adaptability*, ed. P. T. Baker & J. S. Weiner, pp. 521–32. Oxford: Clarendon Press.
Rappaport, A. (1968). *House form and culture*. Englewood Cliffs, NJ: Prentice-Hall.
Rennie, D. W., Howell, B. & Covino, B. G. (1964). Attempt to cold acclimatize human subjects by repeated immersion in cool water. In *Korean sea women*, ed. H. Rahn, S. K. Hong & D. W. Rennie, pp. 75–8. Buffalo, NY: Department of Physiology, State University of New York at Buffalo.
Risemberg, M. F. (1944). Acción de la coca y la cocaína en sujetos habituandos. *Revista de Medicina experimental (Lima)*, **3**, 317–28.
Robinson, H. (1967). *Monsoon Asia: a geographical survey*. New York: Praeger.
Schneider, E. C. & Sisco, D. L. (1914). The circulation of the blood in man at high altitudes. II. The rate of blood flow and the influence of oxygen on the pulse rate and blood flow. *American Journal of Physiology*, **34**, 29–47.
Simoons, F. J. (1960). *Northwest Ethiopian people and their economy*. Madison, Wisc.: University of Wisconsin Press.
Stamp, L. D. (1964). *Africa: a study in tropical development*, 2nd edn. New York: John Wiley & Sons.
Suzuki, M. (1969). Peripheral response to cold. *Journal of the anthropological Society of Nippon*, **77**, 213–23.
Swan, L. W. (1967). Alpine and aeolian regions of the world. In *Arctic and Alpine environments*, ed. H. E. Wright, Jr. & W. H. Osburn, pp. 29–54. Bloomington, Ind.: Indiana University Press.
Szabó, G. (1975). The human skin as an adaptive organ. In *Physiological anthropology*, ed. A. Damon, pp. 39–58. New York: Oxford University Press.
Thomas, R. B. (1973). Human adaptation to a high Andean energy flow system. *Occasional Papers in Anthropology*, **7**. University Park, Pa.: The Pennsylvania State University.
Thomson, A. & Buxton, D. (1923). Man's nasal index in relation to certain climatic conditions. *Journal of the Royal anthropological Institute*, **53**, 92–122.
Weiner, J. S. (1954). Nose shape and climate. *American Journal of physical Anthropology*, **12**, 1–4.
Weitz, C. A. (1969). Morphological factors affecting responses to total body

297

cooling among three populations tested at high altitude. MA Thesis in Anthropology, The Pennsylvania State University.

Williams, D. D., Petajan, J. H & Lee, S. B. (1969). Summer and winter hand calorimetry of children in the village of Old Crow. *International Journal of Biometeorology*, **13**, 11.

Zapata Ortíz, V. (1944). Modificaciones psicológicas y fisiológicas producidas por la coca y la cocaína en los coqueros. *Revista de Medicina Experimental (Lima)*, **3**, 132–62.

# 10. Biological and physiological characteristics of the high-altitude natives of Tien Shan and the Pamirs

M. M. MIRRAKHIMOV

This paper reviews the major results of studies carried out within the IBP framework on human high-altitude adaptation in the USSR with special reference to the physiological characteristics of natives of various Tien Shan and Pamir altitudes.

The first publications on the morphological and physiological features of Tien Shan and Pamir highlanders date back to the end of the last century (Lavrinovich, 1898; Shlomm, 1895; Tretiakov, 1895). Since then many new data have been accumulated (Pavlova, 1937; Filatova, 1961; Mirrakhimov, 1964, 1968; Agadzhanian & Mirrakhimov, 1970; Agadzhanian, Issabaeva & Elfimov, 1973).

In the work described in this article, methods recommended by IBP were used. In those cases where additional methods were used, they are indicated. During physiological studies room temperature was maintained between 18 and 20 °C. Table 10.1 gives certain meteorological data for the high- and low-altitude localities.

Table 10.1. *The main meteorological data for the studied localities in Tien Shan and the Pamirs*

| Locality | Altitude above sea level (m) | Meteorological data | | |
|---|---|---|---|---|
| | | Barometric pressure (mm Hg) | Air temperature (°C) | Relative humidity (%) |
| Tien Shan | | | | |
| Frunze | 760 | 686–702 | +20–+25 | 50 |
| Groznoe | 940 | 678–688 | +18–+20 | 66 |
| Naryn | 2,020 | 595–608 | +14–+26 | 47.3 |
| Susamyr | 2,200 | 593–597 | +16–+22 | 61 |
| The Pamirs | | | | |
| Osh | 1,020 | 683 | +22.7 | 46.5 |
| Kyzyl-Dzhar | 2,300–2,800 | 587–593 | +15–+18 | 35–65 |
| Sary-Tash | 3,200 | 520–526 | +17–+21 | 24–44 |
| Murhab | 3,600 | 495–500 | +10–+15 | 38–45 |
| Murhab suburbs | 4,200 | 445 | +2–+8 | 33 |

299

**Physiological investigations**

*Ventilatory function, chemosensitivity and gas exchange*

Many studies on the respiratory rate of highlanders have shown that the rate is higher in highlanders than in sea-level residents. A slight rise in respiratory rate is considered to be a mechanism producing a smaller tracheo-alveolar oxygen-pressure gradient (Table 10.2). The large chest of highlanders contributes to this increased ventilation.

Pulmonary ventilation ($\dot{V}$), respiratory rate and tidal volume ($V_T$), as well as oxygen consumption ($V_{O_2}$) and ventilatory equivalent ($V_E$), were studied at various altitudes. At 3,600 m $\dot{V}$ exceeds predicted values by 78–84%; the increase in $\dot{V}$ is mainly due to an increase in respiratory rate. Hyperventilation results in a decrease in $V_E$ and $V_{O_2}$ exceeds predicted values.

Alveolar ventilation ($V_{A,CO_2}$) is greater in highlanders than in lowlanders. Since alveolar partial pressures of oxygen and carbon dioxide ($P_{A,O_2}$ and $P_{A,CO_2}$) depend on the $V_{A,CO_2}$ level, it seems natural that highlanders have lower alveolar oxygen and carbon dioxide tensions than lowlanders, although carbon dioxide tension is higher in the former than in subjects undergoing short-term adaptation. The difference in $V_{A,CO_2}$ and alveolar gas tensions in the different groups is due to a lower sensitivity threshold of the respiratory centre to hypoxia in highlanders (Dubinina, Kovaleva & Mirrakhimov, 1967; Akhmedov, 1971). At the same time peripheral chemosensitivity proved to be decreased at 3,200 m, although natives and sojourners responded to a single breath of oxygen by hypoventilation. However, the former had a subsequent moderate hyperventilation, while in the latter this phase was absent. Data obtained during relatively pro-longed oxygen inhalation (5 min) at the same altitude also revealed a hypoventilatory response which was more pronounced in sojourners than in altitude-natives (Mirrakhimov, 1972c).

Studies on the ventilatory response to 5-min oxygen inhalation at 3,200 m and 3,600 m in subjects with a 4 to 24 months' adaptation showed a higher respiratory centre excitability at 3,200 m, and reduced peripheral chemosensitivity and respiratory centre excitability threshold to hypoxia at 3,600 m.

It was shown that at reduced atmospheric pressure, respiratory centre stimulation occurs at lower carbon dioxide tensions, i.e. the respiratory centre sensitivity threshold to carbon dioxide is decreased with a simul-taneous decrease in the ventilatory response to a given $P_{A,CO_2}$ increment. Akhmedov (1971) reports that normally the respiratory-centre sensitivity threshold to carbon dioxide is 21–31 mm Hg (for residents of Moscow and Dushanbe), while in sojourners and natives to 3,600 m it proved to be markedly lower (17.5 mm Hg). Our data obtained at 3,600 m suggest that

Table 10.2. *Certain indices (BTPS) of ventilatory function and alveolar gas tensions in altitude natives*

| Locality and its altitude (m) | N | Respiratory rate ($min^{-1}$) | $\dot{V}$ ($1\ min^{-1}/m^2$) | $\dot{V}_{A,CO_2}$ ($1/m^2$) | $P_{A,O_2}$ (mm Hg) | $P_{A,CO_2}$ (mm Hg) | Authors |
|---|---|---|---|---|---|---|---|
| Tien Shan | | | | | | | |
| Frunze, 760 | 35 | 13.6±0.36 | 4.08±0.11 | 2.66±0.64 | 82.0±0.80 | 38.36±0.59 | Dubinina & Kochenkova (1972) |
| Naryn, 2,020 | 23 | 16.7±0.64 | 4.82±0.17 | 3.25±0.17 | 61.2±0.87 | 34.2±0.64 | Dubinina & Kochenkova (1972) |
| Susamyr, 2,200 | 43 | 17.7±0.34 | 4.66±0.15 | 3.24±0.13 | 65.5±0.79 | 35.9±0.41 | Dubinina & Kochenkova (1972) |
| The Pamirs | | | | | | | |
| Kyzyl-Dzhar, 2,500 | 31 | 15.0±0.60 | 5.40±0.20 | 3.28±0.86 | 64.4±0.83 | 33.7±0.46 | Dubinina et al. (1973) |
| Murhab, 3,600 | 21 | 20.3±0.90 | 6.06±0.60 | 3.31 | 58.0±1.30 | 28.0±1.20 | Akhmedov (1971) |
| Murhab, 3,600 | 39 | 17.9±0.43 | 6.35±0.16 | 4.2±0.15 | 52.1±0.71 | 31.2±0.46 | Dubinina & Kochenkova (1972) |

Table 10.3. *Rebreathing test at 3,200 m*

| Subjects | N | Apnoeic point ($P_{A,CO_2}$ in mm Hg) | | Respiratory centre excitability[a] | |
|---|---|---|---|---|---|
| | | Mean±s.e. | Confidence limits | Mean±s.e. | Confidence limits |
| Permanent residents | | | | | |
| aged 16–19 yr | 17 | 25.6±2.1 | 4.44 | 101.26±3.65 | 7.7 |
| aged 20–40 yr | 13 | 17.8±1.66 | 3.59 | 64.6±1.15 | 2.48 |
| Sojourners with 6–19 months' adaptation | 11 | 21.8±1.94 | 4.26 | 97.29±2.76 | 6.1 |

[a] Respiratory centre excitability ($S$) is the ratio of the increment in $\dot{V}$ ($\Delta V_T$ in ml) to 1 mm Hg increment in $P_{A,CO_2}$ $\left(S = \dfrac{\Delta V_T}{\Delta P_{A,CO_2}}\right)$.

under these conditions ventilation is achieved at lower $P_{A,CO_2}$ values in both sojourners and altitude-natives (Table 10.3).

It is our belief that the decrease in chemosensitivity in highlanders is an adaptive response increasing body tolerance to extreme altitude conditions.

The decrease in $P_{A,CO_2}$ in highlanders as a result of permanent hyperventilation is compensated by a proportional decrease in blood bicarbonate and by metabolic changes which keep the blood pH within normal values. The alkaline reserve is somewhat higher in permanent moderate-altitude residents than in lowlanders, while a decrease in this reserve is noted at higher altitudes (3,600 m) (Mirrakhimov, 1964). These studies also reveal an increased carbonic anhydrase activity with altitude.

It was shown that the actual vital capacity (VC) value at 3,600 m is not different from that at sea level. However, the percentage of total lung volume contributed by the residual volume (RV) is greater in highlanders. Residual volume was 30–31% of total volume in highlanders but only 23% in lowlanders. It must be noted that VC declines and RV increases with age in highlanders. It is assumed that the RV increase is due to the opening of new alveoli. It is also known that alveolar blood supply increases at high altitude, leading to an increase in the rigidity of alveoli.

$\dot{V}_{O_2}$ and basal metabolic rate (BMR) are within normal limits or somewhat lower in natives to low and moderate altitudes. The lower BMR under these conditions is apparently related to decreased thyroid function.

Summarizing the results on ventilation in altitude-natives it can be stated that the major adaptive respiratory and gas exchange responses to hypoxia are:

(*a*) a slight increase in respiratory rate and pulmonary ventilation;
(*b*) an increase in residual volume;
(*c*) a decrease in chemoreceptor sensitivity.

### Cardiovascular function

The cardiac index (CI), the volume of heart output (l.min$^{-1}$)/body surface area was studied at various Tien Shan and Pamir altitudes (up to 4,200 m) by the acetylene and dye-dilution methods. At altitudes below 3,000 m, CI is unchanged, but above 3,000 m, even at rest, there is a marked increase in CI, which is considered to be an adaptive response (Mirrakhimov, 1964; Narbekov, 1970).

Heart rate (HR) is decreased at Tien Shan and Pamir altitudes due to decreased sympathetic tone (Filatova, 1961; Plotnikov, 1963) which, apparently, ensures an adequate cardiac diastolic filling pressure. Studies of linear blood-flow rate at the vascular bed level from the cubital vein to the pulmonary capillaries and further to the ear lobe revealed increased circulation time (Dzhailobaev, 1967; Narbekov, 1970). Circulating blood mass is increased in altitude-natives, mainly through an increase in erythrocytes (Fig. 10.1).

The status of the heart itself in highlanders has been studied by many workers. No definite changes have been observed below 2,000 m, but at somewhat higher altitudes right ventricular hypertrophy, documented by X-ray, ECG and VCG findings, is seen. Right ventricular hypertrophy has

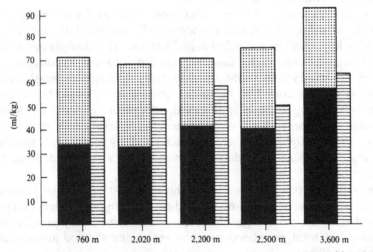

Fig. 10.1. Changes in circulating red blood cell mass (■), plasma (▨), and haematocrit (▤) in natives of various Tien Shan and Pamir altitudes.

been confirmed by a great number of authors and is seen at various Tien Shan and Pamir altitudes (Grinstein, 1966; Dzhailobaev, 1967; Mirrakhimov, 1971). Some highlanders exhibit right atrial hypertrophy. Using VCG recordings, Rudenko (1973) reported signs of left ventricular and left atrial hypertrophy were also found at relatively high localities in the Pamirs (3,600 m, 4,200 m, 4,500 m).

Myocardial hypertrophy is apparently due mainly to alveolar hypoxia and increased pulmonary vascular resistance and reactivity. Indirect measurement of pulmonary arterial pressure by the method of Konopleva, Panichkin & Popov (1971), on the other hand, showed it to be only moderately elevated. At 3,200 m systolic pulmonary arterial pressure was $38.5\pm1.7$ mm Hg, diastolic pressure was $13.1\pm1.2$ mm Hg, and mean pulmonary arterial pressure was $12.6\pm1.3$ mm Hg; at 3,600 m the values were $41.4\pm1.5$ mm Hg, $14.8\pm0.8$ mm Hg and $23.6\pm1.3$ mm Hg, respectively.

Recently myocardial contractility has been intensively studied in highlanders. BCG findings at 3,600 m suggest an increase in myocardial contractility, with higher $H$ and $L$ waves attributable to more asynchronous cardiac performance at relatively high altitudes (Grinstein, 1966). Characteristic of residents of 3,600 m and 4,200 m was, according to left ventricular PCGs, lengthening of the tension phase (AC and IC) and shortening of the E phase (Kudaiberdiev, 1970). This peculiar alteration in myocardial contractility is also observed at somewhat lower altitudes and especially at moderate altitude (2,200 m), which is probably explained by the influence of the annual migration of these people and their livestock to high-altitude summer pastures.

Recent studies with simultaneous left ventricular PCGs and right ventricular apexcardiograms showed evidence of moderate left ventricular myocardial hypodynamics at 3,200 m and 3,600 m. The changes in cardiodynamics may be due to the same regulated myocardial hypodynamics observed at sea level in well-trained athletes. Systemic arterial blood pressure (BP) was found to be unchanged up to 3,000 m, but above 3,000 m arterial hypotension is observed. Moreover, BP does not show the regular elevation with age which is seen in sea-level residents (Mirrakhimov, 1968). Highlanders are characterised by decreased systolic pressure as they age, while diastolic pressure, as a rule, undergoes no significant change.

Pulse wave propagation, not only through elastic but also through muscular type vessels, shows no marked decrease in highlanders of Tien Shan and the Pamirs. A slight decrease was observed only at 4,200 m. Estimations of total peripheral vascular resistance showed an increase only above 3,600 m.

The status of capillary circulation, capillaries and their permeability was studied in highlanders (Mirrakhimov, 1968). The number of nail-bed capillaries was found to be decreased, the capillaroscopic background appeared mostly turbid and a physiological increase in capillary permeability was observed. Tien Shan and Pamir highlanders exhibit elevated cubital vein pressure.

Summarising the results on the cardiovascular system at high altitude it should be emphasised that permanent residence at high altitudes significantly affects both the morphological structure and the functional characteristics of the circulatory system.

## Blood picture

The red blood cell (RBC) counts and haemoglobin levels in subjects living permanently at altitudes up to 1,650 m are within normal limits (Mirrakhimov, 1964; Raimzhanov, Usupova & Barbashova, 1973). At 2,020 m there is a slight increase in both parameters, with a high haematocrit value. RBC counts increase most markedly above 3,500 m (Mirrakhimov, 1971). Evidence for the increase in blood formation is obtained from the increase in circulating blood mass. The average individual erythrocyte size and its haemoglobin content increase with altitude. Reticulocyte counts exceed sea-level values beginning from 2,020 m. The presence of reticulocytosis in highlanders is demonstrated by the significantly greater amount of total circulating reticulocytes (Table 10.4).

Reticulocyte maturation rates are almost twice as high at 2,500 m as at 940 m. Increased reticulocyte formation is also due to enhanced bone marrow haemopoiesis which is maintained by increased erythropoietin production. In altitude-natives of Tien Shan and the Pamirs the haemoglobin content in RBC is normal or slightly elevated. At 3,600 m a moderately decreased blood quotient is observed.

Thus, permanent life at high altitudes produces persistent adaptive changes in blood formation, manifesting themselves in hyperglobulia, increased mean erythrocyte size and predominantly in increased circulating RBC mass. This state is maintained as the result of an enhancement and, perhaps, acceleration of blood formation through adequate erythropoietin production and utilisation.

Relative lymphocytosis is also characteristic of highlanders, and, above 4,000 m, excess segmentation of neutrophils occurs.

Numerous observations have shown characteristic increases in platelets in altitude-natives and extensive studies on blood coagulation and fibrinolytic activity have revealed an increase in both parameters at high altitude (Issabaeva, 1972). A certain prevalence of fibrinolysis possibly ensures a better tissue perfusion in highlanders who are known to exhibit a higher tissue capillarisation.

305

Table 10.4. *Haematological indices (mean±s.e.) in permanent residents of various Tien Shan and Pamir altitudes*

| Haematological indices | Tien Shan | | | | | | | The Pamirs | | | |
|---|---|---|---|---|---|---|---|---|---|---|---|
| | Frunze (760 m) | Groznoe (940 m) | P< | Naryn (2,020 m) | P< | Susamyr (2,200 m) | P< | Kyzyl-Dzhar (2,500 m) | P< | Murhab (3,600 m) | P< |
| Erythrocyte counts (ms) | 4.79 ±0.04 | 4.59 ±0.31 | 0.5 | 4.77 ±0.09 | 0.5 | 4.94 ±0.07 | 0.5 | 4.65 ±0.04 | 0.1 | 5.45 ±0.07 | 0.02 |
| Haemoglobin content (g%) | 15.1 ±0.10 | 14.8 ±0.13 | 0.5 | 15.9 ±0.26 | 0.001 | 16.2 ±0.15 | 0.001 | 15.2 ±0.10 | 0.01 | 18.5 ±0.25 | 0.001 |
| Total circulating Hb (g/m²) | 428.0 ±19.4 | 433.2 ±14.2 | 0.5 | 473.3 ±31.6 | 0.5 | 471.3 ±39.7 | 0.5 | 469.1 ±16.5 | 0.2 | 656.0 ±34.7 | 0.001 |
| Haematocrit (%) | 45.5 ±0.6 | 47.0 ±0.45 | 0.01 | 47.0 ±0.4 | 0.01 | 58.0 ±1.8 | 0.001 | 51.0 ±1.5 | 0.001 | 61.0 ±2.0 | 0.001 |
| Reticulocyte counts (%) | 0.35 ±0.01 | 0.55 ±0.03 | 0.001 | 0.5 ±0.2 | 0.02 | 0.5 ±0.03 | 0.02 | 0.52 ±0.02 | 0.001 | 0.51 ±0.3 | 0.001 |
| Total circulating reticulocytes (ms/m²) (×10¹⁰) | 4.8 | 7.5 | | 6.6 | | 7.5 | | 7.2 | | 9.8 | |
| Reticulocyte maturation (h) | — | 20.2 ±1.289 | | — | | — | | 11.4 ±0.344 | 0.001 | — | |
| Number of subjects | 66 | 32 | | 24 | | 39 | | 85 | | 45 | |

Summarising the results on the physiology of natives to Tien Shan and Pamir altitudes it can be stated that the adaptation process encompasses various physiological functions. The observed changes are oriented both to the maintenance of an adequate oxygen supply and adequate partial carbon dioxide pressure in alveolar air and, consequently, in the blood. Significant changes occur in the pulmonary circulation. Many adaptive responses of natives to Tien Shan and the Pamirs are of a relatively persistent nature.

## Work capacity in altitude natives

Physical work capacity (PWC) in altitude-natives of Tien Shan and the Pamirs has been studied by us for several years (Mirrakhimov, Kudai-berdiev & Schmidt, 1973). Apparently healthy Kirghiz males aged 18 to 25 years, mainly cattle-breeders, were selected for the study. The three groups studied were residents of the villages Groznoe (940 m), Kyzyl-Dzhar (2,500 m) and Murhab (3,600 m), respectively.

PWC was tested on a bicycle ergometer (Elema). The initial work load was 600 Kg min$^{-1}$. Every 2 min it was increased by 200 kg min$^{-1}$. The subject worked until a heart rate of 170–180 min$^{-1}$ was reached or until exhaustion prevented him from continuing. Recordings of heart rate, polycardiograms, tacho-oscillograms and synchronous carotid and femoral artery pulsations, as well as $\dot{V}_{O_2}$ and $\dot{V}$ determinations were made before and during exercise and during a 15 min recovery period.

The total work performed by natives to 2,500 m, 940 m and 3,600 m averaged 33,160 kg (23 min 18 s), 7,895 kg (8 min 48 s) and 4,403 kg (mean 5 min 33 s), respectively. The actual PWC at which heart rate reached 170 (PWC$_{170}$) was highest at 2,500 m and lowest at 3,600 m (Table 10.5). It should be noted that predicted and actual PWC$_{170}$ in all groups was not significantly different. The rather high PWC in moderate-altitude natives was manifest in the cardiovascular responses to increasing work load. For example, maximal heart rate reached 180 at the 6th min of exercise at 3,600 m, 185 at the 10th min at 940 m, and 179 only at the 20th min at 2,500 m.

By calculating the ratio of the sum of pulse counts over 6 min of exercise to the sum of recovery pulse counts, it was possible to assess that, at moderate and low altitudes, exercise was performed with a similar stress on the cardiovascular system, while at the higher altitude the stress was greater.

During graded exercise the rise in $\dot{V}$ was significantly faster at 940 m than at 2,500 m, but was slower than at 3,600 m. At the 4th min of exercise $\dot{V}$ was 5, 2.5 and 4 times the resting value in natives to 3,600 m, 2,500 m and 940 m, respectively. After fifteen minutes' recovery $\dot{V}$

307

Table 10.5. *Physical work capacity in natives to Tien Shan and the Pamirs*

| | Groznoe, 940 m | Kyzyl-Dzhar, 2,500 m | Murhab, 3,600 m |
|---|---|---|---|
| Number of subjects | 22 | 16 | 23 |
| Duration of work | 8 min. 48 s | 23 min. 18 s | 5 min. 33 s |
| Total work performed (kg) | 7.895 | 33.160 | 4.403 |
| Predicted $PWC_{170}$ (kg min$^{-1}$) | 1.123±34.6 | 1.644±93.2 | 618±57.0 |
| Actual $PWC_{170}$ (kg min$^{-1}$) | 1.000±253.9 | 1.800±207.3 | 625±35.6 |
| Predicted $\dot{V}_{O_2}$ max (ml kg$^{-1}$ min$^{-1}$) | 76.2±6.85 | 88.9±6.76 | 54.7±1.89 |
| Actual $\dot{V}_{O_2}$ max (ml kg$^{-1}$ min$^{-1}$) | 33.9±4.43 | 45.7±4.46 | 34.8±5.09 |
| Actual $\dot{V}_{O_2}$ max at heart rate 170 (ml kg$^{-1}$ min$^{-1}$) | 26.9±1.97 | 36.7±2.08 | 30.1±2.32 |
| Oxygen expenditure (ml kg$^{-1}$ min$^{-1}$ per kg body weight) | 0.024±0.008 | 0.018±0.0009 | 0.019±0.003 |

exceeded the base line values by 49%, 15% and 69%, respectively. This suggests that the natives to 940 m and 3,600 m incurred a high oxygen debt with exercise.

The actual $\dot{V}_{O_2}$ max was markedly lower than the predicted values, with the actual values being quite similar at 3,600 m and 940 m. Of interest is the fact that actual $\dot{V}_{O_2}$ max is markedly higher at 2,500 m than at 940 m despite a great difference in mechanical efficiency. Recovery rate (relation of exercise $\dot{V}_{O_2}$ increment to oxygen debt) was highest at 2,500 m and lowest at 3,600 m. The high recovery rate values at 2,500 m were due to a greater duration of exercise. The oxygen cost per unit of work proved to be lower in altitude-natives than in lowlanders.

Exercise tolerance was also assessed from cardiodynamics. At 940 m there was a significant shortening of the cardiac cycle during the first 2 min of exercise (by a mean of 0.274 s), and this shortening continued with increasing work load. The difference in duration between two consecutive cardiac cycles did not exceed an average of 0.02–0.03 s. The average cycle values at 2,500 m and 940 m were similar at the beginning of exercise. It should be noted that gradual shortening of the cardiac cycle was observed only up to 10 min and, starting at a work load of 2,000 kg min$^{-1}$, no further shortening occurred.

The chronology of cardiac cycle phases was significantly different in high-altitude natives (3,600 m) from that in the other two groups. The IC phase shortened both in altitude-natives and in lowlanders at the start of exercise. Maximal IC phase shortening was observed with lowlanders during the first 4 min of work, and thereafter the shortening was gradual. By the end of the tenth minute, the IC phase reached, as a rule, a zero value. In contrast to the 940 m natives, 2,500 m natives had a more rapid IC phase shortening, and a zero value was reached at 6–8 min. At 3,600 m zero values of the IC phase were registered on the average by the sixth minute of exercise.

The left ventricular E phase also shortened during work. At 940 m maximal shortening was observed during the first 4 min of exercise. Thereafter the shortening of the E phase was very gradual. The mean duration of the E phase at a work load of 600 kg min$^{-1}$ was 0.208 s, while at 900 kg min$^{-1}$ it was 0.187 s. Further work-load increase caused an average decrease to 0.172 s. Stabilization of the E phase was not noted at 940 m. At 2,500 m the shortening of the E phase was gradual from the very beginning of work: on the average 0.020 s for every 200 kg min$^{-1}$ (up to 1600 kg min$^{-1}$). At higher work loads the E phase shortened only by 0.010 s. Beginning from 1,800 kg min$^{-1}$, the E phase stabilized, amounting to 0.131±0.007 s. At 3,600 m the main E phase alteration occurred at a work load of 600 kg min$^{-1}$. Residents of this altitude exhibited relatively low values for the E phase at rest and at the end of exercise.

It should be noted that lowlanders performed work at greater cardiac output ($\dot{Q}$) values than moderate-altitude natives. However, the $\dot{Q}$ values of lowlanders were far below those of highlanders, being 619 ml sec$^{-1}$ at the end of exercise.

Summarising the changes in cardiodynamics during graded exercise, the following features may be pointed out. Lowlanders cease to pedal in the initial period of cardiodynamic stabilization (at 1,200 kg min$^{-1}$). Alteration of cardiodynamics is achieved differently in natives of moderate and high altitudes. At 2,500 m the response of cardiodynamics to increasing work load is achieved in three phases: initial response, initial stabilization stage and steady-state cardiodynamics. Steady-state cardiodynamics are manifest at relatively high work loads, when the myocardial hyperdynamics syndrome is at its maximum. At 3,600 m alterations in cardiodynamics occur at low work loads, i.e. there is an early myocardial hyperdynamics syndrome. Our data show a direct correlation between the increase in $\dot{V}_{O_2}$ during exercise and the degree of the enhancement of cardiodynamics.

In the early recovery period (30–60 s) lowlanders show a further E phase shortening, while other parameters (e.g. IC phase) remain at a level typical of the last minutes of work. During this period highlanders had a shortening in both the E and IC phases. The mechanism of this paradoxical phenomenon is related to the abrupt cessation of work and the sharp fall in venous return to the heart. Since in the early recovery period cardiac contractions remain at a level typical of the final work period, it

309

is obvious that blood ejection is very rapid. The more pronounced E phase shortening in highlanders during early recovery suggests evidently a continued enhancement of cardiac function, which plays a definite role in the recovery of physiological functions. During the recovery period, the stabilization phase of IC, E and $\dot{Q}$ began at the tenth minute at 2,500 m and 940 m. At 2,500 m, unlike 940 m, stabilization of the IC and E phases is complete by the fifteenth minute. The AC phase reached a steady state within the first 1–3 min at 2,500 m, and by the tenth minute at 940 m. Delay in the recovery of the cardiac cycle and IC and E phases was characteristic of natives to 3,600 m. Moreover, during fifteen minutes' recovery there was no stabilization stage of cardiodynamics.

Thus, cardiorespiratory responses to graded exercise suggest that residents of 2,500 m have certain advantages. Their relatively high exercise tolerance is related mainly to the functional features of the circulatory system, first of all – the heart. The lower physical endurance at 3,600 m is due to poorer regulation of the cardiorespiratory system.

### Responses of highlanders on return to high altitude after a stay at 760 m

In 1971 we studied young Kirghiz males (twenty subjects) whose mean age was 19 years. Both during base line studies (Frunze, 760 m) and at altitude (3,200 m), the lowlanders (nine subjects) and the highlanders (eleven subjects) had similar dietary and activity patterns. The highland group consisted of natives to 2,000–2,500 m who had lived for the last 1–2 years at 760 m.

Comparison of physiological responses of the two groups showed marked differences. For example, assessment of chemosensitivity at 3,200 m by oxygen inhalation revealed a markedly higher sensitivity in the lowlanders. Despite a smaller decrease in $\dot{V}$ due to oxygen inhalation, the highlanders had a subsequent hyperventilation phase; this can be attributed to the respiratory centre response to blood oxidation due to increased oxyhaemoglobin. Possibly the respiratory centre of the highlander is more sensitive to carbon dioxide although chemoreceptor responses are depressed; despite the altitude-induced hyperventilation in newcomers (mainly due to increased respiratory rate), their partial alveolar oxygen tension was maintained at lower levels than in highlanders.

Marked differences were also found in circulatory function and blood indices during altitude adaptation. Haemodynamic studies (dye-dilution) on days 20 and 40 at 3,200 m revealed greater increases in circulating blood mass and decreases in total vascular resistance in highlanders. It seemed that in highlanders blood played a major role in altitude re-adaptation.

It is noteworthy that the myocardial hypodynamics syndrome that developed in these highlanders, as determined by PCGs, disappeared

310

rather rapidly (within the first 10 days) and was replaced by hyperdynamic events; this correlates with the dynamics of PWC changes.

The data indicate that despite a 1–2-year stay at low altitude, the highlanders retained certain characteristics. These include various physiological functions which became more obvious when they returned to high altitude. This suggests adaptive structural changes in altitude-natives.

We also studied at 760 m several physiological functions in natives to 2,000–3,000 m after a 2-year stay at low altitude (760 m). We found respiratory rate was $15\pm0.28$ per min as against $13.6\pm0.36$ per min in natives to 760 m (controls). $V_{A,CO_2}$ exceeded predicted values by 30% (control 20%); $P_{A,CO_2}$ was $40.25\pm0.71$ mm Hg (control $39.36\pm0.59$), and $P_{A,O_2}$ was $77.1\pm0.68$ (control $81.87\pm0.8$). $V_{O_2}$ and BMR significantly exceeded control values. Therefore, despite a 2-year absence of a hypoxic stimulus, the highlanders retained certain physiological features characteristic of high altitude.

Thus, biological and physiological studies in altitude-natives of Tien Shan and the Pamirs reveal not only the existence of certain particular functional physiological characteristics, but also their stability which, apparently, is related to genetic changes.

**Fertility, pregnancy and placental function**

In women living at moderate and high altitudes menarche is significantly later (1.5 to 2 yr) than it is in lowland women (Toktorbaeva, 1966; Borzhykh, 1971). The onset of menopause is normal and reproductive performance is not impaired. Studies of the dynamics of oestrogen, pregnandiol, and gonadotrophic hormone excretion during several menstrual cycles in highland Tajik women showed that all their cyclic changes in ovarian hormonal functions are similar to those observed in sea-level women (Petranjuk, 1967).

As to fertility, the incidence of primary sterility is higher at high altitude. During pregnancy there is a predisposition to spontaneous abortion. The period between the beginning of sexual life and first pregnancy proves to be prolonged to $2.8\pm0.21$ yr as compared to $2.2\pm0.18$ yr in lowland women; highland women are more likely to have repeated miscarriages. Selected check-ups revealed a smaller total number of births ($5.0\pm0.09$ versus $5.4\pm0.1$ at low altitude) and a smaller number of living children ($4.6\pm0.08$ versus $5.0\pm0.08$) per woman in the highlands compared with women at low altitude. As a rule, fertility is high in women native to the East Pamirs (3,600–4,100 m). At moderate Tien Shan altitudes (1,750 m) delivery is usually at term and is not accompanied by significant complications.

# The biology of high-altitude peoples

Placental function was studied at moderate altitudes (1,750 m) (Lebedeva, 1973). A characteristic feature of arterial umbilical blood in infants at birth was a higher glucose content ($120.3 \pm 3.0$ mg% versus $103.1 \pm 1.8$ mg%). More glucose passed across the placental barrier at moderate altitudes, suggesting a greater functional placental activity; despite an intravenous infusion of the same amount of glucose to mothers at the final phase of delivery, the glucose concentration in umbilical blood at the moment of birth was higher at moderate altitudes than in the controls (760 m). Addition of 40 mg of sigetin producing uterus hyperaemia, caused a more distinct inflow of glucose into umbilical blood and, moreover, utilization of glucose by the foetus was more complete at altitude.

### Morpho-functional features of children native to Tien Shan and the Pamirs altitudes

Anthropological studies showed that the growth and sexual maturation processes are delayed in highland children as compared to lowland children (Miklashevskaya et al., 1973). Increases in erythrocyte and reticulocyte counts, haemoglobin and haematocrit values were found in newborn infants at 2,020 m. At higher altitudes (up to 3,600 m) no further increase in RBC counts was observed, while haemoglobin content rose further. In children up to 5 years of age, RBC counts and haemoglobin content rose with altitude. A smaller rise in RBC counts and haemoglobin concentration in response to altitude hypoxia was observed in children up to 2 years of age. Altitude erythrocytosis and hyperhaemoglobinaemia proved characteristic of under-schoolage (Kudojarov, 1966) and school-age children. Erythrocyte size was greater, although the mean haemoglobin concentration in the individual erythrocytes remained practically unchanged. Exceptions were newborn infants in whom no increase in mean erythrocyte size was observed on the first day after birth (3,600 m).

The hyperglobulinaemia observed in highland children is related to blood formation activation. This is confirmed especially at 3,600 m by the higher reticulocyte counts as well as by higher red cell counts (Afanasenko, 1973).

Newborn infants at 2,020 m have slightly increased leucocyte counts, mainly at the expense of neutrophils, while at 3,600 m decreases are observed. Subsequently (up to 5 years of age), the leucocyte count tends to fall and this may be related to the lower neutrophil count associated with altitude.

At 760 m and 2,020 m there is no difference in heart rate in newborn infants, but at 3,600 m heart rate is decreased and this trend persists through life. From the neonatal period onwards peculiar cardiac cycle alterations (lengthening of diastole, shortening of systole) ensuring a more economic cardiac performance were also found at 3,600 m. Sinus

312

arrhythmia was more pronounced. Complex estimation of ECG records in highland children (especially at 3,600 m) revealed evidence of right and left atrial and right ventricular overload.

The magnitude of right ventricular hypertrophy is similar in newborn infants at all altitudes. During the first months of life but mainly by 6 months, right ventricular hypertrophy is replaced by left ventricular hypertrophy at low and moderate altitudes (up to 2,020 m), while at 3,600 m the former is persistent. At 2,020 m some of the children (8%) older than 6 months retain signs of right ventricular hypertrophy and from 2 years of age highland children also show signs (more often at 3,600 m) of mild left ventricular hypertrophy.

The data suggest that the post-natal adaptation process is quite satisfactory at 2,020 m, whereas at 3,600 m there is a marked stress imposed on the adaptive systems. Thus, post-natal cardiovascular alterations are characterized in highland children by right ventricular overload. It should be mentioned that the major cardiovascular features inherent in adult highlanders are acquired mainly during the first year of post-natal life, but the process of cardiovascular adjustment continues and is complete only at the age of 12–15 years (Dzhunusov, 1970).

### Epidemiology of certain internal diseases at Tien Shan and the Pamirs altitudes

Studies on the incidence of adaptive pathology, congenital cardiac defects, ischaemic heart disease, arterial hypertension, rheumatic heart disease and chronic bronchitis in Tien Shan and the Pamirs revealed certain peculiarities of the occurrence of these diseases among native highlanders. Above 3,000 m permanent residents are characterised by primary altitude pulmonary hypertension (Mirrakhimov, 1971) which sometimes (at 3,600–4,200 m) leads to chronic cor pulmonale with moderate degrees of congestive heart failure. In these subjects there is a marked rise in circulatory blood mass and a decrease in chemosensitivity to hypoxia which decreases lung ventilation. This in turn leads to the enhancement of alveolar hypoxia and pulmonary hypertension, resulting in progressive cor pulmonale and the onset of congestive heart failure (Mirrakhimov, 1972b). This is indicative of the absence of true acclimatization in some altitude-natives.

Congenital defects, as well as patent ductus arteriosus, are much more frequent among highlanders (Afanasenko *et al.*, 1973).

Hypertension and especially ischaemic heart disease are uncommon in Tien Shan and the Pamirs (Dzhailobaev *et al.*, 1973; Mirrakhimov, 1971, 1972a), while the incidence of chronic bronchitis and rheumatic heart disease is about similar at low and high altitudes.

313

# The biology of high-altitude peoples

**References**

Afanasenko, P. P. (1973). In *Human physiology and pathology at high altitude.* Frunze, v. 92, 2nd edn, p. 90.
Afanasenko, P. P., Dzhunusov, M., Baidurina, S. K., Sadybakasov, K. S., Sychev, V. I. & Dzheenbekov, K. D. (1973). In *Human physiology and pathology at high altitude.* Frunze, v. 92, 2nd edn, p. 143.
Agadzhanian, N. A., Issabaeva, V. A. & Elfimov, A. I. (1973). *Chemoreceptors, hemocoagulation and high altitude.* Frunze: Ilim Publishing House.
Agadzhanian, N. A. & Mirrakhimov, M. M. (1970). *High altitude and body resistance.* Moscow: Nauka Publishing House.
Akhmedov, K. Y. (1971). *Respiration of man in altitude hypoxia.* Dushanbe: Donish Publishing House.
Borzhykh, I. V. (1971). Synopsis of Thesis for a Candidate's Degree, Frunze.
Dubinina, J. S., Dzhailobaev, A. D., Kochenkova, E. G. & Schmidt, G. F. (1973). In *Human physiology and pathology at high altitude.* Frunze, v. 92, 2nd edn, p. 26.
Dubinina, J. S. & Kochenkova, E. G. (1972). In *Human adaptation.* Leningrad: Nauka Publishing House.
Dubinina, J. S., Kovaleva, R. I. & Mirrakhimov, M. M. (1967). In *Scientific works of the Kirghiz State medical institute.* Frunze, v. 48, 1st edn, p. 7.
Dzhailobaev, A. D. (1967). Synopsis of Thesis for a Candidate's Degree, Frunze.
Dzhailobaev, A. D., Grinstein, B. J., Dubinina, J. S. & Narbekov, O. N. (1973). In *Human physiology and pathology at high altitude.* Frunze, v. 92, 2nd edn, p. 155.
Dzhunusov, M. G. (1970). Synopsis of Thesis for a Candidate's Degree, Frunze.
Filatova, L. G. (1961). *Studies on the physiology of high-altitude acclimatization in man and animals.* Frunze.
Grinstein, B. J. (1966). Synopsis of Thesis for a Candidate's Degree, Frunze.
Issabaeva, V. A. (1972). In *Human adaptation.* Leningrad: Nauka Publishing House.
Konopleva, L. F., Panichkin, Y. V. & Popov, A. A. (1971). *Cardiology,* **10**, 138.
Kudaiberdiev, Z. M. (1970). Synopsis of Thesis for a Candidate's Degree, Frunze.
Kudojarov, D. K. (1966). Synopsis of Thesis for a Candidate's Degree, Frunze.
Lavrinovich, A. N. (1898). *The Physician,* **19**, 35.
Lebedeva, I. M. (1973). Synopsis of Thesis for a Doctor's Degree, Leningrad-Frunze.
Miklashevskaya, A. N., Solovyeva, V. S., Godina, E. V., Afanasenko, P. P., Dzhunusov, M. G., Dzheenbekov, K. D. & Sadybakasov, K. S. (1973). In *Human physiology and pathology at high altitude.* Frunze, v. 92, 2nd edn, p. 6.
Mirrakhimov, M. M. (1964). *Essays on the influence of mountain climate of Central Asia on the organism.* Frunze: Kirghiztan Publishing House.
Mirrakhimov, M. M. (1968). *The cardiovascular system at high altitude.* Leningrad: Medicina Publishing House.
Mirrakhimov, M. M. (1971). Heart diseases and high altitude. Frunze: Kirghiztan Publishing House.
Mirrakhimov, M. M. (1972a). *Cardiology, Moscow,* **12**, 17.
Mirrakhimov, M. M. (1972b). *Clinical Medicine, Moscow,* **12**, 104.
Mirrakhimov, M. M. (1972c). *Sechenov physiological Journal of the USSR,* **58**, 1816.

Mirrakhimov, M. M., Kudaiberdiev, Z. M. & Schmidt, G. F. (1973). In *Human physiology and pathology at high altitude*. Frunze, v. 92, 2nd edn, p. 118.
Narbekov, O. N. (1970). Synopsis of Thesis for a Candidate's Degree, Frunze.
Pavlova, O. N. (1937). *Transactions of the Uzbek Institute of Experimental Medicine, Tashkent*, 3, 3.
Petranjuk, E. I. (1967). Synopsis of Thesis for a Doctor's Degree, Dushanbe.
Plotnikov, I. P. (1963). Synopsis of Thesis for a Candidate's Degree, Dushanbe.
Raimzhanov, A. R., Usupova, N. Y. & Barbashova, Z. I. (1973). In *Human physiology and pathology at high altitude*. Frunze, v. 92, 2nd edn, p. 105.
Rudenko, R. I. (1973). Synopsis of Thesis for a Candidate's Degree, Frunze.
Shlomm, O. M. (1895). *Proceedings of the sittings of the Fergana Medical Society for 1895–6.*
Toktorbaeva, S. T. (1966). *Proceedings of the IXth Conference of Medical Workers of Kirghizia, Frunze*, 293.
Tretiakov, N. N. (1895). *Military Medical Journal, St. Petersburg*, **183**, 1.

# 11. The adaptive fitness of high-altitude populations

P. T. BAKER

The adaptation of a population to its environment may be measured in a variety of ways. Although adaptation has a very specific meaning in evolutionary theory it is used in practically all the disciplines concerned with man and in each has a slightly different connotation signifying types of adjustments, accommodations, acclimatizations and even simple responses to both the social and natural environment. With this diversity of definitions it is difficult to discuss in a cross-disciplinary framework the adaptation of human populations to the high-altitude environment. Nevertheless, few would disagree that the survival and continuity of a population is a prime indicator of that group's adaptation to an environment and the general health of a group is also widely accepted as an overall indication of success in adjusting to a specific environment. In this chapter I will, therefore, explore the rather limited demographic and health data available on high-altitude peoples.

## Demographic structure of high-altitude peoples
### Total numbers

Many of the high-altitude areas of the world are in the economically least developed areas and for this reason even simple population numbers in relation to altitude are often impossible to obtain. Even in those countries where reasonable census information has been collected the traditional census procedures of analyzing populations by political and socio-economic units makes an analysis by environmental zones difficult. At the beginning of IBP, De Jong (1970) estimated that there might be thirteen to sixteen million people living above 3,300 m on a world-wide basis, but he noted in personal conversation that this was little better than a guess.

The census data for the Andes are generally satisfactory and with access to the original data for each of the Andean countries a reasonably accurate estimate of population size and structure by altitude could be developed. However, at present the only country for which accurate estimates by altitude are available is Peru. The distribution as shown in Fig. 11.1 is based on analysis of the 1961 census (Vásquez, Parlin & Simonson, 1967). By most physiological measures, altitudes above 2,500 m are required to produce evidence of significant hypoxic stress. We will, therefore, arbitrarily designate those living above this altitude as high-altitude popula-

317

*The biology of high-altitude peoples*

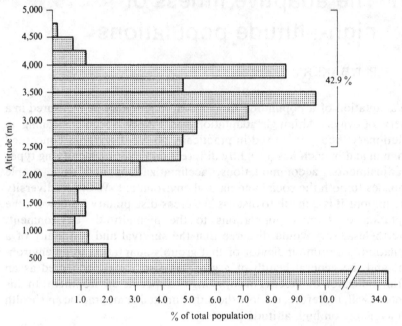

Fig. 11.1. Peruvian population distribution by altitude in 1961. (Modified from Vásquez *et al.*, 1967.)

Table 11.1. *Estimates of population numbers above 2,500 m in the Andes*[a]

| | | Minimum | | Approximation | |
|---|---|---|---|---|---|
| | Survey year | Survey year | 1970 estimate | Survey year | 1970 estimate |
| Argentina | 1960 | — | — | 128,742 | 149,345 |
| Bolivia | 1966 | 1,494,827 | 1,657,874 | 3,314,593 | 3,676,159 |
| Chile | 1963 | — | — | 49,508 | 54,697 |
| Colombia | 1964 | 2,096,470 | 2,536,535 | 4,682,558 | 5,665,577 |
| Ecuador | 1962 | 612,557 | 802,699 | 1,562,469 | 2,047,491 |
| Peru | 1961 | 4,210,285 | 5,773,914 | 4,210,285 | 5,773,914 |
| Total | | | 10,771,022 | | 17,367,183 |

[a] The Minimum columns estimate the total urban population size above 2,500 m. The Approximation columns estimate total population size above 2,500 m by assuming that the proportion of the rural population at high altitude is equal to the urban proportion. No urban areas are located above 2,500 m in Argentina and Chile; the Approximation for these countries is 5% of the rural population.

Sources: Statistical Abstract of Latin America (1972); Encyclopedia Americana (1974); The World Book Encyclopedia (1974); Vásquez *et al.* (1967).

Table 11.2. *Population of high-altitude countries and regions in Asia and Africa*

|  | Year | Population |
|---|---|---|
| Nepal | 1971 | 11,555,983 |
| Tibet | 1971 | 1,300,000 |
| Sikkim | 1971 | 204,760 |
| Bhutan | 1969 | 1,034,774 |
| Province of Ladakh | 1961 | 88,651 |
| Province of Kashmir | 1961 | 1,899,438 |
| Ethiopia (estimate) | 1971 | 26,000,000 |
| Total |  | 42,083,606 |

Sources: U.N. Statistical Yearbook for 1973; The World Book Encyclopedia (1974); Encyclopaedia Britannica (1974).

tions. By this standard, 43% of Peru's population, or more than four million people, lived at high altitude in 1961.

For other Andean countries estimates are necessarily less accurate. In Table 11.1 an attempt to estimate total numbers is made based on the very unorthodox methods of determining the urban populations when altitude of the town is known and extrapolating the rural population distribution from the urban altitude distribution.

For the high-altitude areas of Asia and Africa even this crude estimating technique cannot be used since very few of the population live in settlements large enough to be listed as towns. The total estimated populations for the various countries containing substantial high-altitude populations is shown in Table 11.2. From what little one can deduce from information on these countries, as many as one-quarter of these people may live above 2,500 m. In addition, small permanent populations above this level are to be found in the USSR, Mexico, the United States and possibly in a few other countries. When added together it appears that there are not less than twenty million nor more than thirty million people who are permanent residents at high altitude. This is not a very large percentage of the world's population and is probably a smaller percentage of the world's population than it was 500 years ago. However, it certainly does indicate that a large number of our species has adapted to this seemingly stressful environment.

*Age and sex distribution*

No detailed studies have been made of how the age and sex distributions of high-altitude populations compare with other groups. However, a few small traditional groups have been analyzed in some detail. The structure

319

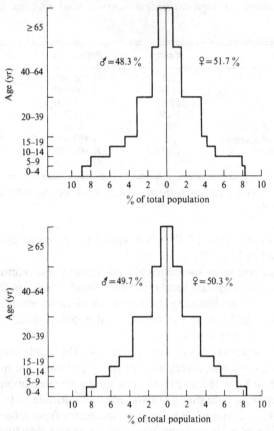

Fig. 11.2. Population structure of the Nuñoa district, population 7,750 (*a*) compared to the total Peruvian population 9,906,746 (*b*). (From Baker, 1976.)

of the population of the District of Nuñoa (4,000 m) compared to the total Peruvian population is shown in Fig. 11.2. The differences are not dramatic but show some interesting differences:

(1) the Nuñoa population is somewhat younger than the general population;

(2) the sex ratio (males/females × 100) is higher in Nuñoa than in all of Peru during infancy and childhood but is low among adults;

(3) there appear to be more older people among the highlanders than in the general population.

The explanations for these differences are generally known for Nuñoa. First, the more youthful profile is caused by the dual processes of a higher birth rate in Nuñoa compared to the country as a whole combined with a high adult emigration (Spector, 1971; Baker & Dutt, 1972). Second, the

320

sex-ratio differences may be explained by a series of phenomena, some of which appear related to altitude responses. A recall fertility questionnaire suggested that the sex ratio at birth was as high as 120 (Hoff, 1968) while an examination of 25 years of birth registrations yielded a sex ratio of 111 (Baker & Dutt, 1972). Both sets of data were subject to error but there was good reason to believe that the actual birth sex ratio was unusually high. On the other hand, male mortality was much higher than female during infancy and remained somewhat higher throughout childhood and early adolescence (Spector, 1971). Thus, the sex ratio would have been near 100 by adulthood if the population were stationary. Instead, emigration rates were high during late adolescence and young adulthood. The emigration was much greater among males than females leaving a very low sex ratio, near 80 for all adults. The high birth sex ratio and high male preferential outmigration appears to have typified the high-altitude areas of Peru (Schadel, 1959; Alers, 1965), northeast Chile (Cruz-Coke *et al.*, 1966), and Bolivia (CEP, 1969; Dutt, 1976a) for at least the last 30 years.

Finally, the high percentage of older individuals in Nuñoa is probably a cultural artifact since people over 50 years of age generally did not know their age but believed it prestigious to provide a very old estimate.

The sex and age structure of the upper Khumbu region in Nepal was reported by Lang & Lang (1971). The structure of this Sherpa population living at an altitude of 3,500 to 5,000 m is shown in Fig. 11.3. The authors note that the structure is not an unusual one for Nepal. However, it will be noted that, as was the case in Nuñoa, females outnumber males during adulthood. In a further analysis of Lang's data, C. A. Weitz (personal communication) noted that the birth sex ratio found was 118 and speculated that as in Nuñoa the male-infant mortality must be considerably higher than the female-infant mortality. He further notes that while there is little permanent outmigration the high percentage of adult males involved in portering for mountain climbing expeditions subjects them to high mortality rates further reducing the adult sex ratio.

The pattern of high sex ratios at birth and high male-infant and child mortality may typify much of high-altitude Nepal as shown by another study. Goldstein (1977) examined three small villages in northwestern Nepal which ranged from 3,500 to 4,000 m and contained 791 people. He found that the sex ratio of children (0–14 yr) was 116 while the sex ratio during the reproductive years (15–44 yr) was only 89. He noted there was virtually no outmigration and only a few male immigrants from Tibet.

While the data are clearly too limited for firm conclusions the available information suggests that at least in some parts of the Andes and Himalayas the numbers of males born relative to females may be significantly higher than one would expect, while the mortality of male infants and

321

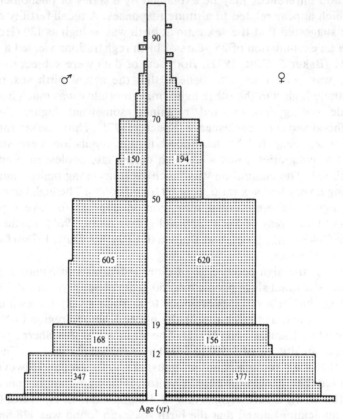

Fig. 11.3. Age and sex distribution of the upper Khumbu area of Nepal. (Modified from Lang & Lang, 1971.)

children is much higher relative to females leaving an adult population with an excess of females. This tendency is exaggerated in many areas by preferential male adult outmigration.

Data on other high-altitude areas are not available for testing these hypotheses and it should be noted that this finding is restricted to populations which were almost totally lacking in modern medical care.

### Fertility and altitude

The idea that high altitude may reduce population fertility is an old one. As Monge (1948) noted, even the early Spanish settlers of the sixteenth century believed that altitude reduced their fertility although they found the high-altitude native quite fertile. As Clegg reports (see Chapter 4) there is good reason to believe that several aspects of basic fecundity are affected by high altitude, but the question of how this affects the net reproduction of high-altitude natives remains a debated point.

A lengthy set of articles was stimulated in recent years by the observation that in the Central Andean countries the number of children under five per woman in the reproductive years declined with altitude. Stycos (1963) first noted that Peru showed a reversed situation to the Latin American norm, so that the child/woman ratio was higher in the urban than the rural setting. This finding based on the 1940 census led him to believe it was caused by the low legal marriage rate among the Amerindian lower class. He later concluded that health differences might also be involved (Stycos, 1965). Heer (1964) and Heer & Turner (1965) disputed the instability of the Amerindian matings and expanding the analysis to Bolivia and Ecuador offered a set of sociological causes. Still later James (1966) pointed out that for the three countries the best relationship to the child/woman ratio among older women was altitude and Heer (1967), in a re-analysis, agreed.

James suggested that the temporary migration of males to the lowlands, the increased neonatal mortality at altitude and a higher spontaneous abortion rate at altitude might explain the results. Other authors (Whitehead, 1968; Bradshaw, 1969) thought the findings might be explained by an under-enumeration of young children in the 1940 census. Of the various explanations several have not been supported by later analysis. Baker & Dutt (1972) examined the 1961 Peruvian census, which was more carefully collected than the 1940 one, and found that the relationship between altitude and the child/woman ratio remained high. Furthermore, they could detect no relationship between this ratio and the adult sex ratio. It thus appears that the use of census data cannot resolve the question and from its analysis one is left with several possible explanations including the possibility that altitude directly affects fertility.

Writing on this question De Jong (1970) suggested that a more detailed approach which considered all of the fundamental variables known to affect reproductive behavior, would be required if definitive conclusions were to be developed. Dutt (1976a), using information generated from a large fertility study in Bolivia (CEP, 1969) plus original data collected by himself in the Bolivian lowlands, attempted to resolve the problem by the altitude-gradient approach. He selected from the larger sample only those women who were lower class and did not appear to be controlling fertility. Three samples were compared from the Departments of La Paz, Cochabamba and Santa Cruz. The samples from La Paz and Cochabamba came from varying altitudes but these averaged about 3,600 m for La Paz and 2,800 m for Cochabamba. The Santa Cruz area was all low altitude. Dutt found completed average fertilities in women over 40 years of age which ranged from 5.27 to above 8.0 depending on residence location. However, he did not find a significant difference in the mean number of live births or total pregnancies according to altitude. The low-altitude women tended to have more pregnancies in their teens and the high-altitude women more in their late 30s and early 40s. While fertility values were generally unrelated to altitude, several aspects of mortality were related. The women in the survey reported fewer abortions at high altitude but significantly increased neonatal and infant mortality. These findings suggested to Dutt (1976c) that much of the negative correlation between altitude and the child/woman ratio in Bolivia may be attibuted to differences in infant and child mortality.

Dutt's findings differ in several details from studies on other Andean populations. Cruz-Coke *et al.* (1966) found an unusually high birth rate of 82.4 per 1,000 in a small group of high-altitude Aymara in Chile and reported that maximum pregnancy risk was very late in the reproductive span.

In the small city of Cerro de Pasco at 4,200 m in Central Peru completed fertility appears to have been about 7.6 births for women who practiced minimal conscious birth control (CISM, 1968). This figure is slightly higher than the estimated 6.7 completed fertility found for the rural area of Nuñoa at 4,000+ m (Hoff, 1968; Hoff & Abelson, 1976). While all of the reported completed fertilities are somewhat below the maxima which have been found for various populations, they are presently providing for rapid population growth and would provide more than replacement at even the highest of expected mortality rates.

These various data clearly indicate that the effect of altitude on the fertility of Andean high-altitude natives is not massive, but they leave unanswered the question of whether the apparent effects of altitude on fecundity are such as to affect fertility if only the stresses of altitude are varied. Some indication of the impact of these stresses was provided by the study of Abelson, Baker & Baker (1974). In this study, recall fertility data were collected on women living in an agricultural valley located on the southern Peruvian coast. The samples studied included three groups – sedente women, women who migrated from low altitude and women who migrated from high altitude – all living under comparable social and health conditions. The birth histories of these women are shown in Table 11.3. While the reproductive performances of the three groups show a general similarity the use of a two factor analysis of variance showed the overall patterns to differ significantly according to place of origin. The fertility pattern of the highland migrants was quite divergent when it was compared to their fertility history prior to migration or compared to the high-altitude Nuñoa women who completed their fertility living in the same altitude zone from which the migrants had originated. While the comparison of the

324

Table 11.3. *Births by 5-year intervals, according to altitude of birthplace and migrant status*

| | Low-altitude born | | | | | | | | | High-altitude born | | | | | |
| | Non-migrant Tambo valley | | | Migrant before migration | | | Migrant after migration | | | Migrant before migration | | | Migrant after migration | | |
| Age group | N | x̄ | s.d. | N | x̄ | s.d. | N | x̄ | s.d. | N | x̄ | s.d. | N | x̄ | s.d. |
|---|---|---|---|---|---|---|---|---|---|---|---|---|---|---|---|
| 15–19.9 | 56 | 0.79 | 1.02 | 36 | 0.44 | 0.65 | 19 | 0.95 | 0.91 | 57 | 0.37 | 0.69 | 23 | 1.08 | 0.99 |
| 20–24.9 | 54 | 1.87 | 1.23 | 18 | 1.28 | 1.13 | 30 | 1.70 | 0.99 | 33 | 1.12 | 1.11 | 39 | 2.00 | 0.75 |
| 25–29.9 | 44 | 1.89 | 1.20 | — | — | — | 34 | 1.65 | 1.07 | 17 | 1.41 | 1.12 | 38 | 1.76 | 1.10 |
| 30–34.9 | 34 | 1.53 | 1.37 | — | — | — | 29 | 1.34 | 0.97 | — | — | — | 31 | 1.84 | 1.03 |
| 35–39.9 | 28 | 1.21 | 1.37 | — | — | — | 24 | 1.00 | 1.14 | — | — | — | 21 | 1.14 | 1.01 |
| 40–44.9 | 16 | 0.75 | 0.86 | — | — | — | 19 | 1.05 | 1.47 | — | — | — | 14 | 0.64 | 1.09 |
| Total | — | 8.04 | — | — | — | — | — | 7.69 | — | — | — | — | — | 8.46 | — |

From Abelson *et al.* (1974).

Table 11.4. *Length of time between parities according to population*

| Parity level[a] | Non-migrant populations | | Migrant populations after migration | |
|---|---|---|---|---|
| | Tambo valley (yr) | Nuñoa (yr) | Low altitude (yr) | High altitude (yr) |
| 1–2 | 3.0 | 4.5 | 3.0 | 3.0 |
| 2–3 | 2.5 | 4.0 | 3.0 | 2.5 |
| 3–4 | 2.5 | 4.0 | 3.0 | 2.5 |
| 4–5 | 3.0 | 5.0 | 3.5 | 2.5 |
| 5–6 | 3.0 | 7.5 | 4.0 | 3.0 |
| 6–7 | 3.5 | — | 5.0 | 3.0 |
| 7–8 | 6.0 | — | — | 5.0 |
| Multiple parity level[a] | | | | |
| 1–6 | 2.80 | 5.00 | 3.25 | 2.70 |
| 1–7 | 3.08 | — | 3.54 | 2.75 |
| 1–8 | 3.50 | — | — | 3.07 |

[a] Refers to the serial order of births; i.e., 1–2 refers to the period between the births of first and second child.
From Abelson *et al.* (1974).

highland migrants with the Nuñoa natives suggested that downward migration may increase completed fertility by about two births, the differences in birth spacing were more striking. As shown in Table 11.4 parity spacing was much greater for any given parturition at high altitude than it was for any of the female samples at low altitude.

Another possibly significant finding of this study was that highland migrant women reported abortions to be only 22/1,000 before migration compared to 58/1,000 after migration. The latter figure is closely comparable to 54/1,000 reported by all the lowland women in the study. Since these are the same women reporting for both high and low altitudes the results strongly suggested that the low abortion rates reported in the previously cited studies of high-altitude Andean natives are not simply the product of poor or faulty recall.

While the limited sample sizes and imperfections of the natural experimental situation used in the Abelson *et al.* (1974) study prevent definitive conclusions, the results are those one would anticipate if life at high altitude did reduce slightly basic fecundity in the high-altitude natives of the Andes.

Reports of fertility for the high-altitude natives of Nepal and Ethiopia differ in many ways from the Andean results. Lang & Lang (1971) reported completed fertility for the Sherpa in Nepal to be 6.0 based on 103 women over age 45 years. Fertility may be rising since C. A. Weitz (personal com-

326

munication) reported that the age of marriage and first pregnancy is now dropping in response to contact with the outside world. While the fertility levels may therefore be quite comparable to values in the Andes, abortion rates appear higher. Lang & Lang (1971) report that their total sample yielded a 55/1,000 live birth rate. Weitz noted that within this sample there was an altitude relationship with abortions rising with the altitude of the community. In the isolated villages of northwestern Nepal, Goldstein (1976) found a lower fertility. He estimated from his small samples that married women might have a completed fertility of between 4 and 5 but because of polyandry practices many women are left unmarried and have a low reproductive performance. While Goldstein's data do not allow a total completed fertility estimate it appears that the figure may be well below four and may vary over time so that reproduction no more than matches loss in the long term.

In the Ethiopian study of Harrison *et al.* (1969) sample sizes were too small for reasonable estimates of birth rates or completed fertility but it appears that reproductive capacity in both the lowland and highland sample may be similar to the high-altitude population in Nuñoa. They reported that birth intervals based on the number of births/years married was in the range of 4.4 to 6.8 at high altitude and 4.2 to 6.8 at low altitude. Both figures seem quite similar to the 5.0 year parity intervals found for Nuñoans in the Andes. While birth intervals appear similar, abortion rates, as in Nepal, appear reversed from the Andean findings. Harrison *et al.* (1969) found only one spontaneous abortion reported in 173 pregnancies at the low-altitude village but twenty-one abortions out of 232 pregnancies in the high-altitude village. These rates which conform to 91/1,000 at high altitude and 6/1,000 at low altitude gave only a one in 1,000 chance of being different from each other by chance when tested by the $\chi^2$ technique.

In summary the fertility data on high-altitude peoples present a complex picture from which obvious conclusions do not emerge. It does seem clear that the native high-altitude peoples of the world have a more than adequate fertility for reproductive needs and indeed in the Andes and parts of Nepal are producing a rapidly growing population. In spite of these findings the studies in the Andes suggest that altitude-related forces tend to limit maximum fertility and increase neonatal mortality. The effects of altitude on spontaneous abortion are not so clear. Clegg has argued convincingly that it should increase abortion rates (see Chapter 4) and the limited abortion recall data from Ethiopia and Nepal support this contention. However, in the Andes all the available information from recall studies shows lower rates of abortion at high altitude. As noted there is good reason to believe that the Andean reports are correct. One is, therefore, left with two rather obvious hypotheses. Either abortion is

327

occurring so early in gestation in the Andes that it goes undetected by women (Clegg, Harrison & Baker, 1970) or the Andean native has genetically adapted to high altitude in a manner which increases fetal retention.

## Effects of altitude on health

The ascent of lowlander to high altitude produces a pronounced reduction in health including for most headaches, nausea, fitful sleep, chills and a decline in many perceptual, cognitive and motor abilities (Ward, 1975). If rapid ascent is attempted above 5,000 m these symptoms are almost universal. For a few individuals the symptoms are more severe and for the relatively few individuals who develop altitude-induced pulmonary edema the symptoms may be fatal.

Most of the symptoms decline within a week or two but given the continuing reduced work capacity (see Chapter 6), the increased probability of thrombosis (Chiodi, 1960) and the frequent persistence or recurrence of symptoms such as Cheyne–Stokes respiration it may be considered an unhealthy environment for lowlanders.

Several Peruvian investigators (Hurtado, 1966; Velásquez, 1966) have pointed out that these findings are all too frequently extrapolated to native high-altitude peoples. Instead they claim that for the native highlanders it may be a healthy environment and for these natives low-altitude environments may be unhealthy. However one views the general health of the various environments, altitude does produce a different pattern of diseases from the one found in the low-altitude areas surrounding the world's high-altitude areas. Some of these differences are specifically related to the reduced oxygen pressure, cold and aridity of the mountains on human biology. Other differences are caused by the environment's effect on disease organisms and their vectors.

### Altitude-specific disease

While several disease states appear to be aggravated by the reduced oxygen pressure found at high altitude only two have been directly attributed to this cause. Of these the best documented one is pulmonary edema. This disease, which occurs relatively infrequently in both lowlanders and highlanders after a rapid ascent to altitude, is frequently fatal if untreated. It is most common in young males according to Ward (1975) and occurs more often if the individual exercises heavily immediately after ascent to altitude. Onset is within 72 hours of ascent. Marticorena *et al.* (1964) noted that among high-altitude natives the highest risk occurred in pre-adolescent males.

The exact causes are not known but are believed to be related to the hypertension of pulmonary arterial circulation normally found at high altitude and some form of arteriole blockage. The clinical findings are usually restricted to fluid build-up in the lungs. The pathology tends to recur in the same individual suggesting idiosyncratic predisposing factors. It is a response not limited to man but may occur in a variety of mammals. Domestic cattle are particularly prone to this response (Monge & Monge, 1966).

A more poorly defined disease specific to reduced oxygen pressure is generally termed chronic mountain sickness (Monge & Monge, 1966). This disease, which is so far only well described for Andean high-altitude natives, is in the opinion of the Monges a breakdown in altitude acclimatization. While the initial symptoms are primarily ones of psychological malfunction, the clinical findings suggest reduced oxygen-transport capabilities. Erythrocyte activity is stimulated and although hemoglobin levels reach from 20.8 to 28.4 g/100 ml the arterial oxygen saturation is below normal native levels (Monge, Lozano & Carcellen, 1964). The only successful treatment for this potentially fatal disease is movement to low altitude.

## Altitude-aggravated disease

The relatively high frequency of thrombosis at high altitude may be seen as either a condition directly caused by reduced oxygen pressure or aggravated by it. While thrombosis is common in lowlanders going to high altitude (Ward, 1975) its frequency and significance to high-altitude natives is unknown. The increased frequency found for lowlanders is presumably related to the elevated red blood cell count and increased viscosity of the blood. Since natives appear to have lower red-cell counts compared to newcomers the problem may not be a serious one (see Chapter 7).

While thrombosis may not be seriously increased in natives the hypoxia of altitude does appear to increase the gravity of any pulmonary disease and to increase greatly the frequency of patent ductus arteriosus. Slight cases of pneumonoconiosis at altitude lead to complications resembling chronic mountain sickness (Monge, 1953) and it may well be that any disease which over a period of time affects normal lung function stimulates an exaggerated erythropoietic activity with concomitant detrimental effects on health. This appears to be confirmed by the observation of Buck *et al.* (1968) that in one high-altitude Peruvian village hemoglobins above 19 g% were associated with respiratory disease including tuberculosis.

Late fetal development and birth at high altitude may stimulate a number of heart defects (Hellriegel, 1967). Certainly it increases the frequency of patent ductus arteriosus. Marticorena *et al.* (1959) reported an incidence

329

of 0.77% in school children born around 4,300 m and Peñaloza *et al.* (1964) estimated the frequency at above 4,000 m to be 14 times greater than it is at sea level.

Finally, it should be noted that any inherited or acquired form of anemia has more serious consequences at high altitude than at sea level. For example, it has been reported that individuals with asymptomatic cases of sickle-cell anemia often develop an acute crisis due to splenic infarction when they travel to high altitude (Aste-Salazar, 1966, see citation in Baker & Dutt, 1972).

### Infectious disease and altitude

While the high radiation, cold and aridity of the high-altitude areas should have a significant effect on the prevalent types of infectious diseases very few studies provide the necessary information for examining this proposition. Of prime value in this regard is the four-community study reported by Buck *et al.* (1968). Their study examined four similar sized Peruvian villages: (1) in the eastern tropical forest, (2) on the slope of the eastern Andean escarpment, (3) on the dry western slope, and (4) on the high *altiplano*. Some of the multiple differences in infectious disease between these villages could be attributed to economic differences and variation in culturally defined behavior patterns. However, some appear significantly or even totally environmentally related. Table 11.5 shows the incidence of some common insect-borne diseases. Buck *et al.* (1968) noted that appropriate mosquitoes for the transmission of malaria and yellow fever are almost totally lacking at high altitude. Lice are possible in all these settings. However, while the heat of the tropical forest settings encourages frequent bathing the cold almost totally precludes it for the

Table 11.5. *Life histories of some infectious diseases among the male populations of four Peruvian villages*

| Locality | History of | | |
|---|---|---|---|
| | Malaria (%)[a] | Yellow fever (%)[a] | Typhus (%)[a] |
| San Antonio (152 m) | 25.3 | 1.1 | 0.0 |
| Cachicoto (730 m) | 13.3 | 1.8 | 0.0 |
| Yacango (1,870 m) | 12.6 | 0.0 | 1.0 |
| Pusi (3,840 m) | 0.7 | 0.0 | 8.4 |

[a] Percentages listed are age-adjusted.
Data from Buck *et al.* (1968).

Table 11.6. Percentages (age-adjusted) with ova and intestinal parasites in single stool examinations in the populations of four Peruvian villages

| Village | Nematodes | | | | Cestodes | | | Trematodes | | E. histolytica | | | Protozoa | | | | | |
|---|---|---|---|---|---|---|---|---|---|---|---|---|---|---|---|---|---|---|
| | Ascaris | Trichuris | Hookworm | Strongyloides | Taenia | H. diminuta | H. nana | S. mansoni | Fasciolidae | Small cysts | Large cysts | Trophozoites | Giardia | Balantidium | Chilomastix | Endolimax | Iodameba | E. coli |
| San Antonio | 79.7 | 76.6 | 49.4 | 1.0 | 0.0 | 0.0 | 0.0 | 0 | 0.6 | 17.5 | 12.2 | 0.6 | 0.8 | 1.8 | 3.6 | 12.6 | 11.3 | 75.1 |
| Cachicoto | 76.6 | 89.7 | 68.3 | 12.9 | 0.5 | 0.7 | 1.4 | 0 | 0.2 | 15.1 | 10.2 | 0.0 | 0.5 | 3.1 | 1.3 | 7.9 | 3.5 | 54.1 |
| Yacango | 1.6 | 1.6 | 0.0 | 0.0 | 0.2 | 0.0 | 8.9 | 0 | 1.4 | 26.5 | 2.5 | 0.0 | 5.8 | 0.0 | 12.4 | 16.9 | 18.5 | 23.6 |
| Pusi | 32.5 | 65.8 | 0.4 | 1.2 | 1.6 | 0.0 | 3.8 | 0 | 0.0 | 28.0 | 5.3 | 0.0 | 3.6 | 1.9 | 8.4 | 12.8 | 13.7 | 76.1 |

From Buck et al. (1968). Health and disease in four Peruvian villages: contrasts in epidemiology. © The Johns Hopkins University Press.

## The biology of high-altitude peoples

high-altitude natives. This partially explains the relatively high frequency of typhus on the *altiplano*.

Differences in the internal parasites also appear as shown in Table 11.6. Since fecal disposal practices were similar in all the villages the major differences can be attributed to environmental characteristics. Considering the known anemia-inducing consequence of hookworm the low infestation in the highlands is fortunate even though the relatively high *Trichuris* infestation may have some anemia-producing effect. The hemoglobin (g%) does not suggest an anemia problem in the *altiplano* village.

Other findings of significance were the almost total lack of arbovirus at high altitude and the relatively low prevalence of diarrhea in the high altitude and dry western slope villages compared to the wetter and hotter eastern slope and tropical forest villages.

Other studies of infectious disease in the Andes support and supplement the study of Buck *et al.* In an analysis of hospital records for a high-altitude hospital in Central Peru, Hellriegel (1967) noted the high frequency of bronchopulmonary related disease including tuberculosis and echino-coccus disease. The latter he attributed to the extensive sheep flocks of the very high-altitude zone. Since sheep are very common even in the pastoral zones dedicated primarily to llama and alpaca one must assume this to be a common problem throughout the high Andes.

In a study of a rural population in the high-altitude zone of Ecuador, Campuzano, Garres & Moldonado (1974) noted that the principal cause for which adults sought medical care was neuralgias of various types, closely followed by respiratory ailments. As Hellriegel noted for Central Peru, Campuzano *et al.* state that chronic bronchitis was common. Almost all consultations for children were related to respiratory disease (45%) or gastro-intestinal disorders (35%).

Beall (1976) analyzed the diagnoses of infants brought for consultation to 'well baby' clinics in cities in the highlands and coastal areas of Southern Peru. Respiratory and gastro-intestinal disease accounted for two-thirds of the illnesses in the highlands and one-half on the coast. In both areas there were slightly more respiratory diseases diagnosed than gastro-intestinal.

Some indication of how the various ailments affected the individual's perception of health was provided by a study in the Peruvian valley previously described in reference to the fertility study of Abelson *et al.* (1974). Among the indices of perceived health measured were the number of days in the preceding month during which the individual felt ill and too ill to work (Dutt & Baker, 1977). A comparison with lowlanders is shown in Fig. 11.4. Since this was primarily a wage-labor society in which the percentage of females employed was lower than males, the male data may be somewhat more indicative of how perceived health affected economic

332

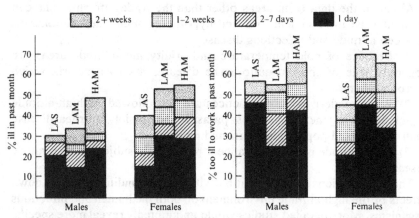

Fig. 11.4. Reported sickness and its effects on work performance in three groups of Peruvians. LAS, low-altitude sedentes; LAM, low-altitude migrants; HAM, high-altitude migrants. (From Dutt & Baker, 1977).

behavior. A further examination of responses to the health questionnaire showed that the young males from high altitude had significantly more symptoms than the low-altitude groups and that both male and female high-altitude migrants suffered significantly more respiratory symptoms than the lowlanders. To what extent this reflected the symptomology of the high-altitude migrants prior to migration could not be determined, but the difference in respiratory symptoms persisted even in migrants who had lived at low altitude for more than 5 years. This result appeared to support Monge's (1934) statement that downward migrants in Peru were particularly prone to active tuberculosis.

Data on infectious disease in the Asian and Ethiopian mountain areas are very fragmentary and no detailed controlled comparisons between high and low altitude are reported. Lang & Lang (1971) in their analysis of the Kunde hospital activities in Nepal note that tuberculosis is common and mostly of the pulmonary variety. Their 1-year record of hospital admissions also suggests that pneumonia may be very common. Leprosy also appears fairly common in contrast to the Andes where it is not reported.

Harrison *et al.* (1969) found that in their two Ethiopian communities several altitude-related differences appeared in a recall analysis of infectious disease. They note that the overall incidence of infectious disease was greater at low altitude and even some specific diseases were significantly less common at high altitude as shown by statistical tests. In contrast to the Andean and Himalayan studies they reported tuberculosis to be of relatively low importance and no difference was found between high-and low-village incidence.

## The biology of high-altitude peoples

Although the data from areas other than the Andes are slim, one can draw at least three conclusions and hypotheses about the relationships between altitude and infectious disease.

(1) Because of cold temperature and aridity high-altitude areas are generally free of the mosquito-borne diseases such as malaria which surround them.

(2) When modern sanitary practices are not followed the high-altitude populations will generally have less parasitic and possibly diarrheal disease than the lowland populations surrounding them.

(3) High-altitude populations appear particularly subject to respiratory disease.

These suggestions seem compatible with both the findings and the known effects of the high-altitude environment on both the human and infectious organisms. More detailed studies would undoubtedly reveal more specific differences particularly in the diseases common to wet tropical areas. It thus appears that the high-altitude areas of the tropical world may, even today, be healthier than the surrounding lowlands and prior to modern medical knowledge must have provided for a population much fitter than that in the surrounding areas.

### Degenerative disease and altitude

If the often suggested long life and health of high-mountain people is indeed the case, a part of it may have been the protection they received from some infectious disease. One would anticipate that another part should be some protection against the degenerative diseases of old age. Mirrakhimov (1971) indeed believes this to be the case and cites the fact that in Kirghizia the mountain people of age 60 years may have a greater life expectancy than the non-mountain people with comparable medical care. No comparable finding has been published for the Andes but in both areas there is evidence that the frequency of cardiovascular disease is very low during middle and old age.

Since many human populations who are not living within modern society have low cholesterol levels (Shaper & Jones, 1959) and low blood pressures (Huizinga, 1972) part of the explanation may be nutritional and psychological in origin. Thus the remoteness and poor economic development of the high-altitude areas may contribute to the apparent cardiovascular health of the high-altitude old. However, several authors have argued that these are insufficient explanations and that there are specific physiological reasons why growth and development at high altitude should provide some protection against at least ischemic heart disease (Poupa, Krofta & Prochazka, 1966; Mirrakhimov, 1971). Basically they suggest that early development under lower oxygen pressures than those encountered by

334

healthy lowlanders has two significant consequences. First, it leads to greater growth in both the numbers and size of capillaries and arterioles in tissue such as the heart. This finding appears widely supported by studies on several animals (Bartlett & Remmers, 1971; Burri & Weibel, 1971; Goldberg, Levy, Siassi & Betten, 1971; Hunter, Barer, Shaw & Clegg, 1974) and on men (Arias-Stella & Castillo, 1966). As an increased capillary bed could very well dampen blood-pressure rises and would certainly reduce the necrotic damage caused by a given arteriole blockage, less cardiovascular disease might very well result. Second, they support Barbashova's (1964) claim that in some manner heart cells develop altered metabolic pathways which reduce aerobic requirements during periods of restricted oxygen flow. Unfortunately, much of the literature on this subject is published only in the Slavic languages and I am unable to evaluate it. In any case, these arguments certainly suggest the need for a close examination of old-age cardiovascular changes in high-altitude people.

Hurtado (1960) suggested quite a while ago that altitude might reduce cardiovascular disease but a survey in the mountain peoples of the United States did not support this finding (Morton, Davids & Lichty, 1964). The difference in the results may have resulted from many causes but clearly the altitudes involved and the permanency of residence was much less in the US study. Hellriegel (1967) commented on the rarity of aneurisms and arteriosclerotic obstructions among the central Andean natives. Lang & Lang (1971) commented on the lack of arteriosclerotic and cardiovascular disease among the Sherpa. Harrison *et al.* (1969) did not obtain quantitative data on cardiovascular disease but commented that in clinical work some cases were encountered. The available information from these areas, combined with Mirrakhimov's (1971) report that cardiovascular disease is very rare in the Tien Shan and Pamir mountains, does indeed suggest that cardiovascular disease is quite uncommon at least among permanent very high-altitude peoples.

A somewhat larger and more systematic body of information exists in relation to blood pressure and cholesterol levels. Marticorena, Severino & Chavez (1967) measured blood pressure on 300 high-altitude natives in central Peru. The results of this study are shown in Table 11.7. Some effects of late adolescent growth appear but there is otherwise no sig-nificant age rise in either systolic or diastolic pressure. A later cross-sectional study of 60- to 80-year olds in the same area showed a possible slight increase in systolic pressure but there was essentially no increase in diastolic pressure. It is worth-while to note that among the ninety indi-viduals studied none showed a systolic pressure above 165 or a diastolic pressure above 95 (Zapata & Marticorena, 1968).

In a study of the more traditional population of Nuñoa at 4,000+ m in

Table 11.7. *Systemic blood pressure of control Peruvian high-altitude natives (3,700 m)*

| Age (yr) | Males | | Females | |
|---|---|---|---|---|
| | Systolic | Diastolic | Systolic | Diastolic |
| 15–20 | 117.2 | 68.0 | 110.9 | 67.0 |
| 21–40 | 118.1 | 73.0 | 116.7 | 73.8 |
| 41–60 | 118.2 | 75.3 | 116.7 | 76.0 |

Data from Marticorena *et al.* (1967).

Table 11.8. *Blood pressure in Ethiopian populations*

| Blood pressure (mm Hg) | Adi-Arkai | | | Debarech | | | Geech | | |
|---|---|---|---|---|---|---|---|---|---|
| | $\bar{x}$ | $N$ | S.E. | $\bar{x}$ | $N$ | S.E. | $\bar{x}$ | $N$ | S.E. |
| | Local males | | | | | | | | |
| Systolic | 119.0 | 71 | 1.20 | 123.2 | 76 | 1.52 | 121.0 | 22 | 2.62 |
| Diastolic | 76.3 | 71 | 1.32 | 76.5 | 76 | 1.29 | 79.9 | 22 | 2.15 |
| | Local females | | | | | | | | |
| Systolic | 115.0 | 35 | 2.41 | 119.2 | 34 | 2.28 | — | — | — |
| Diastolic | 75.9 | 35 | 1.70 | 74.4 | 34 | 1.85 | — | — | — |

Data from Harrison *et al.* (1969).

southern Peru essentially similar results were obtained with no significant hypertension encountered (Baker, 1969). It was, however, noted that at least part of the lack of age increase in blood pressure must be a product of the foods and behavior associated with the traditional life style. When the cross-sectional sample was divided into children who did and did not attend school and adults who showed some or no signs of adopting more modern life styles then significant indications of age increases in blood pressure appeared in the acculturating population segment.

It seems doubtful that significant hypertension occurs in any of the high-altitude peoples who are living in traditional fashion. Buck *et al.* (1968) observed a few hypertensives in their high-altitude village but Lang & Lang (1971) reported none among the Sherpa. Mirrakhimov (1971) claims an almost total absence among the peoples of Pamir and Tien Shan mountains and Harrison *et al.* (1969), although they found slightly higher systolic blood pressures among males in their high-altitude groups, do not report any hypertensive individuals (see Table 11.8).

It also appears that serum-cholesterol levels are very low in at least the

Andean natives. Buck *et al.* reported a mean value of only 122 mg% in their high-altitude village and even among the old the average only rose to 158 mg%. Watt, Picón-Reátegui, Gahagan & Buskirk (1976) reported that among a small group of native soldiers in an urban setting at high altitude the mean was only 150 mg%.

Perhaps the most unexpected finding concerning blood pressure and serum-cholesterol levels was that these values do not alter substantially if the high-altitude native migrates as an adult to low altitude and adopts a style of life which is associated with higher levels of cholesterol and increasing blood pressure with age. In one study highland males who had spent ten or more years in a coastal Peruvian city were examined and found to have blood pressures and cholesterol levels very similar to males who had remained in the high-altitude areas of origin (Watt *et al.*, 1976). In another study both male and female migrants from high-altitude areas were examined and contrasted with lowland natives in a Peruvian coastal valley (Davin, 1975; Baker, 1977). These migrants had been in the valley from 1–20 years. Although blood pressures were slightly higher than the norms found in Nuñoa they were significantly lower than those of the lowland natives in the valley. Furthermore, they demonstrated no age increase and, if anything, appeared to be lower the longer the time since migration. An equally interesting finding was a gradual reduction in systemic blood pressure in lowlanders who remain for many years at high altitude (Galvez, 1966).

It thus appears that the highland native is in many ways protected from cardiovascular disease in middle age provided he has been born and grows up in a traditional society. Whether the same is the case for highland natives developing in modern culture is not known but is critical to testing the hypotheses put forth by the various authors concerning the beneficial effects of hypoxia on cardiovascular disease. A carefully designed study of this problem using selected natural experiment situations seems highly desirable.

## Morbidity and mortality at high altitude

While the information on population structure, fertility and disease is meager in the high mountain areas of the world other than the Andes, this shortage becomes acute in relation to morbidity and mortality. Except for a small amount of data on infant mortality no information is available on areas other than the Andes. Considering how aberrant the findings on Ethiopia have been from those on the Andes and the Asian mountains, and how often the particulars of the Asian mountain people have deviated from those on Andean people any generalizations must by necessity apply only to the Andes.

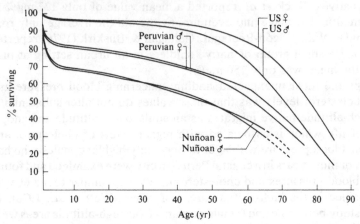

Fig. 11.5. Survivorship estimate for Nuñoa, Peru, compared to ones prepared for the US in 1967 and Peru in 1971. (From Way, 1976.)

Even within this limitation the information on morbidity and mortality must be considered limited in scope. The gross death rate per thousand has considerable significance in terms of judging population growth potential, but considering the complex variation in migration and fertility rates occurring in the Andes it tells us little about the biological effects of altitude on mortality. Mortality rates in the high Andean communities without modern medical care vary from a low of 14.5/1,000 calculated for Vicos, Peru in 1952 (Alers, 1965) to a high of 50.0/1,000 calculated for the area of Lauca in Chile in 1966 (Cruz-Coke *et al.*, 1966). The latter rate was calculated from a very small sample and it seems that a rate of 20–25/1,000 as found in larger studies is a more representative mortality rate. However, these rates reported for such communities as Vicos, Peru, 1963 (Alers, 1965) and Nuñoa, Peru, 1971 (Spector, 1971) must be interpreted in light of the fact that considerable outmigration of adults characterized both districts.

A somewhat more reliable estimate of mortality by age for typical communities in the Andean highlands is provided in the survivorship curve of the Nuñoa District as shown in Fig. 11.5. This curve based on a 1940 and 1961 census was derived from the birth and death registry in the Nuñoa District from 1940 to 1969 (Spector, 1971). Analysis of the net outmigration for the larger province containing Nuñoa (Spector, 1971) and a follow-up study of infants born in the District (McGarvey, 1974) suggested that net outmigration might be as high as 20% over this time span. The effect of this on calculated survivorship was probably greater in young adults than it was on children or older adults.

On the basis of this study it appeared that death rates as shown in Fig.

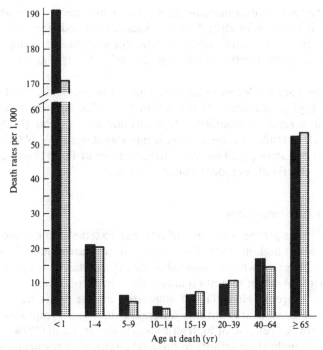

Fig. 11.6. Age- and sex-specific death rates for Nuñoa, Peru, ■, males; ▣, females. (From Baker & Dutt, 1972.)

11.6 were greater in males than females during childhood. This trend may reverse itself slightly during adulthood. Because of the outmigration effect on survivorship the adult loss suggested by Fig. 11.5 may not be as great as indicated, but certainly the total lack of medical care and sanitation leads to a lower survivorship than the rate found for Peru as a whole.

## Neonatal and infant mortality

Given the limitations of the data available it is more instructive to examine in detail mortality during specific life segments. Neonatal and infant mortality is particularly well documented because of the high mortality during this period of life and its general health implications.

Grahn & Kratchman (1963) first observed that neonatal mortality was positively associated with altitude in the United States, and Mazess (1965) observed the same phenomenon in Peru. While this association disappeared in the United States as medical care improved (Frisancho & Cossman, 1970) it has probably persisted in the Andean region in part because medical care tends to be poorer in the high-altitude regions.

## The biology of high-altitude peoples

C. A. Weitz (personal communication) stated that neonatal mortality also tends to increase with altitude in the Sherpa area and Harrison *et al.* (1969) found that infant- and child-loss rates for women averaging about 30-years old was 280/1,000 at high altitude and 150/1,000 at the lower elevation.

Actual rates vary by location and the amount of medical care. Death rates under 1 year old were 73.0/1,000 in urban La Paz, Bolivia, but 123.5/1,000 in a rural community at high altitude near La Paz (Puffer & Serrano, 1973). In Nuñoa the under-1-year rate was about 180/1,000 (Baker & Dutt, 1972) but among infants in the urban center of Puno the rate was 70.3/1,000 (C. M. Beall, personal communication).

### Neonatal and infant morbidity

As discussed in the previous section infants tend to sicken from somewhat different causes at high altitude than they do in the surrounding lowland forest areas. However, when compared to the dry coastal areas to the west and south of the Andean mountain areas, the causes of illnesses are not dramatically different. Haas (1973) suggested on the basis of mother interviews that coastal area infants had more diarrheal symptoms than respiratory with the reverse at high altitude. However, Beall (1976) found that mothers brought their infants to hospital clinics with respiratory and diarrheal pathologies in about equal frequencies in both areas.

Although the symptoms, therefore, appear in similar frequencies their contribution to morbidity is quite different. Fig. 11.7 shows the neonatal and infant death rates from two high-altitude communities compared to the tropical lowland community of Recife, Brazil (Puffer & Serrano, 1973). The very high percentage of respiratory caused deaths in the highlands compared to the lowlands is obvious. That respiratory symptoms precede most infant deaths in the highlands is confirmed by the death registries in Nuñoa, Peru (Spector, 1971). There, of 1,106 deaths before age 1 between 1950 and 1969, over 87% were attributed to respiratory causes. While the cause of death in the Nuñoa registry must be considered unreliable because of the lack of medical expertise the data do suggest a very high frequency of respiratory symptoms at the time of death.

Beall's data from the death registries of Puno (3,860 m) and Tacna (600 m) must be considered more reliable. As shown in Table 11.9 respiratory disease was a much more significant underlying cause of death at high altitude than it was at low altitude. Furthermore, this difference in patterns of morbidity appears to have a differential effect according to the birth weight of the infant. As demonstrated by many studies, including Beall's, birth weights are lower at high altitude than they are at low altitude (see Chapter 4). However, instead of increasing neonatal and infant mortality

340

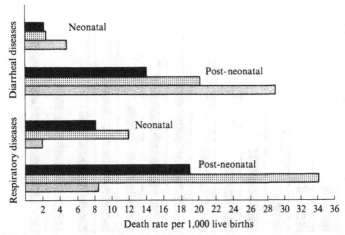

Fig. 11.7. Infant mortality rates in two high-altitude locations (La Paz ■ and Viacha ▨) compared to a low-altitude population in Brazil (Recife ▨). (Data from Puffer & Serrano, 1973.)

Table 11.9. *Infant mortality: summary of municipal infant mortality records from Puno and Tacna, 1971–3*

|  | Puno |  |  | Tacna |  |  |
|---|---|---|---|---|---|---|
| Total N | 479 |  |  | 979 |  |  |
| Age at death (%) |  |  |  |  |  |  |
| 0–7 days | 29 |  |  | 23 |  |  |
| 1 week 1 month | 9 |  |  | 7 |  |  |
| 1 month 1 year | 62 |  |  | 70 |  |  |
| Cause of death (%) | Respiratory | Gastro-intestinal | Other | Respiratory | Gastro-intestinal | Other |
| All | 57.5 | 20.0 | 22.3 | 38.7 | 32.7 | 28.6 |
| 1–7 days | 53.6 | 11.1 | 35.3 | 29.4 | 5.9 | 64.7 |
| 1 week 1 month | 74.4 | 16.2 | 9.4 | 28.5 | 33.3 | 38.2 |
| 1 month 1 year | 55.1 | 27.2 | 17.7 | 42.1 | 41.2 | 16.7 |

From Beall (1976).

this may enhance the probability of survival, at least in the Andes. As shown in Fig. 11.8 survivorship at high altitude is better in low and average birth-weight categories than it is at low altitude. As birth weight becomes higher survivorship becomes progressively better at low altitude than it is at high altitude.

Finally, a very curious finding in the Bolivian case reported by Puffer & Serrano (1973) must be noted. They reported that among the mass

341

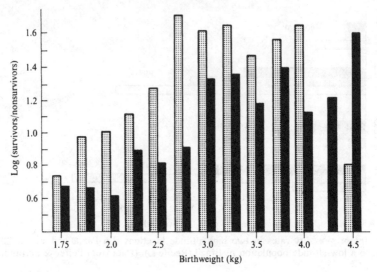

Fig. 11.8. Survivorship rate by birth weight for high- and low-altitude Peruvian infants, from death registers at Puno, 3,860 m, ■, and Tacna, 600 m, ▦.

surveys on the underlying and associated causes of death for fifteen populations surveyed in the western hemisphere, congenital anomalies caused fewer deaths in La Paz than in any other group. Furthermore, in the total number of all congenital anomalies noted in La Taz, infants showed about one-quarter the rate found for the highest incidence area (São Paulo, Brazil) and about one-half the rate of the next lowest (Chaco Province, Argentina). Why congenital anomalies as well as measured abortion rates should be so low in the high-altitude Andean peoples is not at all apparent.

### Childhood mortality and morbidity

Data on death rates for children over 1 year of age are more scarce but it appears that rates are fairly low. In Nuñoa the rate was only 20/1,000 for 1- to 5-year-olds. In Bolivia, the rates were in the same range with a 28/1,000 rate for 1-year-olds and only 6/1,000 for 2- to 4-year-olds. These moderate rates for 1-year-olds suggest a lower incidence of post-weaning deaths than those found in many peasant societies. The causes for these lower rates are probably related to the relatively good nutrition level reported for most of the Andean highlands (see Chapter 8). They may also be related to the relatively long intervals between births of high-altitude Andean populations. The situation may be quite different in Ethiopia where post-weaning deaths in Harrison's high-altitude village were quite numerous.

342

Among the underlying causes of death in these age groups the Nuñoa registry still indicates respiratory disease to be the most frequent cause, but epidemics of scarlet fever and whooping cough account for 25% of the deaths. In the Bolivian study diarrheal disease, respiratory disease and a measles epidemic all contributed about equally to morbidity in the 1-year-olds. In the 2- to 4-year-olds measles ranked as the leading morbidity factor while respiratory disease and diarrhea were somewhat less important.

As in most societies mortality is at its lowest in late childhood and the Nuñoa data (Spector, 1971) shows the 10- to 14-year-olds have the lowest of all mortality rates. The reported rate of only about 3/1,000 compares favorably with industrialized countries.

### Adult mortality and morbidity

Mortality for young adults in the Andean populations is probably in general rather low. In Nuñoa the rate was in the vicinity of 8/1,000 among 20- to 39-year olds and was slightly higher for females than males. The pastoral and agricultural life of most high-altitude Andean natives exposes them to very few risks of accidental death. However, this is changing as increasingly large numbers of people travel over dangerous roads by poorly maintained trucks and buses. This contrasts somewhat with the present situation among the Sherpa in Nepal where mountain climbing deaths appear quite significant for adult mortality.

For women, child-bearing without modern medical care constitutes a significant threat but the data available are too poor for judgments on whether or not child-bearing is more or less risky at altitude. The smaller size of the high-altitude newborn may reduce at least part of the risk and the low levels of various bacterial infections at high altitude may further enhance survival probabilities.

Age determinations are so poor for older individuals in Nuñoa that it is impossible to estimate reasonably the age rates of mortality among the old. The listed causes of death continue to suggest that respiratory problems such as pneumonia and tuberculosis are the most significant causes of death although a significant percentage of deaths are attributed to old age and rheumatism.

Since no other quantitative data on the mortality or morbidity of the old have come to my attention it may be useful to look at the well-documented causes of death among the downward migrants. These were determined from a large hospital death registry on the southern coast of Peru (unpublished data).

Table 11.10 indicates the causes of death for three groups during middle and old age. The three groups were local coastal people (LAN), migrants from middle-altitude areas (MAM), and migrants from high altitude

343

Table 11.10. *Cause of death in three low-altitude groups in Peru (data taken from hospital records)*

| Cause of death | Age: 45–69 yr | | | Age: 70+ yr | | |
|---|---|---|---|---|---|---|
| | Low-altitude natives (%) | Medium-altitude migrants (%) | High-altitude migrants (%) | Low-altitude natives (%) | Medium-altitude migrants (%) | High-altitude migrants (%) |
| | *Males* | | | | | |
| Total cardiovascular | 23.6 | 18.8 | 9.5 | 23.6 | 22.6 | 9.7 |
| Ischemic | 5.7 | 0.0 | 3.2 | 7.3 | 7.5 | 0.0 |
| Cerebrovascular | 7.3 | 3.6 | 6.3 | 3.6 | 7.5 | 0.0 |
| Hypertensive | 1.6 | 0.9 | 0.0 | 3.6 | 1.9 | 0.0 |
| Other forms | 8.9 | 10.9 | 0.0 | 9.1 | 5.7 | 9.7 |
| Total respiratory | 30.1 | 25.4 | 41.1 | 23.5 | 26.4 | 35.5 |
| Tuberculosis | 21.9 | 21.8 | 35.8 | 16.4 | 13.2 | 22.6 |
| Pneumonia | 6.5 | 3.6 | 3.2 | 5.5 | 11.3 | 12.9 |
| Bronchitis and flu | 1.6 | 0.0 | 0.0 | 1.8 | 1.9 | 0.0 |
| All other diseases | 34.1 | 32.7 | 29.5 | 43.6 | 43.4 | 45.2 |
| Accidents | 12.2 | 26.4 | 20.0 | 7.3 | 7.5 | 9.7 |
| N | 123 | 110 | 95 | 55 | 53 | 31 |
| | *Females* | | | | | |
| Total cardiovascular | 22.9 | 23.8 | 16.7 | 30.0 | 23.4 | 22.7 |
| Ischemic | 0.0 | 7.1 | 0.0 | 0.0 | 4.3 | 9.1 |
| Cerebrovascular | 8.6 | 4.8 | 4.2 | 6.7 | 2.1 | 4.5 |
| Hypertensive | 0.0 | 3.4 | 4.2 | 10.0 | 6.4 | 4.5 |
| Other forms | 14.3 | 9.5 | 8.3 | 13.3 | 10.6 | 4.5 |
| Total respiratory | 31.4 | 19.0 | 29.2 | 16.7 | 17.0 | 18.2 |
| Tuberculosis | 22.9 | 19.0 | 16.7 | 13.3 | 10.6 | 9.1 |
| Pneumonia | 5.7 | 0.0 | 12.5 | 0.0 | 6.4 | 9.1 |
| Bronchitis and flu | 2.9 | 0.0 | 0.0 | 3.3 | 0.0 | 0.0 |
| All other diseases | 37.1 | 52.4 | 50.0 | 50.0 | 55.3 | 50.0 |
| Accidents | 8.6 | 4.8 | 4.2 | 3.3 | 4.3 | 9.1 |
| N | 35 | 42 | 24 | 30 | 47 | 22 |

(HAM). Respiratory and cardiovascular diseases were the major causes of death but no single cause was outstanding. For males a clear distinction occurs between high-altitude migrants and lowlanders. The highland male appears much more likely to die from respiratory disease, particularly tuberculosis, than the lowland male. On the other hand, he is much less likely to die of cardiovascular disease.

While high-altitude females also seemed less subject to cardiovascular morbidity than lowlanders they do not seem particularly prone to respiratory morbidity and in fact appear to die from respiratory causes with about the same frequency as lowland women. How representative the morbidity of these migrants is of morbidity in the highlands is doubtful.

Table 11.11. *Fertility, mortality, and estimates of the index of opportunity for selection in Andean communities*

| Variable | Nuñoa | Chapiquina | Huallatire |
|---|---|---|---|
| Mean progeny number, $\bar{x}_s$ | 6.7[a] | 8.5[b] | 7.3[b] |
| Variance, $V_x$ | 9.0 | 5.6 | 9.9 |
| Fertility index, $I_f = V_x/\bar{x}_s$ | 0.200[a] | 0.077[b] | 0.185[b] |
| Proportion offspring surviving to adulthood, $P_s$ | 0.657[a] | 0.751[a] | 0.459[a] |
| Mortality index, $I_m = (I-P_s)/P_s$ | 0.522 | 0.331 | 1.178 |
| $I_{f_s} = I_f/P_s$ | 0.305 | 0.232 | 0.403 |
| Total index, $I = I_m + I_{f_s}$ | 0.827 | 0.563 | 1.581 |

From Garruto & Hoff (1976).
[a] For women over 45 years of age.     [b] For women over 35 years of age.

**Selection and adaptation**

The health difficulties of lowlanders in high-altitude areas and the fact that lowlanders, even with superior technology, have failed biologically to replace native high-altitude populations, has often led to the speculation that high-altitude natives are genetically adapted to their environment and thereby achieve a functional capacity greater than possible for at least most lowlanders (Monge, 1948; Baker, 1969). Some of the physiological evidence collected during the past 10 years tends to support this. The large chest of the Andean native may enhance oxygen transport and is clearly a trait primarily governed by inheritance (Hoff *et al.*, 1972; Beall, Baker, Baker & Haas, 1977). Several hematological characteristics of high-altitude Andean peoples appear to be both adaptive and inherited (see Chapter 7). Even the fact that altitude has a less serious suppressant effect on birth weight among the Andean peoples than it does in North America suggests some basic adaptation (McClung, 1969).

Certainly the opportunity for selection to operate at a high level has been present. Table 11.11 shows the opportunity for selection based on Crow's (1958) index for three Andean populations. It is interesting to note that based on these calculations there is a high opportunity for selection to operate during the post-natal period when the hypoxia effects of high altitude may be presumed to be the most significant. If indeed selection has operated to differentiate the genetic structure of the highland Andean native from the lowlander this will be extremely difficult to discern in South America using local populations, since downwards gene flow has been so great that the present lowlander surrounding the high-altitude populations must share a majority of common genes (Dutt, 1976*b*; Garruto & Hoff, 1976). Harrison *et al.* (1969) noted that a similarly high gene flow exists in Ethiopia. In Nepal the population movements are very recent and casual observation suggests that gene flow has so far had only a modest effect on the highland and lowland populations, which appear to have multiple morphological differences.

A somewhat more specific example of how selection may have differentiated high-altitude and low-altitude populations is provided by a recent study of Beall (1976). She found that optimum birth weights for the survival of newborns in the high-altitude area of Southern Peru was lower than it is at low altitude. Furthermore, the optimal weight distribution for survival conformed very closely to the values actually encountered. The higher and more skewed distribution of birth weights in lowland mothers was less well adapted to survival in the lowlands. Considering that most of the lowland births were to individuals of high-altitude heritage this study suggests that natural selection may have indeed operated on the pre-natal growth rates of high-altitude newborns in Peru.

Despite these multiple indications that at least some populations may have genetically adapted to the environmental conditions of high altitude it must be said that no single identifiable locus in man can be shown to have adaptive value in relation to survival or reproduction at high altitude. It is also clear that in many regards the highland native is functionally affected by the high-altitude environment in the same way as the low-lander. One must presume from our present level of knowledge that most ways in which their general fitness differs from the lowlander coming to altitude is the product of lifelong exposure to the environment.

However, these speculations need further investigation. A survey of the fitness of high-altitude populations supports the earlier noted contention of Peruvian scientists. The highland peoples appear as well adapted to their environment as any lowlanders are to theirs.

### Research perspectives

While in this chapter I have limited the subject of population fitness primarily to demographic and health parameters it is clear that all of the topics covered in this book are in some manner associated with how high-altitude peoples adjust to their environments. The years of research completed since the inception of IBP have obviously enhanced our knowledge on how high-altitude peoples have made these adjustments and the problems they still encounter because of the unique aspects of this environmental zone. Despite the knowledge gained the research has opened new problem areas and also indicated many practical health and welfare problems not previously recognized. Thus from both the scientific and the practical viewpoint it is imperative that research on these peoples not only continue but expand.

346

Preparation of this chapter was aided by NIH Grant No. R01 HDO 9182 awarded by the National Institute of Child Health and Human Development.

## References

Abelson, A. E., Baker, T. S. & Baker, P. T. (1974). Altitude, migration, and fertility in the Andes. *Social Biology*, **21**, 12–27.

Alers, J. O. (1965). Population and development in a Peruvian community. *Journal of Inter-American Studies*, VII, 423–48.

Arias-Stella, J. & Castillo, Y. (1966). The muscular pulmonary arterial branches in stillborn natives of high altitude. *Laboratory Investigations*, **15**, 1951–9.

Aste-Salazar, H. (1966). Unpublished work cited in Monge, C. M. & Monge, C. C., *High-altitude diseases: mechanism and management*. Springfield, Ill.: C. C. Thomas.

Baker, P. T. (1969). Human adaptation to high altitude. *Science, Washington*, **163**, 1149–56.

Baker, P. T. (1976). The evolution of a project: Theory, method and sampling. In *Man in the Andes: A multidisciplinary study of high altitude Quechua*, ed. P. T. Baker & M. A. Little, pp. 1–20. Stroudsburg, Pa.: Dowden, Hutchinson & Ross, Inc.

Baker, P. T. (1977). Migration and biological fitness: a case study in Southern Peru. *American Anthropologist*, in press.

Baker, P. T. & Dutt, J. S. (1972). Demographic variables as measures of biological adaptation: a case study of high altitude human populations. In *The structure of human populations*, ed. G. A. Harrison & A. J. Boyce, pp. 352–78. Oxford: The Clarendon Press.

Barbashova, Z. I. (1964). Cellular level of adaptation. In *Handbook of physiology, Section 4: adaptation to the environment*, ed. D. B. Dill, E. F. Adolph & C. G. Wilber, pp. 37–54. Washington, DC: American Physiological Society.

Bartlett, D. & Remmers, J. (1971). Effects of high altitude exposure on the lungs of young rats. *Respiratory Physiology*, **13**, 116–25.

Beall, C. M. (1976). The effects of high altitude on growth, morbidity and mortality of Peruvian infants. Doctoral Dissertation in Anthropology, The Pennsylvania State University.

Beall, C. M., Baker, P. T., Baker, T. S. & Haas, J. D. (1977). The effects of high altitude on adolescent growth in Southern Peruvian Amerindians. *Human Biology*, **49**, 109–24.

Bradshaw, B. S. (1969). Fertility differences in Peru. A reconsideration. *Population Studies*, **23**, 5–19.

Buck, A. A., Sasaki, T. T., Anderson, R. I., Hitchcock, J. C. & Leigh, G. R. (1968). *Health and disease in four Peruvian villages: contrasts in epidemiology*. Baltimore, Md.: The Johns Hopkins University Press.

Burri, P. H. & Weibel, E. R. (1971). Morphometric estimation of pulmonary diffusion capacity. II. Effect of $P_{O_2}$ on the growing lung. *Respiratory Physiology*, **11**, 247–64.

Campuzano, L., Garres, G. & Moldonado, A. (1974). Motivos de consulta medica mas frecuentes. In *Nutricion y Desarrollo en los Andes Ecuatorianos*, ed. M. Vareo-Teran & J. Vareo-Teran, pp. 336–8. Quito, Ecuador: Artes Graficos.

CEP (1969). *Condicionamientos socio-culturales de la fecundidad en Bolivia*. La Paz, Bolivia: Centro de Estudios de Poblacion y Familia.

347

The biology of high-altitude peoples

Chiodi, H. (1960). Mal de montaña a forma cerebral. Posible mecanismo etiop-
atogénico. *Annales de Faculdad Medicina*, Lima, **43**, 437.

CISM (1968). *Encuesta de fecundidad en la ciudad de Cerro de Pasco*. Lima, Peru:
Centro de Investigaciones Sociales por Muestro and the Ministerio de Trabajo
y Comunidades del Peru.

Clegg, E. J., Harrison, G. A. & Baker, P. T. (1970). The impact of high altitudes
on human populations. *Human Biology*, **42**, 486–518.

Crow, J. F. (1958). Some possibilities for measuring selection intensities in man.
*Human Biology*, **30**, 1–13.

Cruz-Coke, R., Cristoffani, A. P., Aspillaga, M. & Biancani, F. (1966). Evolutionary forces in an environmental gradient in Arica, Chile. *Human Biology*,
**38**, 421–38.

Davin, E. P. (1975). Blood pressure among residents of the Tambo valley. Master's
Thesis in Anthropology, The Pennsylvania State University.

De Jong, G. (1970). Demography and research with high altitude populations.
*Social Biology*, **17**, 114–19.

Dutt, J. S. (1976a). Altitude and fertility: the Bolivian case. Doctoral Dissertation
in Anthropology, The Pennsylvania State University.

Dutt, J. S. (1976b). Population movement and gene flow. In *Man in the Andes:
A Multidisciplinary Study of High Altitude Quechua*, ed. P. T. Baker &
M. A. Little, pp. 115–27. Stroudsburg, Pa.: Dowden, Hutchinson &
Ross.

Dutt, J. S. (1976c). Altitude, fertility, and early childhood mortality: the Bolivian
example. *American Journal of physical Anthropology*, **44**, 175.

Dutt, J. S. & Baker, P. T. (1977). Environment, migration and health in Southern
Peru. *Social Science & Medicine*, in press.

Frisancho, A. R. & Cossman, J. (1970). Secular trend in neonatal mortality in the
mountain states. *American Journal of Physical Anthropology*, **33**, 103–6.

Galvez, J. (1966). Presión arterial en el sujeto de nivel del mar con residencia
prolongada en las grandes alturas. *Archivos del Instituto de Biologia Andina*,
**1**, 238–43.

Garruto, R. M. & Hoff, C. J. (1976). Genetic history and affinities. In *Man in the
Andes: A Multidisciplinary Study of High Altitude Quechua*, ed. P. T. Baker
& M. A. Little, pp. 98–114. Stroudsburg, Pa.: Dowden, Hutchinson &
Ross.

Goldberg, S. J., Levy, R. A., Siassi, B. & Betten, J. (1971). The effects of
maternal hypoxia and hyperoxia upon neonatal pulmonary vasculature.
*Pediatrics*, **48**, 528–33.

Goldstein, M. C. (1976). Fraternal polyandry and fertility in a high Himalayan
Valley in northwest Nepal. *Human Ecology*, **4**, 223–33.

Grahn, D. & Kratchman, J. (1963). Variation in neonatal death rate and birth weight
in the United States and possible relations to environmental radiation, geology
and altitude. *American Journal of Human Genetics*, **15**, 329–52.

Haas, J. D. (1973). Altitudinal variation and infant growth and development in
Peru. Doctoral Dissertation in Anthropology, The Pennsylvania State
University.

Harrison, G. A., Küchemann, C. F., Moore, M. A. S., Boyce, A. J., Baju, T.,
Mourant, A. E., Godber, M. J., Glasgow, B. G., Kopeć, A. C., Tills, D. &
Clegg, E. J. (1969). The effects of altitudinal variation in Ethiopian populations. *Philosophical Transactions of the Royal Society of London*, **256**B,
147–82.

348

Heer, D. M. (1964). Fertility differences between Indian and Spanish speaking parts of Andean countries. *Population Studies*, **18**, 71–84.

Heer, D. M. (1967). Fertility differences in Andean countries: a reply to W. H. James. *Population Studies*, **21**, 71–3.

Heer, D. M. & Turner, E. S. (1965). Areal differences in Latin American fertility. *Population Studies*, **18**, 279–92.

Hellriegel, K. O. (1967). Health problems at altitude. Paper presented at the WHO/PAHO/IBP Meeting of Investigators on Population Biology of Altitude, 13–17 November 1967, Washington, DC.

Hoff, C. J. (1968). Reproduction and viability in a high altitude Peruvian population. *Occasional Papers in Anthropology*, **1**, 85–160. University Park, Pa.: The Pennsylvania State University.

Hoff, C. J. & Abelson, A. E. (1976). Fertility. In *Man in the Andes: A Multidisciplinary Study of High Altitude Quechua*, ed. P. T. Baker & M. A. Little, pp. 128–46. Stroudsburg, Pa.: Dowden, Hutchinson & Ross.

Hoff, C., Baker, P. T., Haas, J., Spector, R. & Garruto, R. (1972). Variaciones altitudinales en el crecimiento y desarrollo fisico del Quechua Peruano. *Revista del Instituto Boliviano de Biologia de la Altura*, IV, 5–20.

Huizinga, J. (1972). Casual blood pressure in populations. In *Human biology of environmental change*, ed. D. S. M. Vorster, pp. 164–9. London: International Biological Programme.

Hunter, C., Barer, G. R., Shaw, J. W. & Clegg, E. J. (1974). Growth of the heart and lungs in hypoxic rodents: A model of human hypoxic disease. *Clinical Science and Molecular Medicine*, **46**, 375–91.

Hurtado, A. (1960). Some clinical aspects of life at high altitudes. *Annals of internal Medicine*, **53**, 247–58.

Hurtado, A. (1966). Natural acclimatization to high altitudes. Review of Concepts. In *Life at high altitudes*, pp. 7–9. Scientific Publication No. 140. Washington, DC: Pan American Health Organization.

James, W. H. (1966). The effect of altitude on fertility in Andean countries. *Population Studies*, **20**, 97–101.

Lang, S. D. R. & Lang, A. (1971). The Kunde hospital and a demographic survey of the upper Khumbu, Nepal. *New Zealand Medical Journal*, **74**, 1–8.

McClung, J. (1969). *Effects of high altitude on human birth: observations on mothers, placentas, and the newborn in two Peruvian populations.* Cambridge, Mass.: Harvard University Press.

McGarvey, S. T. (1974). A follow-up study on the biological and social status of Quechua infants. Master's Thesis in Anthropology, The Pennsylvania State University.

Marticorena, E., Severino, J. & Chavez, A. (1967). Presión arterial sistemica en el nativo de la altura. *Archivos del Instituto de Biologia Andina*, **2**, 18–26.

Marticorena, E., Severino, J., Peñaloza, D. & Hellriegel, K. (1959). Influencia de las grandes alturas en la determinación de la persistencia del canal arterial. Observaciones realizadas en 3500 escolares de altura a 4300 m. Sobre el nivel del mar. Primeros resultados operatorios. *Revista Médica de la Provincia de Yauli*, Ano IV, Nos. 1–2, La Oroya.

Marticorena, E., Tapia, F. A., Dyer, J., Severino, J., Banchera, N., Gamboa, R., Krüger, H. & Peñaloza, D. (1964). Pulmonary edema by ascending to high altitudes. *Diseases of the Chest*, **45**, 273–83.

Mazess, R. B. (1965). Neonatal mortality and altitude in Peru. *American Journal of physical Anthropology*, **23**, 209–14.

# The biology of high-altitude peoples

Mirrakhimov, M. M. (1971). *Heart disease and high altitude.* Frunze: 'Kirghiztan' Publishing House.

Monge, C. C., Lozano, R. & Carcellen, A. (1964). Renal excretion of bicarbonate in high altitude natives with chronic mountain sickness. *Journal of clinical Investigations,* **43,** 2303.

Monge, M. C. (1934). La tuberculosis en el ejercito. *Revista Sanidad Militar,* **7,** 338.

Monge, M. C. (1948). *Acclimatization in the Andes.* Baltimore, Md.: The Johns Hopkins Press. (Republished by Blaine Ethridge Books, Detriot, 1973.)

Monge, M. C. (1953). El problema de la silicosis y el mal de montaña cronico. *Perú Indígena,* Lima, **4,** 7.

Monge, M. C. & Monge, C. C. (1966). *High-altitude diseases.* Springfield, Ill.: C. C. Thomas.

Morton, W. E., Davids, D. J. & Lichty, J. A. (1964). Mortality from heart disease at high altitude: the effects of high altitude on mortality from arteriosclerotic and hypertensive heart disease. *Archives of environmental Health,* **9,** 21–4.

Peñaloza, D., Arias-Stella, J., Sime, F., Recavarren, S. & Marticorena, E. (1964). The heart and pulmonary circulation in children at high altitude. *Pediatrics,* **34,** 568–82.

Poupa, O., Krofta, K. & Prochazka, J. (1966). Acclimation to simulated high altitude and acute cardiac necrosis. *Federation Proceedings,* **25,** 1243–6.

Puffer, R. R. & Serrano, C. V. (1973). *Patterns of mortality in childhood: report of the Inter-American Investigation of Mortality in Childhood.* Scientific Publication No. 262. Washington, DC: Pan American Health Organization.

Schadel, R. (1959). La demografia y los recursos humanos. Plan para el desarollo del sur del Peru. Lima, Peru.

Shaper, A. G. & Jones, K. W. (1959). Serum-cholesterol, diet and coronary heart disease in Africans and Asians in Uganda. *Lancet,* **ii,** 534.

Spector, R. M. (1971). Mortality characteristics of a high altitude Peruvian population. Master's Thesis in Anthropology, The Pennsylvania State University.

Stycos, J. M. (1963). Culture and differential fertility in Peru. *Population Studies,* **16,** 257–70.

Stycos, J. M. (1965). Needed research on Latin American fertility: Urbanization and fertility. *Milbank Memorial Fund Quarterly,* **43,** 299–323.

Vásquez, C. L., Parlin, W. & Simonson, E. (1967). Distribución altitudinal de la población Peruana. Informe preliminar. *Archivos del Instituto de Biologia Andina,* **2,** 150–5.

Velásquez, T. (1966). Acquired acclimatization to sea level. In *Life at high altitudes,* pp. 58–63. Scientific Publication No. 140. Washington, DC: Pan American Health Organization.

Ward, M. (1975). *Mountain medicine.* London: Crosby Lockwood Staples Ltd.

Watt, E. W., Picón-Reátegui, E., Gahagan, H. E. & Buskirk, E. R. (1976). Dietary intake and coronary risk in Peruvian Quechua Indians. *Journal of the American dietetic Association,* **68,** 535–7.

Way, A. B. (1976). Morbidity and postneonatal mortality. In *Man in the Andes: A Multidisciplinary Study of High Altitude Quechua,* ed. P. T. Baker & M. A. Little, pp. 147–60. Stroudsburg, Pa.: Dowden, Hutchinson & Ross.

Whitehead, L. (1968). Altitude, fertility and mortality in Andean countries. *Population Studies,* **22,** 335–46.

Zapata, B. & Marticorena, E. (1968). Presión arterial sistémica en el individuo senil de altura. *Archivos del Instituto de Biologia Andina,* **2,** 220–8.

# Index

# Index

capillary blood vessels: in muscle of HAN, 183, 185, 304–5, 335; in placentae of HAN, 85

carbohydrates: intake of, by Andean HAN, 236; percentage of calorie intake as, 237

carbonic anhydrase, altitude and activity of, 302

cardiac cycle: during exercise and recovery, in groups from different altitudes, 308–10; in HAN children, 312–13

cardiac defects, in HAN newborn, 103

cardiac index, altitude and, 303

cardiovascular disease, altitude and, 334–5, 344

Caucasoid ancestry: in Andes, very small, 55; in Ethiopia, 55; in Himalaya, 51, 54

cell division, hypoxia and rate of, 70, 165, 208

cereals, staple food up to 3,500 m in Andes (rice to 2,000 m, maize 2,000–3,500 m), 221, 222, 223

cestode infestation, at different altitudes in Peru, 331

chemoreceptor sensitivity to oxygen breathing, altitude and, 300, 302, 310

chenopodiums (quinoa and cañihua), in diet of Andean HAN, 221, 222, 224

chest: circumference of, Ethiopian HAN and LAN, 141, 147; circumference/height ratio, in Andean HAN at 4,000 m and above 4,500 m, 125, in Sherpa HAN and Tibetan downward migrant children, 150, 151, and in Tien Shan HAN and Kirov LAN, 158; enlargement of, genetically determined in Andean HAN, 8, 345; enlargement of, not general characteristic of Himalayan HAN, 166; width and depth of, Andean HAN, 119, and Tien Shan HAN and Kirov LAN, 157, 162–3

Chile, 61; estimated population of, above 2,500 m, 318

cholesterol, blood content of, 334; in Andean HAN, and HAN downward migrants, 337

chorionic gonadotrophin, human, response to injection of: in HAN downward migrants, before and after return to high altitude, 75; in HAN and LAN, 76, 77

chromosome abnormalities, none found in Andean HAN, 208

climates, in high-altitude zones, 251; Andes, 33, 253–6; Ethiopia, 258; Himalaya, 34–5, 256–7

clothes at high altitudes, 262–3, 290; Andes, 263–4; Himalaya, 264–5

coca leaf chewing, by Andean HAN, 284; and response of body temperature to exposure to cold, 286–8, 291

cold stress, in Andean HAN, 232, 266–73; influence of drugs on tolerance to, 284–8; interaction of, with hypoxic stress, 281–2; interaction of, with low-humidity stress, 288–9; laboratory studies of 274–9

Colorado, population in high altitude areas of, 4

colour-blindness, rare or absent in Andean HAN, 50

congenital defects: altitude and, 103–4, 313; in Andean HAN, 342

cor pulmonale: chronic, in some HAN (Tien Shan and Pamirs), 313

cortisol: normal production of, in HAN, 78

cosmic radiation, at high altitudes, 56, 69, 101

cultures, altitudinal similarities in? 42

cytochrome, microsomal, in oxygen transport, 89–90

demographic structure of HAN: age and sex distribution, 319–22; total numbers, 317–19

dental development: in Ethiopian HAN and LAN, 145–6; in Sherpa HAN and Tibetan children at low altitude, 154–6; in Tien Shan HAN and Kirov LAN, 160–1

Diego allele: distribution of gene frequency for, in Andean HAN and neighbouring LAN, 58, 59, 60

diet: composition of, in different groups of Andean HAN, 236–44

2,3-diphosphoglycerate content of red cells: in Bohr effect, 201, 208–9; did not increase on transfer to high altitude when alkalosis was prevented, 209; in polycythaemia, 203; in Sherpas and Europeans, 211, 212

diseases, infectious: in babies, Andean HAN and LAN, 105; in Ethiopian HAN and LAN, 138, 139; and food requirements, 233, 245; mosquito-borne, HAN free of, 330, 334; often less common in HAN, 69

divorce rate, in Ethiopia, 27

dogs, hypoxia and foetal haemoglobin in, 94

ductus arteriosus, patent: in HAN, 313, 329–30

Echinococcus infestation, in Andean HAN, 332

Ecuador, 61; estimated population of, above 2,500 m, 318

emphysema, and polycythaemia, 204

energy balance, in Andean HAN, 235

enzymes of oxidative chain, in HAN, 183, 185

eosinophilia: in Ethiopian LAN, indicating nematode infestation, 197

erythropoiesis: in bone marrow, altitude and, 123, 305; in mice, hypoxia and, 191–5

erythropoietic islands: hypoxia and numbers of, in bone marrow and spleen of mice, 191, 193

erythropoietin: increased secretion of, at high altitude, 191–4, 201, 305; in LAN upward migrants, 202

Ethiopia, 18; economic and political structure of, 26–7; estimated population of, 319; ethnic and linguistic diversity of population of, 25–6; exogamy, in 61; genetic distances between HAN and neighbouring LAN in, 55; genetic polymorphisms in, 51–2; geology of, 20–1; migrants between altitudes in, 11, 58; occupied by man longer than other highlands, 51

exogamy, 61

fat: of animals, in diet of Andean HAN, 221, 222, 226; of arm, development of, in Andean HAN, 122; of body, rural–urban differences in, 135; intake of, by Andean HAN, 236, 237–8, (apparent absorption of) 235, (percentage of calorie intake as) 237

fertility, 5; altitude and, 322–8; differential, apparently less important than differential mortality in selection pressure estimate, 56–7; possible

352

# Index

inbreeding, in HAN, 60–1
Incas, in Andes, 28, 30–1, 60, 67; inbreeding in dynasty of, 61
Indian Government: Border Roads Organization of, in Himalaya, 36
insulation, unit of (for clothing), 262
iodine, in diet of HAN, 244
iron: blood content of, in HAN and HAN downward migrants, Andes, 206; intake of, by Andean HAN, 239, 243; intestinal absorption of, promoted by increase in erythropoiesis, 201
irrigation, in Himalaya, 35, 38, 40

Karakorum ranges, Pakistan, 51, 54, 60
Kashmir province, estimated population of, 319
Kirghizia, 77–8

lactic acid, produced in foetus in acute hypoxia of mother, 95
Ladakh province, estimated population of, 319
leprosy, in Nepal, 333
leucocytes, in HAN children, 312
livestock: exposure of HAN to climatic stress in herding of, 265–6; introduced by Spaniards, at first failed to reproduce at high altitudes, 68; lodged on ground floor of Himalayan houses, 261; raising of, in Andes, 34, 117–18, 219, in Ethiopia, 26–7, 138, and in Himalaya, 36, 38–9, 40–1
llama: altitude, and size of placenta in, 83
lung capacity, 163–4, 165–6; effect on, of age at acclimatization to high altitude, 133, 134; residual volume as percentage of, higher in HAN, 302
lung forced vital capacity, in Andean HAN and LAN: chest circumference and, 132–3; compared with US standard, 130–1; and forced expiratory volume, 131–2
lung vital capacity: not affected by altitude, 302; and tidal volume and respiratory volume, in Tien Shan HAN and Kirov LAN, 161, 162, 163
lung oedema, lung ventilation, see pulmonary oedema, pulmonary ventilation
luteinizing hormone: altitude, and production of, 76, 130
lymphocytosis, in HAN, 197, 305

magnesium, in red cells of HAN, 207
malaria: agent of natural selection on sickle-cell haemoglobin, 58; at different altitudes in Peru, 330; eradication of, in Nepal, 39; in Ethiopian HAN and LAN, 138
measles: child mortality from, Bolivian HAN, 343
meat and other animal products, in diet of Andean HAN, 221, 222, 223, 224, 225–6, 244
menarche, altitude and age of, 77–8, 311; in Andes, 128–9; in Himalaya, 154; in Tien Shan HAN and Kirov LAN, 159–60
menstrual cycle, change of altitude and, 73
Mestizos, Peru, 32, 99–100, 118; growth of children of, HAN and LAN, 126–7
methaemoglobin: in normal, anaemic, and polycythaemic HAN, 201
mice: alkalosis in, and sex ratio, 103; effects of

hypoxia in, on congenital defects, 103, on erythropoiesis, 191–4, on foetal haemoglobin, 94, on gametogenesis, 71, 72, on mating behaviour, 74, and on survival and abnormalities of ova, 79–80, 90, 92
migration (gene flow): in Andes, 58, 60, 321, 345–6; in Ethiopia, 11, 58; in Himalaya, 39, 60
mining, in Andes, 220
mitochondria: numbers of, in skeletal muscle of HAN, 183, 185
Mongoloid ancestry, in Himalaya, 51, 54
monogamy, serial, in Andes, 91
morbidity at high altitudes: infant and neonatal, 340–2; childhood, 342–3; adult, 343–5
mortality: death rates, Andes, (by age) 338, (age- and sex-specific) 339, (children) 342–3; differential, apparently more important than differential fertility in selection pressure estimate, 56–7
mortality, infant: birth weight and, HAN and LAN, 341, 342; of first generation Spaniards at Potosi (4,000 m), 67; in HAN and LAN, Andes, 56, 106, 340–1; hypoxia and, in animals, 102; increased in HAN, 101, 106, 244, 399–40; sex and, HAN and LAN, 100–1, 321–2
mountain climbing: oxygen utilization during, by HAN and LAN, 178
mountain sickness, altitude-specific disease, 329
muscle: of arm, development of, in Andean HAN, 122, 124; capillarity of, in HAN, 183, 185, 304–5, 335; number of mitochondria in skeletal, of HAN, 183, 185
mutations: cosmic radiation and, 56, 69; not detected in HAN populations, 56

natural selection, possibility of estimating intensity of: from differential survival and fertility rates, 56–7; from distribution of genetic polymorphisms, 57–8
nematode infestation: at different altitudes, Peru, 331; in LAN, Ethiopia, 197
Nepal, 19, 37; agriculture and stock-raising in, 38–9, 40–1; estimated population of, 319; population distribution in, 37; vegetational zones of, 37, 38
neutrophils: excess segmentation of, at high altitudes, 305; numbers of, in HAN children, 312
newborn: measurements of, in HAN downward migrants, 10, and in HAN and LAN, Andes, 99–100; sex differences in measurements of, 100; weight of, see birth weight
niacin: intake of, by Andean HAN, 239, 240
night vision, altitude and, 18
nitrogen balance, in Andean HAN, 235
Nuñoa district, S. Peru, 31–2, 118; population structure in, 120
nutrition, 5, 163; and growth, Andes, 134–5; of infants, HAN and LAN, 104–5; of Sherpa HAN, and Tibetan children at low altitude, 147, 155

oestrogens: altitude, and concentrations of, in blood of mother and of newborn, 87–8, 102; produced by placenta, 86–7
offspring, numbers of; in Andean HAN and LAN, 90–1, 323–6, 345; in Ethiopian HAN, 327; in

354

offspring (*cont.*)
 Nepal, 327; in Tien Shan, 311; percentage surviving to adult age, Andes, 345
oogenesis, change of altitude and, 71
ova: hypoxia, and survival and abnormalities of, in mice, 79–80, 90, 92
oxygen: altitude and consumption of, 300; cost in, per unit of work, in groups from different altitudes, 308; exercise ventilation and uptake of, in HAN, and in LAN upward migrants, 182–3; in foetal blood, 94; hyperventilatory response to breathing of, altitude and, 300, 302, 310; partial pressure of, at different altitudes, 225–7; transport of, across placenta, 88–90; utilization of, in mountain climbing, by HAN and LAN, 178
oxygen-carrying capacity of blood: effect of hypoxia on, in HAN, and in LAN upward migrants, 184

Pamir Mountains: meteorological data for different altitudes in, 299; peoples of, 4, 51
Pan-American Health Organization, and high-altitude project, 3
parasites, intestinal: in Andes, in HAN, 234, and in populations at different altitudes, 331, 332; in Ethiopian LAN, 146, 163
Peru, estimated percentage of population of, above 2,500 m, 317–19
phosphogluconate dehydrogenase, in HAN and LAN, Ethiopia, 55
phosphorus: blood content of, HAN and HAN downward migrants, Andes, 206; intake of, by Andean HAN, 239, 242–3; ratio of calcium to, in diet of Andean HAN, 242
physical activity, habitual: and aerobic capacity, 175, 177; and hypoxic ventilatory drive, 182; and oxygen utilization in mountain climbing, 178; and ventilatory exchange ratio, 182
placenta: altitude, and infarcts in, (guinea pigs) 82, (humans) 85–6; altitude, and ratio of weight of, to birth weight, humans, 10, 83–4, 87; altitude, and size and function of, (animals) 80–3, (humans) 83, 84, 85, 87; oxygen transport across, 88–90; perfusion of, with various concentrations of oxygen, 84–5; production of hormones by, 86–8
placenta praevia: altitude, and frequency of, 86, 97
plasma proteins, polymorphisms of, 48
plasma volume, in LAN upward migrants, 202
platelets, in HAN, 205, 305
polyandry, in Nepal, 327
polycythaemia: in HAN, 184–5, 190–1, 194–5, 203–4; in hypoxic mice, 191–4; only apparent in Andean HAN in pathological conditions, 199
population, of high-altitude areas of world, 317–19
population genetics, 47
population structure, HAN of Peru compared with total population, 126
potassium: blood content of, in HAN downward migrants and LAN upward migrants, 206–7
potatoes (fresh or dehydrated): in diet of Andean HAN, 221, 223–4; as source of ascorbic acid, 241
Potosi city, Bolivia (4,000 m): first generation Spanish children in, all died at birth or soon after, 67

precipitation: Andes, monthly averages, 253; Ethiopia, of monsoon pattern, 258; Himalaya, 35, 40, (monthly averages) 257
pregnancy, altitude and duration of: in guinea pig, 92; in human, 93
pregnanediol: excretion of, by HAN and LAN women, 77
progesterone, 77; produced by placenta, 86
protein: altitude and requirement for, 234; hypoxia or altitude, and specific dynamic action of, 230; intake of, by Andean HAN, 236, (as percentage of calorie intake), 237
proteinograms: for blood of two ethnic groups of HAN, before and after downward migration, 207–8
protozoan infestations, at different altitudes in Peru, 331
psychomotor development of babies, Andean HAN and LAN, 105, 106
puberty, altitude and age of, 77–8, 312; in Andean HAN and LAN, 129–30; in Ethiopian HAN and LAN, 146; in Tien Shan HAN and Kirov LAN, 159–60; genetic origin of late, in HAN downward migrants, 10
pulmonary oedema, altitude-specific disease, 328–9
pulmonary ventilation: altitude and, 300, 301; exercise and, in groups from different altitudes, 307

Quechua people, Bolivia and S. Peru, 4, 10, 57, 182–3, 207; genetic distances between other tribes and, 54; large chests of, genetically determined, 8; migration of, in 15th century, 59–60

rabbits: hypoxia and foetal haemoglobin in, 94
rats, hypoxia in: and embryonic deaths, 79; and gametogenesis, 71; and litter size and birth weight, 92; and lung development, 133–4; and placenta, 82
red cell antigens, polymorphisms of: in Andes, 48, 49–50; in Ethiopia, 48, 52; in Himalaya, 48, 51; *see also* Diego allele, Rhesus system
red cell proteins, polymorphisms of: in Andes, 48, 49; in Ethiopia, 48, 52; in Himalaya, 48; *see also* glucose 6-phosphate dehydrogenase, haemoglobin
red cells: altitude, and numbers of, 303, 305, 306, 312; numbers of, HAN and LAN, Ethiopia, 105, HAN, LAN, and LAN upward migrants, Himalaya, 198, and LAN upward migrants, Andes, 202; size of, in HAN, 305, 312; volumes of, *see* haematocrit measurements
respiration: altitude, and loss of heat in, 227–9; percentage loss in, of heat produced by LAN (sedentary and active), and HAN (active), 229
respiratory centre, sensitivity threshold to hypoxia of: in HAN, and in LAN acclimatizing to high altitude, 300, 302; in HAN returned to high altitude after period at low altitude, 310
respiratory rate, altitude and, 300, 301
respiratory disease: in HAN, 332, 333, 334; in HAN downward migrants, 344
reticulocytes: altitude, and numbers of, 305, 306, 312; altitude, and maturation rate of, 305, 306

# Index

Rhesus system: distribution of frequency of segments of, from north to south along Andes, 49–50, 58, 59, 60; in Himalaya, 54; as indicator of genetic drift in Andes, 60; possible action of natural selection through variation in resistance of haplotypes of, to salmonella infections, 57–8
riboflavin: intake of, by Andean HAN, 239, 240
rural–urban differences, Andes: in body fat, 135; in skinfold thickness and muscle development, 124; in stature and weight, 123

salmonella infections, variation in resistance of Rhesus haplotypes to, 58
Sardinia, sickle-cell haemoglobin in HAN and LAN in, 58
serum proteins: in Andes, 48, 50; in Ethiopia, 48, 52; in Himalaya, 48
sex: and birth weight in HAN and LAN, Andes, 100; and body measurements of babies, HAN and LAN, 106; and development of arm tissues, Andean HAN and US, 122; and neonatal mortality, HAN and LAN, 100–1; and response to cold stress, 278–9
sex hormones, altitude and, 73–8
sex ratio of births; in HAN and LAN, 103, 321; in mice, alkalosis and, 103
sheep, at high altitude: foetal growth rate in, 92; foetal haemoglobin in, 94; metabolism of foetus in, 95; placenta in, 81, 82, 85
Sherpa people, Nepal, 4, 7; aerobic capacity of, 178, 179; growth and development of, compared with Tibetans at low altitude, 147–55; population structure of, 321, 322
Sikkim, estimated population of, 319
silicosis, and polycythaemia, 204
skeletal age and chronological age: in Andean HAN, 127–8; in Ethiopian HAN and LAN, 144–5; in Sherpa HAN and Tibetan children at low altitude, 153–4, 155
skin colour, of HAN and LAN in same area, 289–90
skinfold thickness: in Andean HAN, 120–1, (rural and urban) 124; in Ethiopian HAN and LAN, 143–4; in Sherpa HAN and Tibetan children at low altitude, 151, 153, 155, 164; in Tien Shan HAN and Kirov LAN, 157, 159, 162
smoking, haematological effects of, 204
socio-economic factors, 66; may obscure effects of altitude, 164–5; may be related to altitude, 69–70
sodium: blood content of, in HAN, HAN downward migrants, and LAN upward migrants, 207
soil erosion, in Nepal, 40
solar radiation, altitude and, 252–3, 289
spermatogenesis, change of altitude and, 70–1, 72–3
survivorship curves, for US, Peru, and Peruvian HAN, 338

Tajikistan, 78, 91
temperature of body: in cold stress, remains higher in extremities of HAN than of LAN, but rectal temperature decreases more in HAN, 274–5, 291; on different parts of skin of HAN, 271–2; of HAN during sleep in different conditions, 266–9, and during normal activities, 269–70; of hands

and feet exposed to cold, in HAN and LAN, 272–3, 277–8; relation of body weight to, 272; sex differences in, 278–9
temperature of surroundings: altitude and, 252; diurnal variation in, increases with altitude; seasonal variation diminishes, 251–2; maximum, at three altitudes, Ethiopia, 258; mean annual, over 40 years, Andes, 256; mean monthly, in Andes at three altitudes, 254, 255–6; in Ethiopia at three altitudes, 258, and in Himalaya, 256–7; within houses, Andes, 259–60
testosterone: altitude, and production and excretion of, 74–6, 130
thalassemia, lacking in HAN, 55
thiamin: intake of, by Andean HAN, 239, 240
thrombo-cythaemia, in LAN upward migrants, 205
thrombo-embolic accidents: rare in HAN, frequent in LAN upward migrants, 205, 329
thyroid, in increased basal metabolic rate of HAN?, 280
Tibet, plateau of, 18–19, 22; estimated population of, 319
Tibetan migrants, Nepal, 4, 10; growth and development of children of, compared with Sherpa HAN, 147–55
Tien Shan Mountains, 19; meteorological data for different altitudes in, 299; peoples of, 4, 51
Tigrean (Tigrinyan) people, Ethiopia, 22, 26, 195
trade, in Himalayan region, 36, 39
trematode infestation, at different altitudes in Peru, 331
tuberculosis, in HAN downward migrants, 333, 344
tubers, staple food in Andes above 3,500 m, 221, 222, 223–4
typhoid fever, variation in resistance of Rhesus haplotypes to, 58
typhus, at different altitudes in Peru, 330, 332

ultraviolet radiation, at high altitudes, 69, 289; and vitamin D requirements of HAN, 233, 245
United States: haematology of HAN and LAN in, 203–4
uric acid content of blood: in HAN and HAN downward migrants, Andes, 206

vascular resistance, altitude and, 304, 310
vasodilation: alcohol and, 285; in extremities of HAN during cold stress? 278; in placenta in hypoxia, 89; in uterus in hypoxia, 86
ventilatory drive, hypoxic; habitual physical activity and, 181, 182
ventilatory exchange ratios, habitual physical activity and, 182
vitamin A: intake of, by Andean HAN, 228, 229, 230, 245
vitamin D: and calcium absorption, 242; ultraviolet radiation at high altitudes, and requirement of, 233, 245

water: effect of lowered boiling point of, at high altitudes, on cooked foods, 226, 240, 245; extra loss of, in respiration at high altitudes, 227–8
weight: of Andean HAN, rural and urban, 123; of

356